Dynamics and Vibration

Dynamics and Vibration
An Introduction

Dr Magd Abdel Wahab

John Wiley & Sons, Ltd

Other Wiley Editorial Offices

John Wiley & Sons Inc., 111 River Street, Hoboken, NJ 07030, USA

Jossey-Bass, 989 Market Street, San Francisco, CA 94103-1741, USA

Wiley-VCH Verlag GmbH, Boschstr. 12, D-69469 Weinheim, Germany

John Wiley & Sons Australia Ltd, 42 McDougall Street, Milton, Queensland 4064, Australia

John Wiley & Sons (Asia) Pte Ltd, 2 Clementi Loop #02-01, Jin Xing Distripark, Singapore 129809

John Wiley & Sons Canada Ltd, 22 Worcester Road, Etobicoke, Ontario, Canada, M9W 1LI

Wiley also publishes its books in a variety of electronic formats. Some content that appears in print may
not be available in electronic books.

Library of Congress Cataloging-in-Publication Data

Wahab, Magd Abdel.
    Dynamics and vibration : an introduction / Magd Abdel Wahab.
        p.   cm.
    Includes bibliographical references and index.
    ISBN 978-0-470-72300-5 (pbk : alk. paper)
    1. Vibration.   2. Structural dynamics.   I. Title.
TA355.W34   2008
620.3–dc22

                            2007041623

British Library Cataloguing in Publication Data

A catalogue record for this book is available from the British Library

ISBN 978-0-470-72300-5

Typeset in 9/12pt Sabon by Aptara Inc, New Delhi, India

To my lord
Allah, Almighty God, the lord of the heavens and the earth
We love you so much

Contents

Preface

This book covers dynamics and vibration modules that could be taught at undergraduate levels to the engineering students at universities in UK and worldwide. The content of the book is suitable for dynamics modules at common Level 1 for civil, mechanical, medical and aerospace engineering students and at Level 2 and Level 3 for mechanical, medical and aerospace engineering students.

The topics presented in the book are:

- motion of particles and rigid bodies with and without reference to masses and forces (kinematics and kinetics) including motion of wheels, gears, linkages and mechanisms;
- balancing of machines including rotating masses and multi-cylinder engines;
- free and forced vibration of systems with a single degree of freedom (a mass, a spring and a damper) including damped and undamped systems and vibration isolators;
- free and forced vibration of systems with two degrees of freedom including vibration absorbers;
- vibration of continuous systems including lateral vibration of cables, longitudinal vibration of bars and lateral vibration of beams using analytical solutions and the finite element method.

Part I of the book covers kinematics and kinetics of particles and rigid bodies and balancing of machines and Part II covers vibration. The chapters in Parts I and II are structured in a similar way: the theoretical background is presented with solved examples and then tutorial sheets provide exercises.

Chapter 1 introduces the kinematics of particles and covers rectilinear motion with constant and non-constant acceleration, and curvilinear motion in the Cartesian, polar and normal-tangential co-ordinate systems. Chapter 2 is concerned with the kinematics of rigid bodies and covers wheels, gears, linkages, slider–crank mechanisms and four-bar linkage mechanisms. Chapter 3 is concerned with the kinetics of particles and includes Newton's laws of universal gravitation and motion, the work and energy; and the impulse and momentum methods. Chapter 4 considers the kinetics of rigid bodies and the concept of mass moment of inertia. The force and acceleration, work and energy, and impulse and momentum methods are applied to wheels, gears, linkages and mechanisms. Chapter 5 considers and discusses the balancing of rotating masses and multi-cylinder engines.

The theoretical background of the free vibration of undamped and damped systems with one degree of freedom is presented in Chapter 6. Chapter 7 covers forced vibration of an undamped and damped system with a single degree of freedom and introduces the concept of vibration isolation. Chapter 8 describes the solution of free and forced vibration of undamped systems with two degrees of freedom and provides an introduction to vibration absorbers and damped systems. Chapter 9 presents the vibration of continuous systems, including the lateral vibration of cables, longitudinal vibration of bars and lateral vibration of beams. Finite element analysis and its application to vibration problems is described in Chapter 10.

Appendix A describes a software Dynamic Analysis for Mechanical Application (DAMA) in which simulations of mechanisms and vibrating systems have been implemented. These simulations are used to enhance student's learning and understanding of moving and vibrating mechanical systems. Sixteen simulations have been developed:

- A slider–crank mechanism
- A four-bar linkage mechanism
- A simple gear train
- A compound gear train
- An engine gearbox
- A sliding bar
- A rotating bar
- A balancing of rotating masses
- A balancing of a multi-cylinder engine
- A system with a single degree of freedom
- A system with two degrees of freedom (two masses and two springs)
- A system with two degrees of freedom (a rigid bar supported by two springs)
- Lateral vibration of a cable or a string
- Longitudinal vibration of a bar
- Lateral vibration of a beam
- Whirling of shafts

The LabView visual programming language has been used to develop the simulations and to animate and visualize the motion and vibration of the systems. A similar layout has been designed for each simulation using separate frames for input parameters, output parameters, animation and a figure that illustrates the input parameters. The simulations were designed in such a way that any change in the input parameters is immediately reflected in the animation, output plots and output parameters.

Appendix D provides a summary of the formulae from the chapters. A supplementary appendix containing solutions to the tutorial questions is available to lecturers and instructors on a web site.

Acknowledgements

First of all thanks to Allah, almighty God, the most gracious, the most merciful, the great creator, without whom there would be no life on Earth and who taught us all the sciences: 'Read in the name of your Lord Who created. He created man from a clot. Read and your Lord is Most Honourable, Who taught (to write) with the pen, Taught man what he knew not.' Quran (96:1–5)

I would like to acknowldge the Fund for the Strategic Development of Learning and Teaching (FSDLT) at the University of Surrey and the Higher Education Funding Council for England (HEFCE) for funding the e-learning project Dynamic Analysis for Mechanical Applications (DAMA), within which the simulations presented in this book have been developed. Special thanks go to Professor David Airey, Pro-Vice Chancellor of Teaching and Learning at the University of Surrey, for supporting the project and leading the University of Surrey to excellence in teaching and professional training.

I would like to thank Dr Djelloul Mahboub, who has been working as a post-doctoral researcher on the DAMA project (2003–5). My thanks also goes to my PhD students who have assisted me in teaching dynamics courses and laboratory classes over the last few years: Mr Libardo Vanegas-Useche, Mr Irfan Hilmy, Mr Salah Rashwan, Mr Ahmed Ashhab, Mr Tauqeer Ali, Mr Seun Adediran and Mr Alex Graner Solana. Special thanks go to Mr Irfan Hilmy for his contribution in developing some of the simulations presented in DAMA. Thanks are due to my colleagues who previously taught dynamics at the University of Surrey: Dr John Thomas, Dr Tony Cartwright and Professor Jingzhe Pan.

Special thanks are due to the editorial board at John Wiley & Sons: commissioning editor Dr Vivien Ward, publishing assistant Ms Flick Williams, project editors Mr David Barnard and Ms Emma Cooper, and marketing manager Mr Sam Crowe; and to the copyeditor at Mitcham Editorial Services, Ms Shena Deuchars.

And, last but not least, special thanks to my late parents for their moral support throughout my life and to my children: Leila, Mona, Younes and Fatima Abdel Wahab, for the nice and joyful moments they give me in this life.

List of symbols

List of symbols is given for each chapter in the following format, when appropriate:

Symbol = meaning (SI Units)

Chapter 1

a = linear acceleration (m/s^2)
a_n = normal acceleration (m/s^2)
a_r = radial acceleration (m/s^2)
a_t = transverse acceleration (m/s^2)
$a_x = \ddot{x}$ = linear acceleration in the x-direction, $\ddot{x} = \dfrac{d^2 x}{dt^2}$ (m/s^2)

$a_y = \ddot{y}$ = linear acceleration in the y-direction, $\ddot{y} = \dfrac{d^2 y}{dt^2}$ (m/s^2)

g = the acceleration of gravity (m/s^2)
L = connecting rod length (m)
s = linear displacement (m)
r = radius, radial displacement (m)
$\dot{r}$ = rate of change in radial displacement, radial velocity $\dot{r} = \dfrac{dr}{dt}$ (m/s)

$\ddot{r}$ = rate of change in radial velocity, $\ddot{r} = \dfrac{d\dot{r}}{dt}$ (m/s^2)

t = time (s)
u = initial linear velocity (m/s)
v = final linear velocity, magnitude of velocity (m/s)
v_o = initial speed of a projectile (m/s)
v_n = normal velocity (m/s)
v_r = radial velocity (m/s)
v_t = transverse velocity (m/s)
$v_x = \dot{x}$ = linear velocity in the x-direction, $\dot{x} = \dfrac{dx}{dt}$ (m/s)

$v_y = \dot{y}$ = linear velocity in the y-direction, $\dot{y} = \dfrac{dy}{dt}$ (m/s)

$\dot{v}$ = rate of change in tangential velocity (m/s^2)
x_o = initial linear displacement in the x-direction (m)
y_o = initial linear displacement in the y-direction (m)
$\dot{x}_o$ = initial linear velocity in the x-direction (m/s)

$\dot{y}_o$ = initial linear velocity in the y-direction (m/s)

θ = angle, angular displacement (rad)

$\dot{\theta}$ = angular velocity, $\dot{\theta} = \dfrac{d\theta}{dt}$ (rad/s)

$\ddot{\theta}$ = angular acceleration, $\ddot{\theta} = \dfrac{d\dot{\theta}}{dt}$ (rad/s^2)

ρ = radius of curvature (m)

Chapter 2

a = offset distance in slider-crank mechanism (m)

$a_{B/A}$ = relative acceleration between B and A (m/ s^2)

$a_{nB/A}$ = relative normal acceleration between B and A (m/ s^2)

a_D = magnitude of acceleration of point D (m/ s^2)

$a_{tB/A}$ = relative transverse acceleration between B and A (m/ s^2)

b = length (m)

$dr_{B/A}$ = relative displacement between B and A (m)

C , C_1, C_2, C_3, C_4, C_5, C_6 = constants

G = gear ratio

L = length of connecting rod (m)

L_1 = length of crank link (m)

L_2 = length of coupler link (m)

L_3 = length of follower link (m)

L_4 = length of fixed frame (m)

nt = number of teeth in a gear

r = radial displacement, crank's radius (m)

r' = radius of a gear on the layshaft (m)

$v_{B/A}$ = relative velocity between B and A (m/s)

ω = angular speed (rad/s)

ψ = angular position of coupler link (rad)

φ = angular position of follower link (rad)

$\dot{\psi}$ = angular velocity of coupler link (rad/s)

$\dot{\varphi}$ = angular velocity of follower link (rad/s)

$\ddot{\psi}$ = angular acceleration of coupler link (rad/s^2)

$\ddot{\varphi}$ = angular acceleration of follower link (rad/s^2)

Chapter 3

a = linear acceleration (m/s^2)

m = mass (kg)

e = coefficient of restitution

F = force (N)

F_s = static friction force (N)

F_d = dynamics friction force (N)

F_x = force in the x-direction (N)
F_y = force in the y-direction (N)
F_{EM} = force of attraction between the Earth and the Moon (N)
F_{Es} = force of attraction between the Earth and the Sun (N)
h_1 = initial vertical height (m)
h_2 = final vertical height (m)

G = Universal Constant of Gravitation = $6.673 \times 10^{-11} \left(\dfrac{m^3}{kg.s^2} \right)$

KE = kinetic energy (N.m)
k = constant, stiffness of a spring (N/m)
M, M_E = mass of the Earth (kg)
M_M = mass of the moon (kg)
M_s = mass of the sun (kg)
N = normal force (N)
R_{EM} = distance between the earth and the moon (m)
R_{ES} = distance between the earth and the sun (m)
R = radius of Earth (m)
R_M = radius of moon (m)
T = tension force (N)
v = linear velocity (m/s)
v_1 = initial linear velocity (m/s)
v_2 = final linear velocity (m/s)
s_1 = initial displacement (m)
s_2 = final displacement (m)
SE = strain energy (N.m)
x_1 = initial extension (m)
x_2 = final extension (m)
PE = potential energy (N.m)
g = acceleration of gravity (m/s^2)
g_M = Moon's gravitational acceleration (m/s^2)
W_e = work done by external forces (N.m)
μ_s = coefficient of the static friction
μ_d, μ = coefficient of the dynamic friction
$\int F dt$ = linear impulse of a force F (N.s)
$\int P dt$ = linear impulse in deformation phase (N.s)
$\int R dt$ = linear impulse in restitution phase (N.s)

Chapter 4

I = mass moment of inertia (kg.m^2)
I_G = mass moment of inertia about the centre of gravity G (kg.m^2)
L = length (m)
k_G = radius of gyration (m)

R = radius (m)

k = stiffness of a spring (N/m)

x_i, y_i = x and y co-ordinates of particle i (m)

x_G, y_G = x and y co-ordinates of the centre of gravity G (m)

$\ddot{x}_i, \ddot{y}_i$ = x and y accelerations of particle i (m/s²)

$\ddot{x}_G, \ddot{y}_G$ = x and y accelerations of the centre of gravity G (m/s²)

$x_{i/G}, y_{i/G}$ = x and y relative co-ordinates between particle i and the centre of
                    gravity G (m)

$\ddot{x}_{i/G}, \ddot{y}_{i/G}$ = x and y relative accelerations between particle i and the centre of gravity
                    G (m/s²)

u = deformed length of a spring (m)

M = bending moment (N.m)

M_G = bending moment about the centre of gravity G, external couple (N.m)

M_O = bending moment about point O (N.m)

V = volume (m³)

dV = element's volume (m³)

dA = area of element's cross section (m²)

T = tension force (N)

R_x = reaction force in the x-direction (N)

R_y = reaction force in the y-direction (N)

R_B = reaction force at point B (N)

P = force (N)

ρ = density (kg/ m³)

U_θ = work done by a couple (N.m)

v_G = linear velocity of the centre of gravity G (m/s)

W_e = work done by external forces (N.m)

KE = kinetic energy (N.m)

SE = strain energy (N.m)

PE = potential energy (N.m)

$\int M_G dt$ = angular impulse of moment M_G (kg.m²/s)

H_G = angular momentum at the centre of gravity G (kg.m²/s)

H_A = angular momentum at point A (kg.m²/s)

L_G = linear momentum at the centre of gravity G (N.s)

Chapter 5

a = length (m)

m_i = mass of particle i (kg)

m = mass (kg)

F = force, magnitude of force (N)

F_x = force in the x-direction (N)

F_y = force in the y-direction (N)

F_A = force acting at point A (N)

L = length (m)
R_A = reaction force at point A (N)
R_B = reaction force at point B (N)
ω = angular speed (rad/s)
n = ratio of the length of the connecting rod to the radius of the crack.
r = radial position, crank's radius (m)
M = bending moment (N.m)
M_x = bending moment about x-axis (N.m)
M_y = bending moment about y-axis (N.m)
M_A = bending moment about point A (N.m)
z = distance along z-axis, z co-ordinate (m)
θ = angle, angular position (rad)

Chapters 6 and 7

A, B = constants
A = cross-section area (m^2)
x = displacement (m)
$\dot{x}$ = linear velocity in the x-direction (m/s)
$\ddot{x}$ = linear acceleration in the x-direction (m/s^2)
ω = angular frequency, forcing frequency (rad/s)
τ = period of oscillation, time for one cycle (s)
f = external force function (N)
f' = frequency of oscillation (Hz)
F_x = force in the x-direction (N)
m = mass (kg)
k = stiffness (N/m)
ω_n = fundamental angular frequency (rad/s)
f_n = fundamental natural frequency (Hz)
δ_o = static deflection, axial displacement (m)
g = acceleration of gravity (m/s^2)
x_o = initial displacement (m)
$\dot{x}_o$ = initial linear velocity (m/s)
C = amplitude of vibration (m)
ψ = phase angle (rad)
τ_n = period of motion (s)
F = force, magnitude of force (N)
E = Young's modulus (N/m^2)
L = length (m)
I = area moment of inertia (m^4)
u = displacement, deflection (m)
ρ = density (kg/ m^3)

P = force, load (N)

σ = axial stress (N/m^2)

ε = axial strain

M_o = bending moment about O (N.m)

I_o = mass moment of inertia about point O (kg.m^2)

θ = angle, angular displacement (rad)

T = torque (N.m)

I_B = mass moment of inertia about point B (kg.m^2)

J = mass moment of inertia of a rotor or disc (kg.m^2)

G = shear modulus (N/m^2)

I_p = polar moment of inertia (m^4)

c = viscous damping coefficient (N.s/m)

ζ = damping ratio

c_c = critical damping coefficient (N.s/m)

D = amplitude (m)

ω_d = damped natural frequency (Hz)

τ_d = damped period of motion (s)

δ = logarithmic decrement

F_o = magnitude of a harmonic force (N)

x_c = complimentary or transient solution for x (m)

x_p = particular or steady state solution for x (m)

χ = steady-state amplitude (m)

ϕ = phase angle (rad)

Y = excitation amplitude (m)

y = harmonic displacement function (m)

$\overline{F}$ = impulse force (N.s)

Δt = time interval (s)

F_t = transmitted force (N)

T_r = force transmissibility or transmission ratio of an isolator

T_d = displacement transmissibility

Chapter 8

m_1, m_2 = masses of body 1 and body 2 (kg)

k_1, k_2 = stiffness of spring 1 and spring 2 (N/m)

k_f = stiffness of front spring (N/m)

k_r = stiffness of rear spring (N/m)

x_1 = linear displacement of mass 1 in the x-direction (m)

x_2 = linear displacement of mass 2 in the x-direction (m)

$\ddot{x}_1$ = linear acceleration of mass 1 in the x-direction (m/s^2)

$\ddot{x}_2$ = linear acceleration of mass 2 in the x-direction (m/s^2)

F_{x1} = force acting on mass 1 in the x-direction (N)

F_{x2} = force acting on mass 2 in the x-direction (N)

y_C = vertical displacement (y_C) at the centre of gravity C (m)

θ_C = angular displacement, rotation at the centre of gravity C (rad)

f_1 = external force function acting at degree of freedom 1 (N)

f_2 = external force function acting at degree of freedom 2 (N)

δ_o = static deflection (m)

f_{yC} = external vertical force function at the centre of gravity C (N)

$M_{\theta C}$ = external bending moment $M_{\theta C}$ at the centre of gravity C (N.m)

δ_A, δ_B = static defections at points A and B (m)

y_A, y_B = vertical displacement at points A and B (m)

I_C = mass moment of inertia at point C (kg.m^2)

P = force exerted on a mass by a spring (N)

M = mass matrix

S = stiffness matrix

D = degree of freedom vector

$\ddot{D}$ = acceleration vector

F = force vector

D_1, D_2 = displacement time response of degree of freedoms 1 and 2 (m)

D_{m1}, D_{m2} = maximum values or amplitudes of D_1 and D_2 (m)

ω_1 = fundamental angular frequency (rad/s)

ω_2 = second angular frequency (rad/s)

r_1 = amplitude ratio for fundamental mode

r_2 = amplitude ratio for second mode

a, b, c = constants of the quadratic formula

f_1 = fundamental natural frequency (rad/s)

f_2 = second natural frequency (rad/s)

T = torque (N.m)

J = mass moment of inertia of a disc (kg.m^2)

G = shear modulus (N/m^2)

I_p = polar moment of inertia (m^4)

F_o = magnitude of a harmonic force (N)

A_1, A_2 = amplitudes of the steady state responses of D_1 and D_2 (m)

c_1, c_2 = viscous damping coefficients (N.s/m)

D = is the velocity vector

ω_{d1} = fundamental damped angular frequency (rad/s)

ω_{d2} = second damped angular frequency (rad/s)

Chapter 9

a = constant, length (m)

L = length (m)

ρ' = mass per unit length (kg/m)

F = force (N)

$f(x,t) =$ transverse force per unit length (N/m)

$y(x,t) =$ vertical co-ordinate y, transverse displacement (m)

$\Delta x =$ infinitesimal length (m)

$\ddot{y} =$ acceleration in the y-direction (m/s^2)

$\Delta F =$ small change in force (N)

$c =$ constant

$Y(x) =$ displacement function depends only on x (m)

$T(t) =$ displacement fucntion depends only on t(m)

$\omega_n =$ angular frequency for the nth mode (rad/s)

$f_n =$ natural frequency for the nth mode (Hz)

$Y_n(x) =$ mode shape function for the nth mode of a cable (m)

$Y_o =$ initial dispalcement function in the y-direction (m)

$\dot{Y}_o =$ initial velocity function in the y-direction (m/s)

$A_n, B_n, C_n, D_n =$ constants for the nth mode

$\delta =$ initial deflection (m)

$E =$ Young's modulus (N/m^2)

$A =$ cross-section area (m^2)

$I =$ area moment of inertia (m^4)

$u_x =$ axial displacement function (m)

$\rho =$ density (kg/ m^3)

$\sigma =$ axial stress (N/m^2)

$\varepsilon_x =$ axial strain

$m =$ mass (kg)

$\ddot{u}_x =$ acceleration in the x-direction (m/s^2)

$A, B, C, D =$ constants

$U_n(x) =$ mode shape function for the nth mode of a bar or beam (m)

$U(x) =$ displacement function depends only on x (m)

$U xo =$ initial dispalcement function in the x-direction (m)

$\dot{U}_{xo} =$ initial velocity function in the x-direction (m/s)

$F_o =$ initial force (N)

$U_o =$ initial axial displacement (m)

$u_y(x,t) =$ vertical displacement function in the y-direction (m)

$M(x, t) =$ bending moment function (N.m)

$V(x, t) =$ shear force function (N)

$f_y =$ force in the y-direction (N)

$M_C =$ bending moment about point C (N.m)

$\ddot{u}_y =$ acceleration in the y-direction (m/s^2)

$\beta_n =$ constant for the nth mode

$M =$ mass of a disc (kg)

$u_y =$ whirl amplitude (m)

$\omega =$ angular speed (rad/s)

$\zeta =$ damping ratio

Chapter 10

u_x = axial displacement function (m)
E = Young's modulus (N/m^2)
A = cross-section area (m^2)
I = area moment of inertia (m^4)
ρ = density (kg/ m^3)
U_i , U_j = nodal displacements at nodes i and j (m)
N_i , N_j = shape functions for displacement at nodes i and j
σ_x = axial stress (N/m^2)
ε_x = axial strain
KE = kinetic energy (N.m)
m_l = mass per unit length (kg/m)
v_x = velocity in the x-direction (m/s)
U_i, U_j = nodal velocities for nodes i and j (m/s)
SE = strain energy (N.m)
$[m]$ = mass matrix
$[k]$ = stiffness matrix
$\{U\}$ = nodal displacement vector, nodal degrees of freedom vector
$\{\dot{U}\}$ = nodal velocity vector
$\{\ddot{U}\}$ = nodal acceleration vector
$\{F\}$ = force vector
$\{U_m\}$ =nodal amplitude vector, eigen vector
λ = eigenvalue
u_y = displacement function in the y-direction (m)
φ_i , φ_j = nodal rotation at node i and j (rad)
N_i^r, N_j^r = shape functions for rotation at nodes i and j
M = bending moment (N.m)
dV = element's volume (m^3)
dA = element's cross section area (m^2)

PART I

Dynamics

CHAPTER 1

Kinematics of Particles

1.1 INTRODUCTION

Dynamics is a branch of mechanics that deals with motion and its effect on a body. Unlike statics, which deals with bodies at rest, dynamics takes into account the effect of velocities and accelerations on the forces acting on bodies. In general, dynamics can be divided into two main categories, kinematics and kinetics. In kinematics, the motion of a body is analyzed without studying the forces acting on it; in kinetics, the effect of forces and masses on the motion of a body is studied. Vibration is regarded as a branch of dynamics and, more specifically, as a branch of kinetics, since forces and masses are taken into account in vibration analysis. However, very often vibration is presented in textbooks as an independent topic, separated from other dynamics topics. In this textbook, introductions to kinematics, kinetics and vibration are presented and integrated in one edition.

The study of dynamics has been possible since accurate measurement of time was made available. One of the early inventors in dynamics was Galileo Galilei (1564–1642), who studied pendulums and falling bodies. The major contribution to dynamics was made by Isaac Newton (1642–1727). Newton formulated three important and fundamental laws of motion and the law of universal gravitation, on which recent technologies and development in dynamics are based. The importance of Newton's laws became significant in the development of high-speed machines and engines that required the application of dynamics principles. Nowadays, the applications of Newton's laws and dynamics principles are widespread and include the motion analysis and design of automobiles, aircraft, space craft, machinery, motors, robotic devices, missiles, rockets, pumps, machine tools, electrical devices.

This chapter presents the kinematics of particles, divided into two main parts. Rectilinear motion is analyzed in the case of motion with constant acceleration and in the case of motion with non-constant acceleration. Curvilinear motion is analyzed in three different co-ordinate systems: the Cartesian or rectangular co-ordinate system, the polar or cylindrical co-ordinate system and the normal-tangential or normal-intrinsic co-ordinate system.

1.1.1 Terminology of kinematics

- **Kinematics** is the study of the relationships between displacement, velocity, acceleration and time without any reference to forces or weights acting on the system.
- A **particle** can be defined as a point or a body of zero dimensions. However, this definition is not useful for engineering applications because most engineering components have a finite size. A more useful definition is that a particle is a body of finite dimensions, in which all parts undergo the same motion. It follows that any point in the body represents the motion of the body as a whole and that particles can translate but

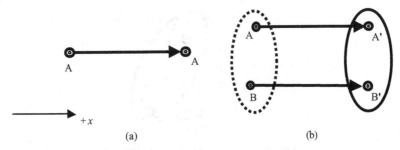

Figure 1.1: Rectilinear translation of (a) a particle and (b) a rigid body

cannot rotate. The motion of a particle in a rectilinear translation and in a curvilinear translation are shown in Figures 1.1(a) and 1.2(a), respectively.

- A **rigid body** can be defined as a body that has finite dimensions and in which all points keep the same distance between them before and after motion but do not necessarily undergo the same motion. It follows that a rigid body can translate and rotate. The motion of a rigid body in a rectilinear translation, a curvilinear translation and a rotation are shown in Figures 1.1(b), 1.2(b) and 1.3, respectively. The distance between points A and B (AB), before motion, is equal to the distance between points A′ and B′ (A′B′), after motion. However, points A and B might have different motion, i.e. AA′ has a different path from BB′, in the case of combined translation and rotation (see Figure 1.4).
- An **elastic body** is a body that can undergo elastic deformation. This means that the distance between the points in the body before and after motion are not equal. In Figure 1.3, if the distance AB is not equal to A′B′ then the body is considered to be an elastic body and undergoes elastic deformation.

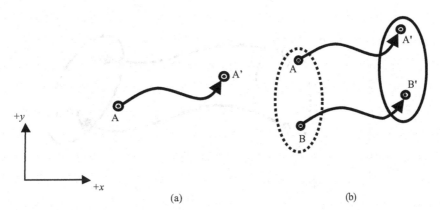

Figure 1.2: Curvilinear translation of (a) a particle and (b) a rigid body

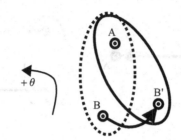

Figure 1.3: Rotation of a rigid body

- **Translation** is a type of motion, in which all points in the body move with the same displacement and in the same direction. There are two types of translation: rectilinear translation (also called rectilinear motion) and curvilinear translation (also called curvilinear motion). In a rectilinear motion, all points in the body move together in parallel and straight lines as shown in Figure 1.1. Rectilinear motion is considered as a one-dimensional translation, i.e. only one axis is required to define the motion. In Figure 1.1, points A and B are both translated with the same displacement in the x direction.
- In a **curvilinear motion**, all points in the body move together in curved paths as shown in Figure 1.2. Curvilinear motion can be considered as a two-dimensional plane motion or as a three-dimensional motion. Figure 1.2 shows a two-dimensional curvilinear motion, where points A and B are translated in both x and y directions.
- **Rotation** is a type of motion in which all points in the body rotate around a point, called a centre of rotation. In Figure 1.3, point B rotates about the centre of rotation A. The distances AB and AB′ are equal. A rigid body can undergo both translation and rotation. In Figure 1.4, point B moves in a curvilinear translation similar to point A and rotates to B′.

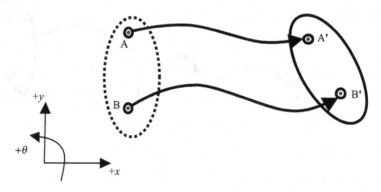

Figure 1.4: Combination of translation and rotation of a rigid body

1.1.2 Origin, co-ordinate systems and sign conventions

In order to measure motion, an origin, a co-ordinate system and a sign convention are required. The origin is a reference point, from which motion is measured. The co-ordinate system comprises reference axes that indicate the positive directions through sign conventions. The origin, the sign convention and the co-ordinate system depend on the type of motion being analyzed. For a rectilinear motion, a one-dimensional co-ordinate system is enough to describe motion as shown in Figure 1.5.

Figure 1.5: A one-dimensional co-ordinate system (rectilinear motion)

For a curvilinear motion in a plane (i.e. two-dimensional motion), three co-ordinate systems can be defined:

- The Cartesian, or rectangular, co-ordinate system (x, y) shown in Figure 1.6(a) is basically a combination of two one-dimensional rectilinear co-ordinate systems perpendicular to one another.
- The polar, or cylindrical, co-ordinate system (r, θ) shown in Figure 1.6(b) has the co-ordinate r in the radial direction and the co-ordinate θ in the transverse direction.

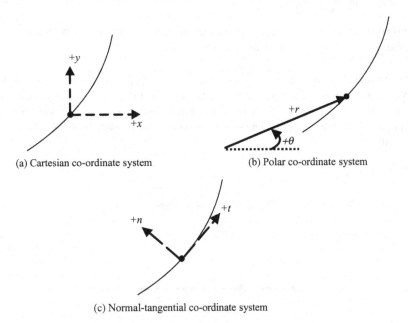

(a) Cartesian co-ordinate system

(b) Polar co-ordinate system

(c) Normal-tangential co-ordinate system

Figure 1.6: Two-dimensional co-ordinate systems (curvilinear motion)

- The normal-tangential, or normal-intrinsic, co-ordinate system (n, t) shown in Figure 1.6(c) has the co-ordinate n in the normal direction and the co-ordinate t in the transverse direction to the body motion path.

Cartesian and polar co-ordinate systems have their origin fixed to the ground or to the Earth, while the normal-tangential co-ordinate system has its origin fixed to and moving with the body under consideration.

1.2 RECTILINEAR MOTION

Consider a particle in a rectilinear motion, which is moving along a straight line from A to A' as shown in Figure 1.7. The displacement (s) is calculated from the origin at point O. A one-dimensional co-ordinate system is considered as the x-axis and a sign convention is defined such that the positive motion is in the positive x direction.

The co-ordinate of point A at any instant of time defines its displacement from the origin O along the x-axis. If the particle moves from A to A', a displacement equivalent to Δs, in a period of time Δt, its average velocity is estimated as $v_{av} = \frac{\Delta s}{\Delta t}$. The instantaneous velocity, which is equal to $v = \frac{ds}{dt}$, approaches the average velocity when Δt is an infinitesimally small time interval $(\Delta t \to 0)$, i.e.

$$v = \lim_{\Delta t \to 0} \frac{\Delta s}{\Delta t} = \frac{ds}{dt} \tag{1.1}$$

It should be noted that velocity is a vector, which has magnitude and direction. The magnitude of velocity is a scalar and is known as speed.

If the instantaneous velocity of the particle in Figure 1.7 changes from v at A to $v + \Delta v$ at A', the average acceleration, which is defined as the rate of change of velocity, over the time interval Δt is $a_{av} = \frac{\Delta v}{\Delta t}$. Again, the instantaneous acceleration is equal to $a = \frac{dv}{dt}$ and approaches the average acceleration when Δt is an infinitesimally small time interval $(\Delta t \to 0)$, i.e.

$$a = \lim_{\Delta t \to 0} \frac{\Delta v}{\Delta t} = \frac{dv}{dt} \tag{1.2}$$

Substituting Equation (1.1) into Equation (1.2), gives:

$$a = \frac{d^2s}{dt^2} \tag{1.3}$$

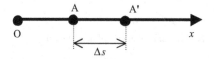

Figure 1.7: Motion of a particle

Eliminating dt between Equations (1.1) and (1.2) $\left(\dfrac{dv}{dt} \times \dfrac{ds}{ds} = \dfrac{ds}{dt} \times \dfrac{dv}{ds} = v\dfrac{dv}{ds}\right)$, yields:

$$a = v\frac{dv}{ds} \tag{1.4}$$

Similarly to displacement and velocity, acceleration is a vector, which has magnitude and direction.

Equations (1.1) to (1.4) are considered to be the basic differential equations of motion for particles in rectilinear motion. By integrating these equations, kinematic problems involving time, displacements, velocities and accelerations can be solved.

1.2.1 Motion with constant acceleration

In many practical engineering applications, bodies undergo motion with constant acceleration. Constant acceleration could be due to the application of external forces or free fall close to the Earth's surface. Consider again the particle shown in Figure 1.7; if it moves a displacement Δs from A to A' in a time Δt, with a constant acceleration a from an initial velocity u to a final velocity v, then from Equation (1.2), the velocity can be integrated as:

$$\int_{u}^{v} dv = \int_{0}^{t} a\,dt = a\int_{0}^{t} dt$$

Performing the integration gives:

$$v = u + at \tag{1.5}$$

And from Equation (1.4), the acceleration can be integrated as:

$$\int_{0}^{s} a\,ds = \int_{u}^{v} v\,dv \Rightarrow a\int_{0}^{s} ds = \int_{u}^{v} v\,dv$$

And again, performing the integration gives:

$$v^2 = u^2 + 2as \tag{1.6}$$

Similarly, substituting Equation (1.5) into Equation (1.1) and performing the integration over displacement and time, the following equation is obtained:

$$s = ut + \frac{1}{2}at^2 \tag{1.7}$$

And from Equations (1.5) and (1.7), the displacement s can be written as:

$$s = ut + \frac{1}{2}at^2 = \frac{t}{2}(2u + at) = \frac{t}{2}(u + u + at) = \frac{t}{2}(u + v)$$

Which gives the average displacement as:

$$s = \frac{1}{2}(u + v)t \tag{1.8}$$

Example 1.1 Rectilinear motion: constant acceleration I

A car travels with an initial velocity of 60 km/hr and accelerates on a straight road to 110 km/hr in 28 seconds as shown in Figure 1.8. Determine the acceleration of the car during this time interval and the distance travelled.

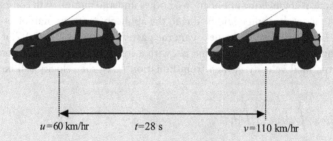

u=60 km/hr t=28 s v=110 km/hr

Figure 1.8: Representation of Example 1.1

Solution

The initial velocity in m/s is:

$$u = 60\,\text{km/hr} = 60 \times \frac{1000}{60 \times 60} = 16.667\,\text{m/s}$$

The final velocity in m/s is:

$$v = 110\,\text{km/hr} = 110 \times \frac{1000}{60 \times 60} = 30.5556\,\text{m/s}$$

To calculate the acceleration, using Equation (1.5) gives:

$$v = u + at \Rightarrow 30.5556 = 16.667 + a \times 28 \Rightarrow a = 0.496\,\text{m/s}^2$$

To calculate the distance s, using Equation (1.6) gives:

$$v^2 = u^2 + 2as \Rightarrow (30.5556)^2 = (16.667)^2 + 2 \times 0.496 \times s$$

$$\Rightarrow s = 661.143\,\text{m} = 661\,\text{m}$$

Example 1.2 Rectilinear motion: constant acceleration II

A particle moves along a straight line with a constant velocity of 40 m/s when it passes a point A until it reaches point B, which is 100 m away from A as shown in Figure 1.9. From point B on, it travels with a constant deceleration of 8 m/s^2, until it reaches a turning point C, then it returns to point A. Determine the time and the distance travelled from point B to the turning point C, and the time that the particle takes to return from point C to point A and its velocity when it reaches A.

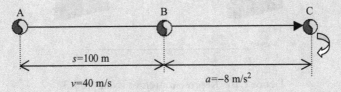

Figure 1.9: Representation of Example 1.2

Solution

At the turning point, the final velocity $v = 0$, the initial velocity $u = 40$ m/s and the acceleration $a = -8$ m/s^2. Using Equation (1.6), the maximum distance (from point B to C) is obtained as:

$$v^2 = u^2 + 2as \Rightarrow 0 = (40)^2 + 2 \times (-8) \times s_{max} \Rightarrow s_{max} = 100\,\text{m}$$

The time to turning point C (from point B to C) is obtained from Equation (1.5) as:

$$v = u + at \Rightarrow 0 = 40 - 8 \times t \Rightarrow t = 5\,\text{s}$$

The distance from A to the turning point C = AB + BC = 100 + 100 = 200 m

To calculate the time to travel back to A, using the initial velocity (point C) $u = 0$, acceleration $a = -8$ m/s^2, the distance $s = -200$ m and Equation (1.7), the time is obtained as:

$$s = ut + \frac{1}{2}at^2 \Rightarrow -200 = 0 + \frac{1}{2} \times (-8) \times (t)^2 = 7.07\,\text{s} = 7.1\,\text{s}$$

And from Equation (1.5), the particle's velocity when it returns to point A is calculated as:

$$v = u + at \Rightarrow v = 0 - 8 \times 7.07 \Rightarrow v = 56.57\,\text{m/s}$$

Example 1.3 Rectilinear motion: constant acceleration III

The acceleration of a high-speed train, shown in Figure 1.10, during the interval of time from 5 s to 10 s is 2 m/s^2 and at time 5 s its velocity is 180 km/hr. Use the equations of rectilinear translation with constant acceleration to determine:

a) the train velocity in km/hr at time 10 s;
b) the train displacement in meters from time 5 s to time 10 s.

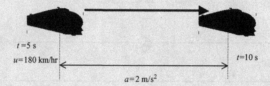

Figure 1.10: Representation of Example 1.3

Solution

a) The train velocity in m/s at $t = 5$ s is:

$$u = 180 \, \text{km/hr} = 180 \times \frac{1000}{60 \times 60} = 50 \, \text{m/s}$$

To calculate the velocity at $t = 10$ s in the time interval (5 s to 10 s), Equation (1.5) is applied as follows:

$$v = u + at \Rightarrow v = 50 + 2 \times (10 - 5) \Rightarrow v = 60 \, \text{m/s} = 60 \times \frac{60 \times 60}{1000}$$

$$= 216 \, \text{km/hr}$$

b) The train displacement after 5 s (i.e. at $t = 10$ s), with initial velocity $u = 50$ m/s and final velocity $v = 60$ m/s, is obtained from Equation (1.6) as:

$$v^2 = u^2 + 2as \Rightarrow (60)^2 = (50)^2 + 2 \times 2 \times s \Rightarrow s = 275 \, \text{m}$$

1.2.2 Motion involving non-constant acceleration

In some cases, bodies move with non-constant accelerations due to external forces and particular constraints. Displacements, velocities and accelerations may vary as functions of time as shown in the example below, where the displacement is expressed as a function of time:

$$s = 3t^3 + 2t^2 + t \tag{1.9}$$

Or, possibly, acceleration could be expressed as a function of displacement:

$$a = 3s^3 + 2s^2 \tag{1.10}$$

In such cases, the basic differential equations of motion for particles in rectilinear motion should be used and appropriate differentiation or integration should be performed (Equations (1.1) to (1.4)).

Example 1.4 Rectilinear motion: displacement as a function of time I

The motion of a particle is described as:

$$s = t^3 - 10.5t^2 + 30t + 4$$

where s is the displacement in meters and t is the time in seconds.

Determine the total distance travelled between $t = 0$ and $t = 4$ s.

Solution

Displacements at start $t = 0$ and end $t = 4$ s are given by:

$$t = 0 \Rightarrow s = 0^3 - 10.5 \times 0^2 + 30 \times 0 + 4 = 4\,\text{m}$$
$$t = 4 \Rightarrow s = 4^3 - 10.5 \times 4^2 + 30 \times 4 + 4 = 20\,\text{m}$$

The time to the turning point is obtained by equating the velocity to zero, i.e. $v = 0$ (see Figures 1.11 and 1.12) as:

$$v = \frac{ds}{dt} = 3t^2 - 21t + 30 = 0 \Rightarrow t^2 - 7t + 10 = 0$$

from which

$$(t - 2)(t - 5) = 0 \Rightarrow t = 2s, t = 5s$$

Only $t = 2$ s is possible in the time interval $0 \leq t \leq 4$ s.

The displacement at $t = 2$ s is calculated as:

$$s = 2^3 - 10.5 \times 2^2 + 30 \times 2 + 4 = 30\,\text{m}$$

The total distance travelled is $30 - 4 + 10 = 36$ m.

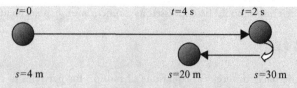

Figure 1.11: Representation of Example 1.4

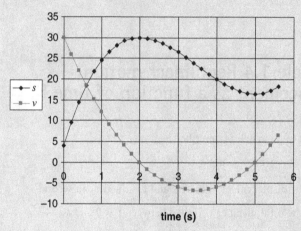

Figure 1.12: Graph of Example 1.4

Example 1.5 Rectilinear motion: acceleration as a function of displacement

A rocket, shown in Figure 1.13, travels with an upwards acceleration of $a = 5.9 + 0.019s$, where a is in m/s^2 and s is the vertical distance in meters. Determine the vertical distance travelled by the rocket when its velocity is 500 m/s. Use initial velocity $v = 0$ at $s = 0$.

Solution

Since the acceleration is a function of displacement, using Equation (1.4) to relate acceleration to velocity and displacement gives:

$$a = v\frac{dv}{ds} = (5.9 + 0.019s) \Rightarrow vdv = (5.9 + 0.019s)ds$$

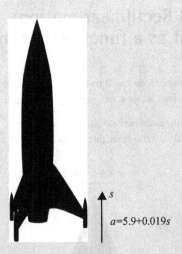

$a = 5.9 + 0.019s$

Figure 1.13: Representation of Example 1.5

integrating $\int v\,dv = \int (5.9 + 0.019s)\,ds$, gives:

$$\frac{v^2}{2} = 5.9s + 0.019\frac{s^2}{2} + C$$

Using the initial condition $v = 0$ at $s = 0 \Rightarrow C = 0$, the following equation is obtained:

$$v^2 = 11.8s + 0.019s^2$$

when $v = 500$ m/s, it gives:

$$500^2 = 11.8 \times s + 0.019 \times s^2$$

which can be written as:

$$0.019s^2 + 11.8s - 250\,000 = 0$$

Solving for s using the quadratic formula gives:

$$s = \frac{-11.8 \pm \sqrt{11.8^2 - 4 \times 0.019 \times (-250\,000)}}{2 \times 0.019} = +3330.13 \text{ or } -3951.17$$

Since the rocket travels upwards, only positive displacement is acceptable and the travelled distance is:

$$s = 3330.13\,\text{m} = 3.33\,\text{km}$$

Example 1.6 Rectilinear motion: displacement as a function of time II

During the interval of time from $t = 0$ to $t = 20$ s, the boat's displacement, shown in Figure 1.14, is given by $s = 8t + 6t^2 - 0.02t^3$. Determine:

a) the boat's velocity and acceleration at $t = 5$ s;
b) the maximum velocity of the boat during this interval of time.

$$s = 8t + 6t^2 - 0.02t^3$$

Figure 1.14: Representation of Example 1.6

Solution

a) Differentiating the displacement with respect to time gives the velocity as:

$$v = \frac{ds}{dt} = (8 + 12t - 0.06t^2)$$

Differentiating again with respect to time, the acceleration is obtained as:

$$a = \frac{dv}{dt} = (12 - 0.12t)$$

For $t = 5$ s, the velocity and acceleration are:

$$v = 8 + 12 \times 5 - 0.06 \times 5^2 = 66.5 \, \text{m/s}$$
$$a = 12 - 0.12 \times 5 = 11.4 \, \text{m/s}^2$$

b) Since the maximum velocity takes place at $a = 0$, the time at maximum velocity is calculated as:

$$a = \frac{dv}{dt} = (12 - 0.12t) = 0 \Rightarrow t = 10\,\text{s}$$

And the maximum velocity is given by:

$$v = 8 + 12 \times 10 - 0.06 \times 10^2 = 122\,\text{m/s}$$

1.3 CURVILINEAR MOTION

Curvilinear motion is involved when a particle moves in a two-dimensional plane or a three-dimensional space rather than along a straight line. If the particle is restricted to move only in the x-y plane, then the motion is called a plane curvilinear motion. A three-dimensional motion is considered as an extension of plane motion by superimposing an additional one-dimensional rectilinear co-ordinate system. This section is devoted to two-dimensional curvilinear motion. As mentioned in Section 1.1, three co-ordinate systems can be used: Cartesian (x, y), polar (r, θ) and normal-tangential (n, t). Figure 1.6 shows how the position of a body can be specified in the three different co-ordinate systems. The (x, y) and (r, θ) are fixed whilst the (n, t) system is attached to the body. All three co-ordinate systems can be used to deal with any two-dimensional curvilinear motion problem and should lead to the same answer. However, a specific co-ordinate system could have a certain advantage, e.g. in providing a simplified solution for a particular problem.

1.3.1 Cartesian co-ordinate system

The Cartesian co-ordinate system has two rectilinear co-ordinate axes, x and y, perpendicular to one another at a right angle as shown in Figure 1.15. The particle displacement is defined by two co-ordinates x and y, which are measured from the origin in the x and y directions, respectively. Sign conventions that indicate the positive directions are defined in both x and y co-ordinate axes. The components of velocity and acceleration in the x and y directions are obtained by differentiating the displacement components in the x and y directions with respect to time. The magnitude and direction of velocity and acceleration are determined using the vector summation of the x and y components of velocity and acceleration, respectively.

Derivation of velocity and acceleration components

Consider the Cartesian co-ordinate system in Figure 1.15. The position of a particle moving with a plane curvilinear motion is given by x and y co-ordinates. The velocity components in the x and y directions are denoted as $\dot{x}$ (or v_x) and $\dot{y}$ (or v_y), respectively. Similarly, the acceleration components in the x and y directions are denoted as $\ddot{x}$ (or a_x) and $\ddot{y}$ (or a_y), respectively.

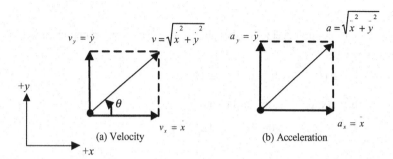

Figure 1.15: Plane curvilinear motion in the Cartesian co-ordinate system

If a particle moves in the x direction a displacement of Δx over a time interval Δt, the instantaneous velocity in the x direction is given by:

$$v_x = \lim_{\Delta t \to 0} \frac{\Delta x}{\Delta t} = \frac{dx}{dt} = \dot{x} \tag{1.11}$$

Similarly, if a particle moves in the y direction a displacement of Δy over a time interval Δt, the instantaneous velocity in the y direction is:

$$v_y = \lim_{\Delta t \to 0} \frac{\Delta y}{\Delta t} = \frac{dy}{dt} = \dot{y} \tag{1.12}$$

The total velocity magnitude is obtained through vector summation of the two velocity components, Equations (1.11) and (1.12):

$$v = \sqrt{\left(\frac{dx}{dt}\right)^2 + \left(\frac{dy}{dt}\right)^2} = \sqrt{\dot{x}^2 + \dot{y}^2} \tag{1.13}$$

If the velocity of the particle in the x direction varies by Δv_x over a time interval Δt, the instantaneous acceleration in the x direction is given by:

$$a_x = \lim_{\Delta t \to 0} \frac{\Delta v_x}{\Delta t} = \frac{dv_x}{dt} = \frac{d^2x}{dt^2} = \ddot{x} \tag{1.14}$$

Similarly, if the velocity of the particle in the y direction varies by Δv_y over a time interval Δt, the instantaneous acceleration in the y direction is given by:

$$a_y = \lim_{\Delta t \to 0} \frac{\Delta v_y}{\Delta t} = \frac{dv_y}{dt} = \frac{d^2y}{dt^2} = \ddot{y} \tag{1.15}$$

The total acceleration magnitude is obtained through vector summation of the two acceleration components, Equations (1.14) and (1.15):

$$a = \sqrt{\left(\frac{d^2x}{dt^2}\right)^2 + \left(\frac{d^2y}{dt^2}\right)^2} = \sqrt{\ddot{x}^2 + \ddot{y}^2} \tag{1.16}$$

Figure 1.15 summarizes the velocity and acceleration components in the Cartesian co-ordinate system. It should be noted that the resultant velocity direction is always tangential to the instantaneous path of the particle. Thus, the instantaneous path of the particle $\tan\theta = \frac{dy}{dx}$ is equivalent to the resultant velocity direction as follows:

$$\tan\theta = \frac{dy}{dx} = \frac{dy}{dt} \times \frac{dt}{dx} = \frac{v_y}{v_x} \tag{1.17}$$

Example 1.7 Curvilinear motion: Cartesian co-ordinate system I

A particle moves in the x-y plane such that:

$$x = 2 + 4t + 6t^2 - 8t^3, \quad y = 3 + 9t + 20t^2 - 4t^3$$

where x and y are the displacement components (in meters) in the x and y directions, respectively, and t is the time in seconds.

When $t = 2$ s, determine:

a) the velocity of the particle in the x and y directions and its total magnitude;
b) the acceleration of the particle in the x and y directions and its total magnitude;
c) the instantaneous direction of the particle motion.

Solution

a) Differentiating x and y with respect to time to derive velocity components, gives:

$$\dot{x} = 4 + 12t - 24t^2, \quad \dot{y} = 9 + 40t - 12t^2$$

When $t = 2$ s

$$\dot{x} = 4 + 12 \times 2 - 24 \times 2^2 = -68 \, \text{m/s}, \quad \dot{y} = 9 + 40 \times 2 - 12 \times 2^2 = 41 \, \text{m/s}$$

The total magnitude of velocity is obtained from Equation (1.13) as:

$$v = \sqrt{\dot{x}^2 + \dot{y}^2} = \sqrt{(-68)^2 + (41)^2} = 79.40 \, \text{m/s}$$

b) Differentiating $\dot{x}$ and $\dot{y}$ with respect to time to derive acceleration components, gives:

$$\ddot{x} = 12 - 48t, \quad \ddot{y} = 40 - 24t$$

When $t = 2$ s

$$\ddot{x} = 12 - 48 \times 2 = -84\,\text{m/s}^2, \quad \ddot{y} = 40 - 24 \times 2 = -8\,\text{m/s}^2$$

The total magnitude of acceleration is obtained from Equation (1.16) as:

$$a = \sqrt{\ddot{x}^2 + \ddot{y}^2} = \sqrt{(-84)^2 + (-8)^2} = 84.38\,\text{m/s}^2$$

c) From Equation (1.17) and Figure 1.16, the instantaneous direction is given by:

$$\tan\theta' = \frac{\dot{y}}{\dot{x}} = \frac{41}{-68} = -0.6029 \Rightarrow \theta' = -31.09°$$

$\theta = 180 - 31.09 = 148.91°$ anticlockwise as shown in Figure 1.16.

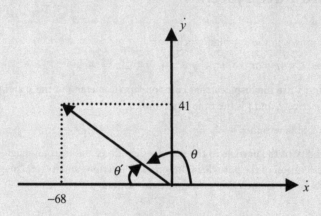

Figure 1.16: Components of Example 1.7

Example 1.8 Curvilinear motion: Cartesian co-ordinate system II

A rocket travels vertically up to an altitude of 52 m, as shown in Figure 1.17, then travels along a path such that $y = 52 + 13\sqrt{x}$, where x and y are in meters. The vertical component of the rocket velocity is constant and equal to 220 m/s. Determine the magnitude of the rocket acceleration.

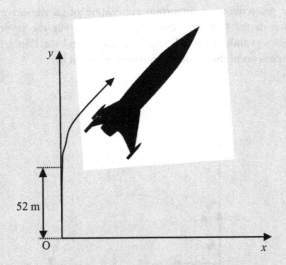

Figure 1.17: Representation of Example 1.8

Solution

Re-arrange the co-ordinate relationship $y = 52 + 13\sqrt{x}$ so that:

$$(y - 52)^2 = 169x$$

Differentiate with respect to time in order to derive velocity, gives:

$$2(y - 52)\dot{y} = 169\dot{x}$$
$$2\dot{y}y - 104\dot{y} = 169\dot{x}$$

Differentiating again with respect to time to derive acceleration, gives:

$$2\ddot{y}y + 2\dot{y}^2 - 104\ddot{y} = 169\ddot{x}$$

Since $\dot{y}$ is constant, $\ddot{y} = 0$ and $\ddot{x}$ becomes:

$$\ddot{x} = \frac{2\dot{y}^2}{169}$$
$$\ddot{x} = \frac{2 \times 220^2}{169} = 572.78 \, \text{m/s}^2$$

The total magnitude of acceleration is obtained from Equation (1.16) as:

$$a = \sqrt{\ddot{x}^2 + \ddot{y}^2} = \sqrt{(572.78)^2 + (0)^2} = 572.78 \, \text{m/s}^2$$

Motion of a projectile

The motion of a projectile is an important application of curvilinear motion that can be easily analyzed in the Cartesian co-ordinate system. For the projectile shown in Figure 1.18, if aerodynamic drag and earth's curvature and rotation are neglected, the acceleration components in the x and y directions are given by:

$$\ddot{x} = 0 \tag{1.18}$$

$$\ddot{y} = -g \tag{1.19}$$

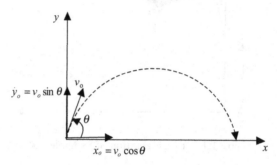

Figure 1.18: Motion of a projectile

Therefore, only a constant acceleration in the y direction, which is equal to g and acts downwards, takes place. The velocity components are obtained by integrating Equations (1.18) and (1.19) over time using initial conditions of $t = 0$, $\dot{x} = \dot{x}_o$ and $\dot{y} = \dot{y}_o$ to give:

$$\dot{x} = \dot{x}_o \tag{1.20}$$

$$\dot{y} = \dot{y}_o - gt \tag{1.21}$$

where $\dot{x}_o$ and $\dot{y}_o$ are the initial velocity components in the x and y directions, respectively. Integrating Equations (1.20) and (1.21) over time and using initial conditions of $t = 0$, $x = x_o$ and $y = y_o$ gives:

$$x = x_o + \dot{x}_o t \tag{1.22}$$

$$y = y_o + \dot{y}_o t - \frac{1}{2}gt^2 \tag{1.23}$$

where x_o and y_o are the initial co-ordinates of the projectile. Applying Equation (1.6), for motion with constant acceleration, in the y direction yields (use $a = -g$ and $s = (y - y_o)$):

$$\dot{y}^2 = \dot{y}_o^2 - 2g(y - y_o) \tag{1.24}$$

If the velocity of the projectile v_o makes an angle θ with the horizontal, the velocity components are obtained as:

$$\dot{x}_o = v_o \cos\theta \tag{1.25}$$

$$\dot{y}_o = v_o \sin\theta \tag{1.26}$$

At the maximum height (y_{max}), the velocity component of the projectile in the y direction vanishes, i.e. $\dfrac{dy}{dt} = \dot{y} = 0$.

Example 1.9 Curvilinear motion: projectile I

A racing motorcycle jumps off a ramp of 0.8 m height, which has a slope of 30° with the horizontal as shown in Figure 1.19. If the motorcycle remains in the air for 1.2 seconds during the jump, determine:

a) the initial speed, v_o;
b) the horizontal distance (L);
c) the maximum height (h).

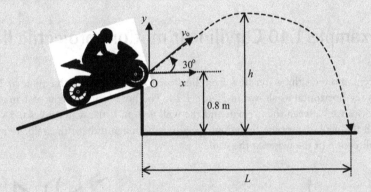

Figure 1.19: Representation of Example 1.9

Solution

From Equation (1.26), the velocity component in the y direction is:

$$\dot{y}_o = v_o \sin 30$$

a) To calculate the initial speed (v_o), substitute the velocity component in the y direction into Equation (1.23), taking the origin at point O, i.e. $y = -0.8$ m, $y_o = 0$ and using $t = 1.2$ s gives:

$$y = y_o + \dot{y}_o t - \frac{1}{2}gt^2 \Rightarrow -0.8 = 0 + v_o \sin 30 \times 1.2 - \frac{1}{2} \times 9.81 \times 1.2^2$$

The initial speed is calculated as $v_o = 10.44$ m/s.

b) From Equation (1.22), for $x = L$, $x_o = 0$ (origin at O), $v_o = 10.44$ m/s, $\dot{x}_o = v_o \cos 30$ (from Equation (1.25)) and $t = 1.2$ s, the horizontal distance (L) is obtained as:

$$x = x_o + \dot{x}_o t \Rightarrow L = 0 + 10.44 \times \cos 30 \times 1.2 = 10.85 \, \text{m}$$

c) To calculate the maximum height (h), substitute the velocity component in the y direction into Equation (1.24) and use $\dot{y} = 0$, $y = (h - 0.8)$ m and $y_o = 0$ to yield:

$$\dot{y}^2 = \dot{y}_o^2 - 2g(y - y_o) \Rightarrow 0 = (10.44 \times \sin 30)^2 - 2 \times 9.81 \times (h - 0.8 - 0)$$

The maximum height is $h = 2.19$ m

Example 1.10 Curvilinear motion: projectile II

A fireman directs the water flow from his hose towards a wall at an angle of 20° with the horizontal as shown in Figure 1.20. The hose is at a height of 1 m and the distance between the fireman and the wall is 9 m. If the water takes 0.5 s to reach the wall, determine the initial speed v_o, the magnitude of the velocity and the direction of the water at the wall.

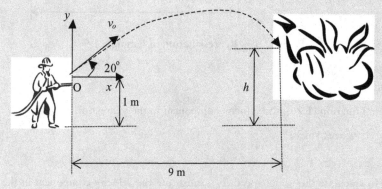

Figure 1.20: Representation of Example 1.10

Solution

Taking the origin at point O, i.e. $x_o = 0$ (at the hose), and $x = 9$ m (at the wall), and using Equations (1.22) and (1.25) gives:

$$x = x_o + \dot{x}_o t \Rightarrow x = x_o + v_o \cos \theta \times t \Rightarrow 9 = 0 + v_o \times \cos 20 \times 0.5$$

From which, the initial speed is calculated as:

$$v_o = 19.155 \, \text{m/s}$$

The water velocity in the y direction at the wall is obtained from Equation (1.21):

$$\dot{y} = \dot{y}_o - gt \Rightarrow \dot{y} = 19.155 \times \sin 20 - 9.81 \times 0.5 = 1.646 \, \text{m/s}$$

The water velocity in the x direction at the wall is obtained from Equation (1.20):

$$\dot{x} = \dot{x}_o = 19.155 \times \cos 20 = 18 \, \text{m/s}$$

The total velocity magnitude is:

$$v = \sqrt{\dot{x}^2 + \dot{y}^2} = \sqrt{(18)^2 + (1.646)^2} = 18.075 \, \text{m/s}$$

The angle that the water at the wall makes with the horizontal is calculated from Equation (1.17):

$$\theta = \tan^{-1} \frac{\dot{y}}{\dot{x}} = \tan^{-1} \frac{1.646}{18} = 5.22°$$

1.3.2 Polar co-ordinate system

The polar co-ordinate system is another popular co-ordinate system that can be used to define the displacement, velocity and acceleration of a body in a two-dimensional plane. The polar co-ordinate system has two co-ordinate axes, a radius r and an angle θ, as shown in Figure 1.21. The particle displacement is defined by r, measured from the origin, and θ, measured anti-clockwise. Sign conventions that indicate the positive directions are defined in both the r and θ co-ordinate axes. The components of velocity and acceleration are obtained by differentiating the displacement components in the x and y directions with respect to time and using a co-ordinate transformation method to resolve them into the r and θ directions. The magnitude and direction of velocity and acceleration are determined using the vector summation of the r and θ components of velocity and acceleration, respectively.

Consider the polar co-ordinate system in Figure 1.21; the position of a particle moving with plane curvilinear motion is given by the r and θ co-ordinates. The velocity and acceleration components in the radial, r, and transverse, θ, directions are expressed in terms of (r, θ) and their first and second derivatives with respect to time.

In order to develop expressions for the radial and transverse components of the velocity and acceleration in polar co-ordinate system, the Cartesian co-ordinates x and y are first expressed in terms of r and θ. Next, by differentiating them with respect to time, the velocity components $(\dot{x}, \dot{y})$ and the acceleration components $(\ddot{x}, \ddot{y})$ can be obtained. Finally, by resolving these derivatives into the r and θ directions, the velocity component in the radial direction (v_r), the velocity component in the transverse direction (v_t), the

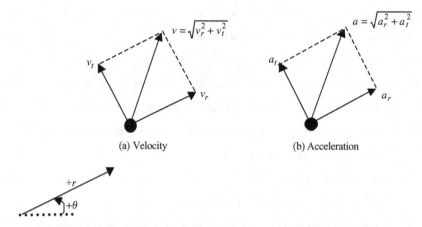

(a) Velocity   (b) Acceleration

Figure 1.21: Plane curvilinear motion in the polar co-ordinate system

acceleration component in the radial direction (a_r) and the acceleration component in the transverse direction (a_t) can be calculated. This process is called co-ordinate transformation, through which one co-ordinate system can be related to another.

The motion of the body being analyzed is unaffected by the chosen co-ordinate system. Using different co-ordinate systems simply expresses the same motion in different ways.

In Figure 1.22, the positions x and y of any point can be written in terms of r and θ as:

$$x = r \cos \theta \tag{1.27}$$
$$y = r \sin \theta \tag{1.28}$$

(Consider differentiation of the product of two functions, thus if $x = f(r, \theta)$ then $\frac{dx}{dt} = \theta \frac{df(r)}{dt} + r \frac{df(\theta)}{dt}$ and $\frac{df(\theta)}{dt} = \frac{df(\theta)}{d\theta} \times \frac{d\theta}{dt}$.) Differentiating Equations (1.27) and (1.28) with

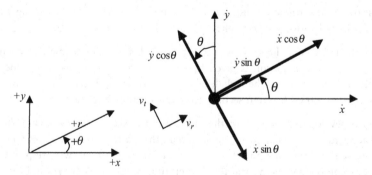

Figure 1.22: Resolving velocity components in the polar co-ordinate system

respect to time, the Cartesian velocity components are obtained as:

$$\dot{x} = \dot{r}\cos\theta - r\dot{\theta}\sin\theta \tag{1.29}$$
$$\dot{y} = \dot{r}\sin\theta + r\dot{\theta}\cos\theta \tag{1.30}$$

where $\dot{r} = \frac{dr}{dt}$ is the rate of change in radial displacement, i.e. radial velocity and $\dot{\theta} = \frac{d\theta}{dt}$ is the rate of change in the angle θ, i.e. angular velocity.

Resolving $\dot{x}$ and $\dot{y}$ in the radial direction, as shown in Figure 1.22, yields:

$$v_r = \dot{x}\cos\theta + \dot{y}\sin\theta \tag{1.31}$$

Substituting Equations (1.29) and (1.30) into Equation (1.31) and simplifying, gives:

$$v_r = \dot{r} \tag{1.32}$$

Similarly, resolving x and y in the transverse direction, as shown in Figure 1.22, yields:

$$v_t = \dot{y}\cos\theta - \dot{x}\sin\theta \tag{1.33}$$

And again by substituting Equations (1.29) and (1.30) into Equation (1.33) and simplifying, gives:

$$v_t = r\dot{\theta} \tag{1.34}$$

In order to derive the acceleration components in the Cartesian co-ordinate system, differentiating Equations (1.29) and (1.30) with respect to time, gives:

$$\ddot{x} = \frac{d\dot{x}}{dt} = (\ddot{r}\cos\theta - \dot{r}\dot{\theta}\sin\theta) - \dot{r}\dot{\theta}\sin\theta - r(\ddot{\theta}\sin\theta + \dot{\theta}^2\cos\theta) \tag{1.35}$$

$$\ddot{y} = \frac{d\dot{y}}{dt} = (\ddot{r}\sin\theta + \dot{r}\dot{\theta}\cos\theta) + \dot{r}\dot{\theta}\cos\theta + r(\ddot{\theta}\cos\theta - \dot{\theta}^2\sin\theta) \tag{1.36}$$

where $\ddot{r} = \frac{d\dot{r}}{dt}$ is the rate of change in radial velocity and $\ddot{\theta} = \frac{d\dot{\theta}}{dt}$ is the rate of change in angular velocity, i.e. angular acceleration.

Resolving x and y in the radial direction as shown in Figure 1.23, similar to the radial velocity case, gives:

$$a_r = \ddot{x}\cos\theta + \ddot{y}\sin\theta \tag{1.37}$$

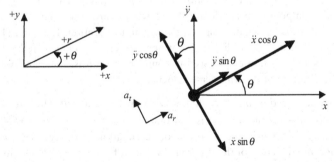

Figure 1.23: Resolving acceleration components in the polar co-ordinate system

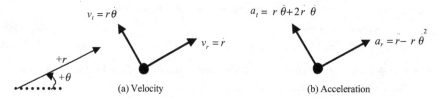

Figure 1.24: Velocity and acceleration components in the polar co-ordinate system

Resolving x and y in the transverse direction as shown in Figure 1.23, similarly to the transverse velocity case, gives:

$$a_t = \ddot{y}\cos\theta - \ddot{x}\sin\theta \qquad (1.38)$$

Substituting Equations (1.35) and (1.36) into Equation (1.37), the acceleration in the radial direction is obtained as:

$$a_r = \ddot{r} - r\dot{\theta}^2 \qquad (1.39)$$

Similarly, substituting Equations (1.35) and (1.36) into Equation (1.38), the acceleration in the transverse direction is obtained as:

$$a_t = r\ddot{\theta} + 2\dot{r}\dot{\theta} \qquad (1.40)$$

Figure 1.24 shows a summary diagram for the velocity and acceleration components in the polar co-ordinate system.

The total velocity magnitude is obtained through vector summation of the radial and transverse velocity components (Equations (1.32) and (1.34)) as:

$$v = \sqrt{v_r^2 + v_t^2} \qquad (1.41)$$

Similarly, the total acceleration magnitude is obtained through vector summation of the radial and transverse acceleration components (Equations (1.39) and (1.40)) as:

$$a = \sqrt{a_r^2 + a_t^2} \qquad (1.42)$$

The radial acceleration, a_r, contains two terms, $\ddot{r}$ and $-r\dot{\theta}^2$. The rate of change of radial velocity, $\ddot{r}$, i.e. $\ddot{r} = \frac{d}{dt}(\dot{r}) = \frac{d}{dt}(v_r)$, is shown in Figure 1.25(a). The term $-r\dot{\theta}^2$ is called the centripetal component of acceleration and acts towards the origin of the polar co-ordinate system (it has a negative sign in the radial acceleration, Equation (1.39)). The centripetal component of the radial acceleration represents the change in the direction of the transverse velocity, $r\dot{\theta}$, as can be seen in Figure 1.25(b). It is concerned only with the change in the direction of the transverse velocity.

The transverse acceleration, a_t, also contains two terms, $r\ddot{\theta}$ and $2\dot{r}\dot{\theta}$. The rate of change of transverse velocity when the radius r is constant, $r\ddot{\theta}$, i.e. $\frac{d}{dt}v_t = \frac{d}{dt}(r\dot{\theta}) = .r\frac{d(\dot{\theta})}{dt}$, is shown in Figure 1.25(c). The Coriolis component of acceleration, $2\dot{r}\dot{\theta}$, represents a change in the magnitude and direction of the transverse velocity, from $r_1\dot{\theta}$ to $r_2\dot{\theta}$, as can be seen in Figure 1.25(d). The Coriolis component of the transverse acceleration is concerned with the change in both magnitude and direction of the transverse velocity.

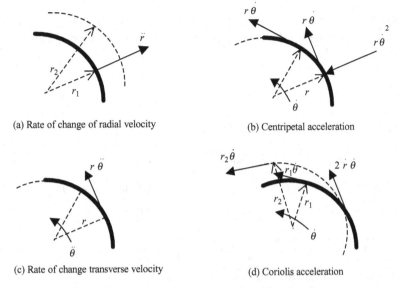

(a) Rate of change of radial velocity

(b) Centripetal acceleration

(c) Rate of change transverse velocity

(d) Coriolis acceleration

Figure 1.25: Components of acceleration

Example 1.11 Curvilinear motion: polar co-ordinate system I

A hydraulic cylinder of length $L_1 = 200$ mm rotates, anti-clockwise, around point O, as shown in Figure 1.26. The length of the piston rod (L_2) varies according to the action of oil pressure in the cylinder and is given by $L_2 = 2t^3$, where L_2 is in

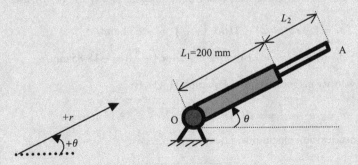

Figure 1.26: Representation of Example 1.11

mm and t is in seconds. If the cylinder angular position varies as $\theta = 0.5 \cos \frac{\pi t}{4}$ (where θ is in radians), calculate the magnitudes of the velocity and the acceleration at A when $t = 2$ s.

Solution

The co-ordinates of point A are given by:

$$r = 200 + 2t^3, \quad \theta = 0.5 \cos \frac{\pi t}{4}$$

Differentiating with respect to time to derive velocities and accelerations, gives:

$$\dot{r} = 6t^2, \quad \dot{\theta} = -0.5 \times \frac{\pi}{4} \sin \frac{\pi t}{4} = -\frac{\pi}{8} \sin \frac{\pi t}{4}$$

$$\ddot{r} = 12t, \quad \ddot{\theta} = -\frac{\pi}{8} \times \frac{\pi}{4} \cos \frac{\pi t}{4} = -\frac{\pi^2}{32} \cos \frac{\pi t}{4}$$

When $t = 2$ s, the radius r becomes:

$$r = 200 + 2 \times 2^3 = 216 \, \text{mm}$$

And the radial and angular velocities and their rates are calculated as:

$$\dot{r} = 6 \times 2^2 = 24 \, \text{mm/s}, \quad \dot{\theta} = -\frac{\pi}{8} \sin \frac{\pi \times 2}{4} = -\frac{\pi}{8} \, \text{rad/s}$$

$$\ddot{r} = 12 \times 2 = 24 \, \text{mm/s}, \quad \ddot{\theta} = -\frac{\pi^2}{32} \cos \frac{\pi \times 2}{4} = 0$$

Radial and transverse velocities, from Equations (1.32) and (1.34), are:

$$v_r = \dot{r} = 24 \, \text{mm/s}, \quad v_t = r \dot{\theta} = 216 \times \frac{-\pi}{8} = -84.823 \, \text{mm/s}$$

Radial and transverse accelerations, from Equations (1.39) and (1.40), are:

$$a_r = \ddot{r} - r\dot{\theta}^2 = 24 - 216 \times \left(-\frac{\pi}{8}\right)^2 = -9.31 \, \text{mm/s}^2$$

$$a_t = r \ddot{\theta} + 2\dot{r} \dot{\theta} = 216 \times 0 + 2 \times 24 \times \left(-\frac{\pi}{8}\right) = -18.85 \, \text{mm/s}^2$$

Total velocity magnitude, from Equation (1.41), is:

$$v = \sqrt{v_r^2 + v_t^2} == \sqrt{(24)^2 + (-84.823)^2} = 88.15 \, \text{mm/s}$$

Total acceleration magnitude, Equation (1.42), is:

$$a = \sqrt{a_r^2 + a_t^2} = \sqrt{(-9.31)^2 + (-18.85)^2} = 21.02 \, \text{mm/s}^2$$

Example 1.12 Curvilinear motion: polar co-ordinate system II

A ball moves in a two-dimensional plane following the path shown in Figure 1.27. At the position shown, the ball has a radial velocity $\dot{r} = 1.5$ m/s and angular velocity $\dot{\theta} = 25$ °/s. The rate of change in radial velocity ($\ddot{r}$) is 2.5 m/s² and the angular acceleration ($\ddot{\theta}$) is 45 °/s². At the position shown, determine the velocity and acceleration components in the Cartesian co-ordinate system.

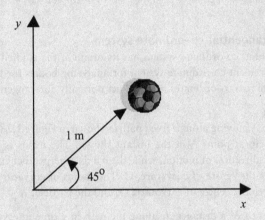

Figure 1.27: Representation of Example 1.12

Solution

Using Equations (1.32) and (1.34), the radial and transverse velocities are calculated as:

$$v_r = \dot{r} = 1.5\,\text{m/s}$$
$$v_t = r\dot{\theta} = 1 \times \left(25 \times \frac{2\pi}{360}\right) = 0.4363\,\text{m/s}$$

Referring to Figure 1.22 and resolving the radial and transverse velocities in the x and y directions, the velocity components in the Cartesian co-ordinate system are:

$$\dot{x} = v_r \cos\theta - v_t \sin\theta = 1.5 \times \cos 45 - 0.4363 \times \sin 45 = 0.752\,\text{m/s}$$
$$\dot{y} = v_r \sin\theta + v_t \cos\theta = 1.5 \times \sin 45 + 0.4363 \times \cos 45 = 1.37\,\text{m/s}$$

The radial and transverse accelerations, Equations (1.39) and (1.40), are:

$$a_r = \ddot{r} - r\dot{\theta}^2 = 2.5 - 1 \times \left(25 \times \frac{2\pi}{360}\right)^2 = 2.3096 \, \text{m/s}^2$$

$$a_t = r\ddot{\theta} + 2\dot{r}\dot{\theta} = 1 \times \left(45 \times \frac{2\pi}{360}\right) + 2 \times 1.5 \times \left(25 \times \frac{2\pi}{360}\right) = 2.094 \, \text{m/s}^2$$

Referring to Figure 1.23 and resolving the radial and transverse acceleration in the x and y directions, the acceleration components in the Cartesian co-ordinate system are:

$$\ddot{x} = a_r \cos\theta - a_t \sin\theta = 2.3096 \times \cos 45 - 2.094 \times \sin 45 = 0.152 \, \text{m/s}^2$$

$$\ddot{y} = a_r \sin\theta + a_t \cos\theta = 2.3096 \times \sin 45 + 2.094 \times \cos 45 = 3.11 \, \text{m/s}^2$$

1.3.3 Normal–tangential co-ordinate system

The normal-tangential co-ordinate system has its origin at the studied body and moves with it. It is a convenient co-ordinate system for analyzing bodies for which the motion path is known. The two co-ordinates, n and t, act normally and tangentially to the path, respectively.

Consider a particle A moving along a fixed path as shown in Figure 1.28. The co-ordinate system has its origin at point A at the instant shown. The t axis is tangential to the particle path in the direction of motion, while the n axis is perpendicular to the t axis and is directed towards the centre of curvature, O. The radius of curvature, ρ, is defined as the distance between the centre of curvature, O, and the particle, A.

If particle A moves to A′ a distance ds along the path in a time interval dt as shown in Figure 1.29, the particle velocity v is given by:

$$v = \frac{ds}{dt} = \dot{s} \tag{1.43}$$

which has a direction tangential to the path as shown in Figure 1.30.

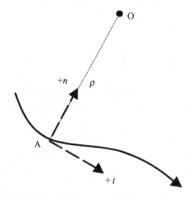

Figure 1.28: Normal-tangential co-ordinate system

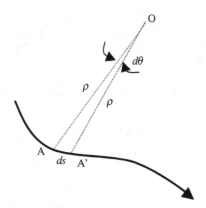

Figure 1.29: Curvilinear motion in the normal-tangential co-ordinate system

If the particle velocity changes by an amount dv in a time interval dt, the tangential acceleration a_t can be obtained as:

$$a_t = \frac{dv}{dt} = \dot{v} \tag{1.44}$$

which again acts tangentially to the path, as shown in Figure 1.30. To derive an expression for the normal acceleration, consider again Figure 1.29; if particle A moves to A′, the angle $d\theta$ between OA and OA′ is approximated as $d\theta = \tan^{-1}\frac{ds}{\rho}$. For a small angle $d\theta$, $\tan^{-1}\frac{ds}{\rho} \approx \frac{ds}{\rho}$, therefore $d\theta$ is given by:

$$d\theta = \frac{ds}{\rho} \Rightarrow ds = \rho d\theta \tag{1.45}$$

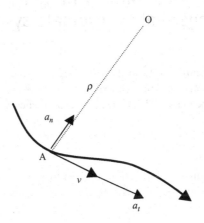

Figure 1.30: : Velocity and acceleration in the normal-tangential co-ordinate system

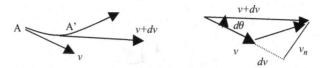

Figure 1.31: Normal acceleration in the normal-tangential co-ordinate system

Differentiating Equation (1.45) with respect to time, gives:

$$\frac{ds}{dt} = \rho \frac{d\theta}{dt} \Rightarrow v = \rho \dot{\theta} \Rightarrow \dot{\theta} = \frac{v}{\rho}$$

(1.46)

From the vector analysis in Figure 1.31, if point A moves to A', the angular displacement $d\theta$ is approximated by $\sin d\theta = \frac{v_n}{(v+dv)}$ and, for infinitesimally small $d\theta$, $\sin d\theta \approx d\theta$ and the product $dv \cdot d\theta$ is neglected so that the normal velocity is given by $v_n = v d\theta$. Therefore, the normal acceleration a_n is given by:

$$a_n = v \frac{d\theta}{dt}$$

(1.47)

Substituting Equation (1.46) into Equation (1.47), gives:

$$a_n = \frac{v^2}{\rho}$$

(1.48)

The total acceleration magnitude is obtained through vector summation of the normal and tangential acceleration components (Equations (1.44) and (1.48)) as:

$$a = \sqrt{a_t^2 + a_n^2}$$

(1.49)

Example 1.13 Curvilinear motion: normal-tangential co-ordinate system I

A motorcycle (see Figure 1.32) travels with a velocity of 2 m/s and acceleration of 0.2 m/s² when it starts the curve described by $y = 0.4\, x^2$, where y and x are in meters. Determine the magnitude of the acceleration when $x = 3$ m.

Solution

The velocity (in the tangential direction) is given by $v = 2$ m/s and the tangential acceleration is given by $a_t = 0.2$ m/s². To calculate the normal acceleration, the

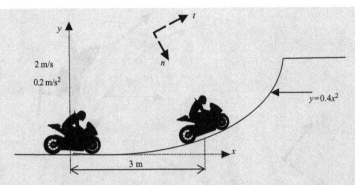

Figure 1.32: Representation of Example 1.13

radius of curvature is required. The radius of curvature is calculated as (from any calculus textbook):

$$\rho = \frac{\left[1 + \left(\frac{dy}{dx}\right)^2\right]^{3/2}}{\left|\frac{d^2y}{dx^2}\right|} = \frac{\left[1 + (0.8x)^2\right]^{3/2}}{|0.8|}$$

For $x = 3$ m, the radius of curvature is:

$$\rho = \frac{\left[1 + (0.8 \times 3)^2\right]^{3/2}}{|0.8|} = 21.97\,\text{m}$$

From Equation (1.48), the normal acceleration is:

$$a_n = \frac{v^2}{\rho} = \frac{2^2}{21.97} = 0.182\,\text{m/s}^2$$

The acceleration magnitude is (Equation (1.49)):

$$a = \sqrt{a_t^2 + a_n^2} = \sqrt{0.2^2 + 0.182^2} = 0.27\,\text{m/s}^2$$

Example 1.14 Curvilinear motion: normal–tangential co–ordinate system II

A jet plane travels along the path shown in Figure 1.33 with a velocity of 125 m/s and an acceleration of 24 m/s^2 that makes an angle of 45° with the tangential direction. Determine the plane's tangential acceleration and the radius of curvature ρ.

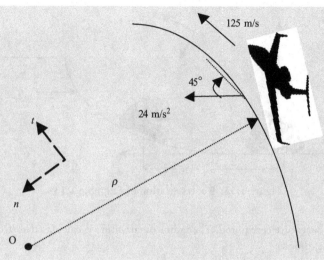

Figure 1.33: Representation of Example 1.14

Solution

The (tangential) velocity is given as $v = 125$ m/s and the tangential and the normal acceleration are obtained by resolving the acceleration in t and n directions as:

$$a_t = 24 \times \cos 45 = 16.97\,\text{m/s}^2$$
$$a_n = 24 \times \sin 45 = 16.97\,\text{m/s}^2$$

From Equation (1.48), the radius of curvature is calculated as:

$$a_n = \frac{v^2}{\rho} \Rightarrow 16.97 = \frac{125^2}{\rho} \Rightarrow \rho = 920.7\,\text{m}$$

1.4 Tutorial sheet

1.4.1 Rectilinear motion: constant acceleration

Q1.1 A particle starts motion with an acceleration of 0.3 m/s^2 as shown in Figure 1.34. After a certain distance, it runs with a constant velocity of 22 m/s. The particle decelerates in 1 minute before coming to a stop. If the total travel distance is about 11 km, use the equation of rectilinear translation with constant acceleration to determine:

a) the distance and time in the acceleration phase;

[0.807 km, 1 min 13 s]

b) the acceleration and the distance in the deceleration phase;

[−0.367 m/s^2, 0.66 km]

c) the total travelled time.

[9 min 27 s]

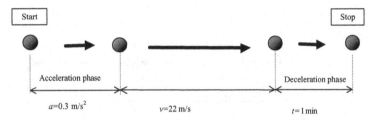

Figure 1.34: Representation of Question 1.1

Q1.2 A motorcycle, shown in Figure 1.35, starts its motion with an initial acceleration of 4 m/s^2. After travelling 200 m, determine:

a) the velocity of the motorcycle;

[40 m/s]

b) the time for which it has been travelling.

[10 s]

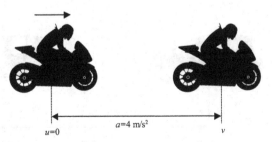

Figure 1.35: Representation of Question 1.2

Q1.3 A particle starts to move at point A with a constant acceleration of 3 m/s^2 until it reaches point B with a speed of 60 m/s, as shown in Figure 1.36. Between points B and C, it decelerates with a constant deceleration of 6 m/s^2 for 5 s. Determine:

a) the distance travelled between A and C;

[825 m]

b) the time taken to travel from A to C;

[25 s]

c) the velocity at C.

[30 m/s]

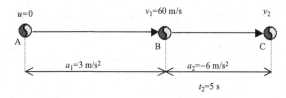

Figure 1.36: Representation of Question 1.3

Q1.4 A car travels with an initial velocity of 60 km/hr and accelerates at 6000 km/hr^2 along a straight road to 110 km/hr as shown in Figure 1.37. Determine:

a) the length of time in which the acceleration took place;

[30 s]

b) the distance travelled during that time.

[708 m]

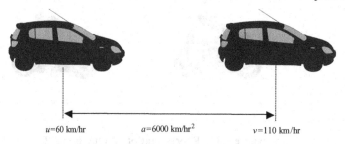

Figure 1.37: Representation of Question 1.4

Q1.5 A particle moves along a straight line with a constant velocity of 40 m/s when it passes a point A until it reaches point B after 3 s, as shown in Figure 1.38. From point B, it travels with a constant deceleration of 10 m/s^2, until it reaches a turning point C, then it returns to point A. Determine:

a) the distance travelled by the particle in the constant velocity phase AB;

[120 m]

b) the time and the distance travelled from point B to the turning point C;

[4 s, 80 m]

c) the time that the particle takes to return from point C to point A and its velocity when it reaches A;

[6.3 s, 63.25 m/s]

d) the total time elapsed and the total distance travelled by the particle.

[400 m, 13.3 s]

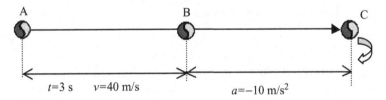

Figure 1.38: Representation of Question 1.5

Q1.6 A truck accelerates from 20 km/hr to 100 km/hr in 12 seconds as shown in Figure 1.39. Determine:

a) the truck's acceleration;

[24 000 km/hr^2]

b) the distance travelled during that time.

[200 m]

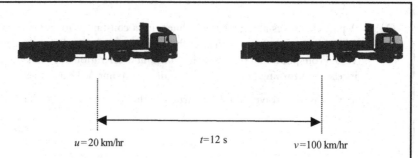

Figure 1.39: Representation of Question 1.6

1.4.2 Rectilinear motion: non–constant acceleration

Q1.7   The motion of a particle is described as $s = t^3 - \frac{15}{2}t^2 + 18t + 5$ where s is the displacement in metres and t is the time in seconds. The particle stops twice in the time interval between 1 s and 4 s. For each of these two turning points, determine:

a) the time in seconds;

[2 s, 3 s]

b) the displacement in meters;

[19 m, 18.5 m]

c) the acceleration in m/s^2.

[−3 m/s^2, 3 m/s^2]

d) Calculate the displacement at $t = 4$ s and sketch the particle path (plot s against t).

[21 m]

Q1.8   A rocket, shown in Figure 1.40, travels with an upwards acceleration of $a = 5.8 + 0.021s$, where a is in m/s^2 and s is the vertical distance in meters. Determine the rocket's velocity after travelling 3 km, using initial velocity $v = 0$ at $s = 0$.

[473.1 m/s]

Figure 1.40: Representation of Question 1.8

Q1.9   A particle is moving with a velocity $v = 4s^2 + 3$, where v is in m/s and s is the displacement in meters.

a) Show that $a = v\frac{dv}{ds}$ and derive an expression for the particle acceleration in terms of the displacement.

$$[a = 32s^3 + 24s]$$

b) If the particle travels 3 m, determine the particle velocity and acceleration.

$$[39 \text{ m/s}, 936 \text{ m/s}^2]$$

Q1.10 A particle decelerates with a rate $a = -12t$, where t is the time in seconds. If the particle has an initial velocity of 20 m/s and a zero initial displacement, determine the distance travelled before it comes to a stop.

$$[24.35 \text{ m}]$$

Q1.11 A car travels along a straight road with a velocity equal to $v = t^2 + t/2$, where v is in m/s and t is the time in seconds. The car starts from rest, i.e. at time $t = 0$, the initial velocity $v_o = 0$ and the initial position $s_o = 0$. When $t = 3$ s, determine:

a) the position of the car;

$$[11.25 \text{ m}]$$

b) the acceleration of the car.

$$[6.5 \text{ m/s}^2]$$

Q1.12 When a projectile is fired into a fluid tank as shown in Figure 1.41, it decelerates with a rate $a = -0.5v^3$, where a is in m/s^2 and v is the velocity in m/s. If the projectile's initial velocity is $v_o = 100$ m/s, determine its velocity after 3 seconds.

[0.577 m/s]

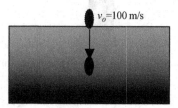

Figure 1.41: Representation of Question 1.12

1.4.3 Curvilinear motion: Cartesian co-ordinate system

Q1.13 A particle moves in the x-y plane with the following displacements:

$$x = 2t^3 - 10t^2 + 2t + 2 \quad \text{and} \quad y = 3t^3 - 4t^2 + 5t + 3,$$

where x and y are in the Cartesian co-ordinate directions, expressed in meters, and t is the time in seconds.
When $t = 4$ s, determine:

a) the velocity components in the Cartesian co-ordinate system and its total magnitude;

[18 m/s, 117 m/s, 118.38 m/s]

b) the acceleration components in the Cartesian co-ordinate system and its total magnitude.

[28 m/s^2, 64 m/s^2, 69.86 m/s^2]

Q1.14 A weather balloon moves in the x-y plane so that its co-ordinates measured from a reference point, as shown in Figure 1.42, are given by $x = 8t$ and $y = 0.03x^2$, where t is the time in seconds and x and y are in meters. When $t = 3$s, determine:

a) the magnitude of the velocity;

[14.03 m/s]

b) the direction of the balloon's motion;

[55.22°]

c) the magnitude of the acceleration.

[3.84 m/s²]

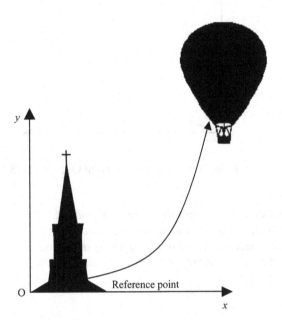

Figure 1.42: Representation of Question 1.14

Q1.15 A helicopter takes off from the origin O, as shown in Figure 1.43, and its flight path is defined such that $x = 3t^2$ and $y = 0.06t^3$, where x and y are in meters and t is in seconds. When $t = 10$ s, determine:

a) the distance between the helicopter and the origin O;

[305.9 m]

b) the magnitude of the helicopter's velocity;

[62.64 m/s]

c) the magnitude of the helicopter's acceleration.

[7 m/s²]

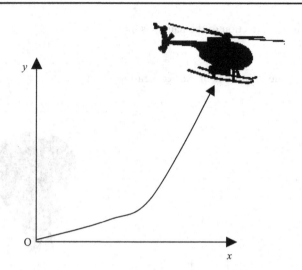

Figure 1.43: Representation of Question 1.15

Q1.16 A rocket travels vertically up to an altitude of 50 m, as shown in Figure 1.44, then travels along a path such that $y = 50 + 13\sqrt{x}$, where x and y are in meters. The vertical component of the rocket velocity is constant and equal to 200 m/s. When $y = 100$ m, determine:

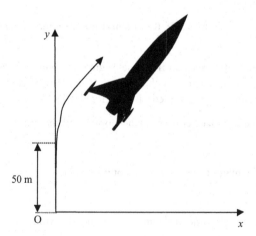

Figure 1.44: Representation of Question 1.16

a) the distance between the rocket and the origin O;

[101.1 m]

b) the magnitude of the rocket's velocity;

[232.39 m/s]

c) the magnitude of the rocket's acceleration.

[473.37 m/s^2]

Q1.17 A racing motorcycle jumps off a ramp of 1 m height, which has a slope of 30° with the horizontal as shown in Figure 1.45. If the initial velocity of the motorcycle when he jumps off the ramp is 9 m/s during the jump, determine:

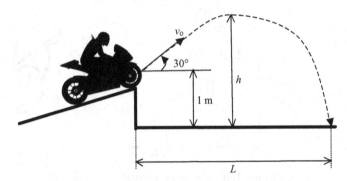

Figure 1.45: Representation of Question 1.17

a) the time it took during the jump;

[1.1 s]

b) the horizontal distance (L);

[1.4 m, 0.023 m]

c) the maximum height (h).

[2.03 m]

Q1.18 A fireman directs the water flow from his hose towards a wall at a speed of 15 m/s and at an angle of 20° with the horizontal as shown in Figure

1.46. If the hose is at a height of 1 m and the distance between the fireman and the wall is 9 m, determine:

a) the time that the water takes to reach the wall;

[0.64 s]

b) the magnitude of the velocity and the direction of the water at the wall;

[14.14 m/s, −4.66°]

c) the height of the water at the wall (*h*).

[2.27 m]

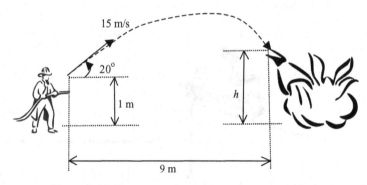

Figure 1.46: Representation of Question 1.18

Q1.19 The stream of water released from a water sprinkler, shown in Figure 1.47, has a velocity of 4 m/s and makes an angle of 60° with the horizontal. The water hits the ground at a point A on a hill with its co-ordinates defined as $y = 0.012x^2$, where x and y are in meters. Determine:

a) the time that the water takes to strike point A;

[0.7 s]

b) the co-ordinates of point A.

[1.4 m, 0.023 m]

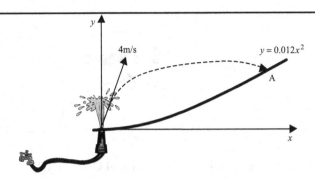

Figure 1.47: Representation of Question 1.19

1.4.4 Curvilinear motion: polar co-ordinate system

Q1.20 A particle moves in the polar co-ordinate system with the following displacements:

$$r = 3t^2 + 3, \quad \theta = \frac{t}{2}\sin\frac{\pi t}{4}$$

where r is the radius in metres, θ is the angle measured anti-clockwise in radians and t is the time in seconds. When $t = 2$ s, determine:

a) the angular velocity and angular acceleration;

[0.5 rad/s, −0.617 rad/s^2]

b) the velocity components in the polar co-ordinate system and the total magnitude of velocity;

[12 m/s, 7.5 m/s, 14.15 m/s]

c) the acceleration components in the polar co-ordinate system and the total magnitude of acceleration.

[2.25 m/s^2, 2.75 m/s^2, 3.55 m/s^2]

Q1.21 A truck travels along a circular road of radius 50 m, as shown in Figure 1.48, with a speed of 15 m/s. Determine:

a) the angular velocity $\dot{\theta}$;

[0.3 rad/s]

b) the radial acceleration component.

[−4.5 m/s^2]

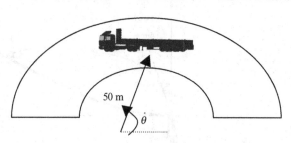

Figure 1.48: Representation of Question 1.21

Q1.22 A ball moves in a two-dimensional plane following the path shown in Figure 1.49. At the position shown, the ball has a radial velocity $\dot{r} = 1$ m/s and angular velocity $\dot{\theta} = 20$ °/s. The rate of change in radial velocity ($\ddot{r}$) is 3 m/s² and the angular acceleration ($\ddot{\theta}$) is 50 °/s². At the position shown, determine:

a) the velocity components in the Cartesian co-ordinate system;

[0.52 m/s, 1.1 m/s]

b) the acceleration components in the Cartesian co-ordinate system.

[1.17 m/s², 3.49 m/s²]

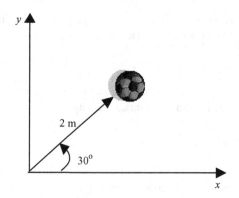

Figure 1.49: Representation of Question 1.22

Q1.23 A train travels along a circular track of radius 180 m, as shown in Figure 1.50, with an angular velocity of 0.03 rad/s and angular acceleration of -0.001 rad/s^2. Determine:

a) the magnitude of the train's velocity;

[5.4 m/s]

b) the magnitude of the train's acceleration.

[0.242 m/s^2]

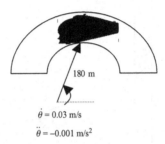

$\dot{\theta} = 0.03$ m/s

$\ddot{\theta} = -0.001$ m/s^2

Figure 1.50: Representation of Question 1.23

Q1.24 An amusement park ride of radius 3 m rotates about its centre, O, in the horizontal plane with a constant angular velocity of 0.2 rad/s as shown in Figure 1.51. For a seat A, on its edge, determine:

a) the radial and transverse velocity components;

[0, 0.6 m/s]

b) the radial and transverse acceleration components.

[-0.12 m/s^2, 0]

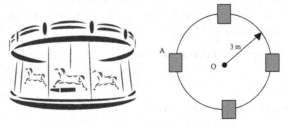

Figure 1.51: Representation of Question 1.24

Q1.25 A ship is anchored at a horizontal distance of 866 m from the road as shown in Figure 1.52. If a searchlight on the ship makes an angle of 30° with the horizontal line and is turned on a car travelling with a constant speed of 20 m/s, determine the angular velocity of the searchlight at the position shown in Figure 1.52.

[1 °/s]

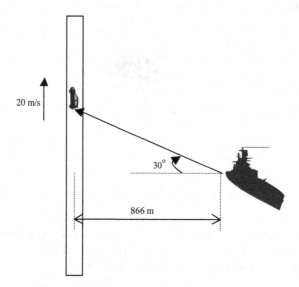

Figure 1.52: Representation of Question 1.25

1.4.5 Curvilinear motion: normal–tangential co-ordinate system

Q1.26 A skier moves along a curved path with a velocity of 5.5 m/s and acceleration of 2.2 m/s^2 at the position shown in Figure 1.53. The path can be described as $y = 0.04x^2$, where y and x are in meters. At the position shown, when $x = 15$ m, determine:

a) the radius of curvature;

[47.64 m]

b) the normal acceleration;

[0.635 m/s^2]

c) the magnitude of the total acceleration.

$$[2.29 \text{ m/s}^2]$$

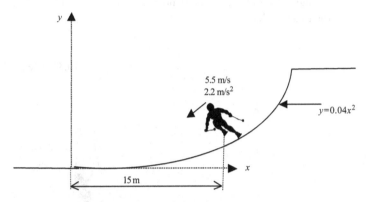

Figure 1.53: Representation of Question 1.26

Q1.27 A race car travels along a curved path of radius 100 m as shown in Figure 1.54. At the position shown, the car has a velocity of 9 m/s and a constant acceleration of 2 m/s². Determine:

a) the normal acceleration of the car;

$$[0.81 \text{ m/s}^2]$$

b) the time required to reach this velocity if the car starts from rest.

$$[4.5 \text{ s}]$$

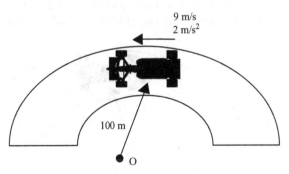

Figure 1.54: Representation of Question 1.27

Q1.28 A jet plane, shown in Figure 1.55, travels along a path described as $y = 0.3x^2$, where y and x are in kilometers. At the instant shown, $x = 6$ km and the plane has a velocity of 220 m/s and a tangential acceleration of 0.5 m/s^2, determine the plane's normal acceleration and the magnitude of the total acceleration.

[0.56 m/s^2, 0.75 m/s^2]

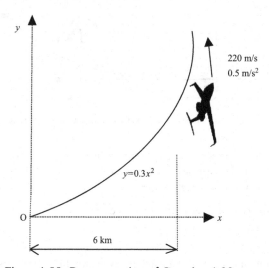

Figure 1.55: Representation of Question 1.28

Q1.29 A race car travels along a curved path of radius 250 m as shown in Figure 1.56. If its speed increases from 14 m/s to 30 m/s in 4 seconds, determine:

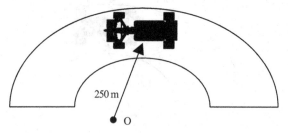

Figure 1.56: Representation of Question 1.29

a) the tangential acceleration of the car;

[4 m/s^2]

b) the normal acceleration of the car when its speed is 25 m/s.

[2.5 m/s^2]

Q1.30 A train has a velocity of 22 m/s and an acceleration of 15 m/s^2 that makes an angle of 60° with the tangential direction at the position shown in Figure 1.57. At the instance shown, determine:

a) the tangential acceleration;

[7.5 m/s^2]

b) the radius of curvature.

[37.26 m]

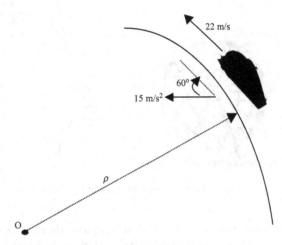

Figure 1.57: Representation of Question 1.30

Q1.31 A Ferris wheel, shown in Figure 1.58, of radius 11 m rotates about its centre in the vertical plane so that the speed of its passengers increases by $\dot{v} = 1.1t$, where v is in m/s and t is the time in seconds. Consider the situation where the wheel starts from rest, i.e. at time $t = 0$, the initial velocity $v_o = 0$ and the initial angular position $\theta_o = 0$. When $t = 3$ s, determine:

a) the passengers' velocity;

[4.95 m/s]

b) the normal and tangential components of the passengers' acceleration;

[2.23 m/s^2, 3.3 m/s^2]

c) the wheel's angular velocity;

[0.45 rad/s]

d) the wheel's angular position.

[25.8°]

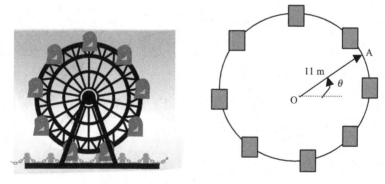

Figure 1.58: Representation of Question 1.31

Q1.32 A satellite is positioned at an altitude, h, from the earth's surface as shown in Figure 1.59. The satellite rotates around the earth with a constant speed of 5000 m/s and has a normal acceleration of 2 m/s^2. If the diameter of the earth is 12 713 km, determine the altitude h.

[6143.5 km]

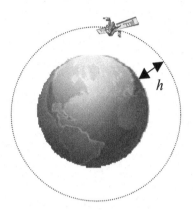

Figure 1.59: Representation of Question 1.32

Q1.33 In Figure 1.59, if the altitude, h, is 7000 km, determine the satellite's normal acceleration (for a constant of speed 5000 m/s).

$[1.87 \text{ m/s}^2]$

CHAPTER 2

Kinematics of Rigid Bodies

2.1 INTRODUCTION

As mentioned in Chapter 1, a rigid body can undergo both translation and rotation. In analysis of two-dimensional motion, also called planar motion, a rigid body can have three types of motion: translation, rotation and general plane motion (a combination of translation and rotation). In addition to the equations of motion of translation derived for particles in Chapter 1, which are also applicable to rigid bodies, relationships for angular quantities are required to describe the kinematics of a rigid body. A rigid body can be defined as a body that contains a system of particles keeping a constant distance between them before and after motion. This means that a rigid body does not undergo any elastic deformation or change in its original shape. This assumption is ideal since most materials can deform under the action of external forces. However, if the deformation is very small compared to the size of the actual body or the overall movements of the body, the assumption is broadly acceptable and yields very accurate results. Analyzing rigid bodies is necessary for the design of cams, gears, linkages and mechanisms. The analysis of displacement, velocity and acceleration of these systems is useful in designing their geometry and mechanical parts. This analysis is also required for later application of Newton's second law in Chapter 4, in order to calculate the force acting on machine elements. This chapter considers the analysis of motion of rigid bodies. First, we derive the kinematics equations for rotational motion. Then the analysis of rigid body kinematics is divided into two main sections, the kinematics of wheels and gears and the kinematics of linkages and mechanisms.

2.2 RIGID BODY MOTION

2.2.1 Translation

When every line in a rigid body after motion remains in its original position before motion, the body undergoes only translation, no rotation. Similarly to a particle, a rigid body can undergo rectilinear translation and curvilinear translation. As all points in the rigid body undergo the same translation, it follows that any point in the body can describe the motion of the whole body. The equations of motion that have been derived for a particle in Chapter 1 are applicable to a rigid body and, therefore, there is no need to repeat them here.

2.2.2 Rotation

Rotation of a rigid body about a fixed axis of rotation is its angular motion about that axis. All points in the rigid body rotate about the centre of rotation at the same angle and in the same time. Consider a rigid body rotating about point O (on the axis of rotation, which is perpendicular to the x-y plane). As shown in Figure 2.1, the angular position is defined by the angle θ measured anticlockwise from a reference axis. The angular displacement

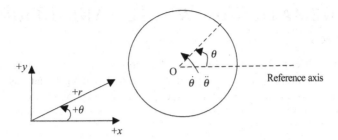

Figure 2.1: Angular motion

is the change in the angular position and is denoted as $d\theta$. The angular velocity is defined as the rate of change in the angle θ, i.e.

$$\dot{\theta} = \frac{d\theta}{dt} \tag{2.1}$$

The angular velocity is a vector that has a magnitude measured in radians per second (rad/s) or revolutions per minute (rev/min), where 1 revolution $= 2\pi$. Its direction is the same as the angle θ, i.e. along the axis of rotation.

Similarly, the angular acceleration is the rate of change in angular velocity and is given by:

$$\ddot{\theta} = \frac{d\dot{\theta}}{dt} = \frac{d^2\theta}{dt^2} \tag{2.2}$$

The angular acceleration is a vector that has magnitude and direction. Its magnitude is often expressed in rad/s^2 (radians per second squared).

In a similar way to Equations (1.5) to (1.8) for rectilinear motion with constant acceleration, the following equations can easily be derived for rotation (by replacing s by θ, v by $\dot{\theta}$ and a by $\ddot{\theta}$):

$$\dot{\theta} = \dot{\theta}_o + \ddot{\theta}t \tag{2.3}$$

$$\dot{\theta}^2 = \dot{\theta}_o^2 + 2\ddot{\theta}\theta \tag{2.4}$$

$$\theta = \dot{\theta}_o t + \frac{1}{2}\ddot{\theta}t^2 \tag{2.5}$$

$$\theta = \frac{1}{2}(\dot{\theta}_o + \dot{\theta})t \tag{2.6}$$

where $\dot{\theta}_o$ is the initial angular velocity. Equations (2.3) to (2.6) are very useful in analyzing rigid body rotation with constant angular acceleration.

2.3 KINEMATICS OF WHEELS AND GEARS

2.3.1 Wheels

Consider the wheel in Figure 2.2 rotating about point O. The position of a point A is defined by the radius r and the angular position θ. Recalling Equations (1.32) and (1.34), since r is constant, $\dot{r} = 0$ and the velocity components of point A in the radial and transverse directions are:

$$v_r = \dot{r} = 0 \tag{2.7}$$
$$v_t = r\dot{\theta} \tag{2.8}$$

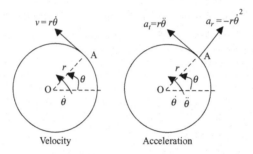

Figure 2.2: Velocity and acceleration components of wheel

Therefore, the total velocity magnitude is equal to the transverse velocity $v = v_t = r\dot{\theta}$. Similarly, recalling Equations (1.39) and (1.40) and substituting the radial and transverse components, the acceleration of point A becomes:

$$a_r = -r\dot{\theta}^2 \tag{2.9}$$
$$a_t = r\ddot{\theta} \tag{2.10}$$

Example 2.1 Wheels and gears: car wheel

A car travels with a constant speed of 110 km/hr and has tyres of radius 355 mm as shown in Figure 2.3. When the driver applies the brakes, the car has a constant deceleration of 2.5 m/s². Consider the moment just after applying the brakes and determine:

a) the angular velocity of the car's wheels;
b) the angular acceleration of the car's wheels;
c) the radial and transverse acceleration of a point A, 300 mm from the wheel's centre.

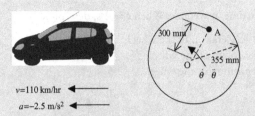

Figure 2.3: Representation of Example 2.1

Solution

a) The transverse velocity of the wheels is equal to the car velocity:

$$v = v_t = 110 \times \frac{1000}{3600} = 30.556 \, \text{m/s}$$

Using Equation (2.8), the angular velocity of the wheel is calculated as:

$$v_t = r\dot{\theta} \Rightarrow 30.556 = \frac{355}{1000} \times \dot{\theta} \Rightarrow \dot{\theta} = 86.073 \, \text{rad/s}$$

b) The transverse acceleration of the wheels is equal to the car acceleration:

$$a_t = -2.5 \, \text{m/s}^2$$

From Equation (2.10), the angular acceleration of the wheels is calculated as:

$$a_t = r\ddot{\theta} \Rightarrow -2.5 = \frac{355}{1000} \times \ddot{\theta} \Rightarrow \ddot{\theta} = -7.042 \, \text{rad/s}^2$$

c) From Equation (2.9), the radial component of the acceleration at point A ($r = 0.3$ m) is:

$$a_r = -r\dot{\theta}^2 = -0.3 \times 86.073^2 = -2222.6 \, \text{m/s}^2$$

And from Equation (2.10), the transverse component of the acceleration at point A is:

$$a_t = r\ddot{\theta} = 0.3 \times (-7.042) = -2.11 \, \text{m/s}^2$$

Example 2.2 Wheels and gears: wheel on cord

In Figure 2.4, a cord is warped around a wheel of radius 200 mm and is initially at rest, i.e. at $t = 0$, $\theta = \dot{\theta} = \ddot{\theta} = 0$. When an upwards force F is applied, the wheel rotates about its centre point O and has an acceleration of $a = 6t$, where a is in m/s² and t is the time in seconds. When $t = 2$ s, determine:

a) the angular acceleration of the wheel;
b) the angular velocity of the wheel;
c) the angular position of the wheel.

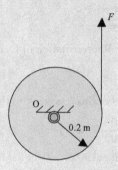

Figure 2.4: Representation of Example 2.2

Solution

a) From Equation (2.10), the angular acceleration of the wheel is calculated as:

$$a_t = r\ddot{\theta} \Rightarrow 6t = 0.2 \times \ddot{\theta} \Rightarrow \ddot{\theta} = 30t \,\text{rad/s}^2 \qquad (E2.2a)$$

At $t = 2$ s, the angular acceleration is:

$$\ddot{\theta} = 30 \times 2 = 60 \,\text{rad/s}^2$$

b) To calculate the angular velocity, Equation (2.2) is integrated over time, i.e.

$$\ddot{\theta} = \frac{d\dot{\theta}}{dt} \Rightarrow \ddot{\theta}dt = d\dot{\theta} \Rightarrow \dot{\theta} = \int \ddot{\theta}dt \qquad (E2.2b)$$

Substituting Equation (E2.2a) into Equation (E2.2b) and integrating over time gives:

$$\dot{\theta} = \int \ddot{\theta}dt = \int 30t \, dt = 15t^2 + C$$

Using the initial condition $t = 0$ and $\dot{\theta} = 0$, thus $C = 0$ and the angular velocity becomes:

$$\dot{\theta} = 15t^2 \qquad (E2.2c)$$

At $t = 2$s, the angular velocity is:

$$\dot{\theta} = 15 \times 2^2 = 60 \,\text{rad/s}$$

c) To calculate the angular position, Equation (2.1) is integrated over time, i.e.

$$\dot{\theta} = \frac{d\theta}{dt} \Rightarrow \dot{\theta} \, dt = d\theta \Rightarrow \theta = \int \dot{\theta} \, dt \qquad \text{(E2.2d)}$$

Substituting Equation (E2.2c) into Equation (E2.2d) and integrating over time yields:

$$\theta = \int \dot{\theta} dt = \int 15t^2 \, dt = 5t^3 + C$$

Again, from the initial condition $t = 0$ and $\theta = 0$, thus $C = 0$ and the angular position becomes:

$$\theta = 5t^3$$

At $t = 2s$, the angular position is:

$$\theta = 5 \times 2^3 = 40 \text{ rad or } \frac{40}{2\pi} = 6.37 \text{ revolutions.}$$

Example 2.3 Wheels and gears: sphere

A sphere, shown in Figure 2.5, has a radius of 0.3 m and starts to rotate about its axis from rest, i.e. at $\theta = 0$, $\dot{\theta} = 0$, with an angular acceleration equal to $\ddot{\theta} = 2\theta + 2$, where $\ddot{\theta}$ is in rad/s^2 and θ is in radians. When $\theta = 3$ radians, determine the angular velocity of the sphere.

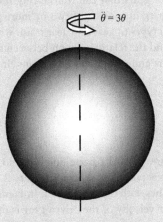

Figure 2.5: Representation of Example 2.3

Solution

From Equation (2.2), the angular acceleration can be written as $\ddot{\theta} = \frac{d\dot{\theta}}{dt} \Rightarrow \ddot{\theta} = \frac{d\dot{\theta}}{dt} \times \frac{d\theta}{d\theta} = \dot{\theta}\frac{d\dot{\theta}}{d\theta}$ which can be expanded (using $\ddot{\theta} = 2\theta + 2$) as:

$$\ddot{\theta} = \dot{\theta}\frac{d\dot{\theta}}{d\theta} \Rightarrow \ddot{\theta}\,d\theta = \dot{\theta}\,d\dot{\theta} \Rightarrow (2\theta + 2)d\theta = \dot{\theta}\,d\dot{\theta}$$

Integrating both sides over the angular position θ and angular velocity $\dot{\theta}$ and using the initial condition, at zero angular position $\theta = 0$ the angular velocity is zero $\dot{\theta} = 0$, gives:

$$\int \dot{\theta}\,d\dot{\theta} = \int (2\theta + 2)d\theta \Rightarrow \frac{\dot{\theta}^2}{2} = \theta^2 + 2\theta$$

At $\theta = 3$ rad, the angular velocity is calculated as:

$$\frac{\dot{\theta}^2}{2} = 3^2 + 2 \times 3 \Rightarrow \dot{\theta} = 5.48\,\text{rad/s}$$

2.3.2 Gears

Gears are used to transmit rotational motion between shafts through engaging gear teeth. As gears mesh with each other, force, torque and motion are transmitted between them. If two engaged gears are of different sizes, motion can be magnified or reduced from one shaft to another. A system of gears that can transmit motion through shafts is called a gear train. In general, a gear train consists of two or more gears that are engaged with each other. There are several types of gear and gear train, which are distinguished by the gear and shaft orientation and the relative motion between them. This section considers spur gears, simple gear trains and compound gear trains. It also introduces the concept of an engine gearbox.

Spur gear

In a spur gear, the shafts are parallel to one another and the gear teeth are straight and parallel to the shaft's orientation, as shown in Figure 2.6. The driver gear, which is connected to the motor, meshes with the driven gear. In Figure 2.6, where the gears have external contact, the angular velocity of the driven gear is opposite to that of the driver gear.

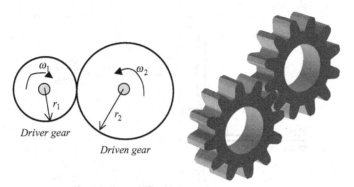

Figure 2.6: Spur gear

The driver gear has a pitch radius r_1 and rotates with an angular speed ω_1. The driven gear has a pitch radius r_2 and rotates with an angular speed ω_2. Their transverse velocities (see Equation 2.8) at the contact point are identical and the following expression is satisfied:

$$r_1\omega_1 = r_2\omega_2 \tag{2.11}$$

The gear ratio, G, is defined as the ratio between the angular speed of the driver gear and the angular speed of the driven gear:

$$G = \frac{\omega_1}{\omega_2} = \frac{r_2}{r_1} \tag{2.12}$$

If $G > 1$, then $\omega_1 > \omega_2$ and the speed is reduced, while if $G < 1$, then $\omega_1 < \omega_2$ and the speed is magnified. The gear ratios can also be written in terms of the number of teeth of the gears as:

$$G = \frac{nt_2}{nt_1} \tag{2.13}$$

where nt_1 and nt_2 are the number of teeth in the driver and the driven gears, respectively.

Simple gear train

In a simple gear train, all gears are directly engaged with each other as shown in Figure 2.7. Each gear is mounted on a separate shaft and the gear ratios are quite simple to obtain, in a similar way to Equation (2.12):

$$\text{Gear ratio for gear 2} = G_2 = \frac{\omega_1}{\omega_2} = \frac{r_2}{r_1} \tag{2.14}$$

$$\text{Gear ratio for gear 3} = G_3 = \frac{\omega_1}{\omega_3} = \frac{r_3}{r_1} \tag{2.15}$$

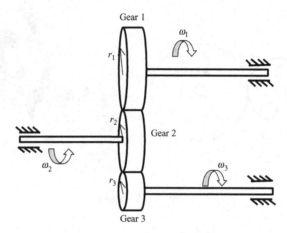

Figure 2.7: A simple gear train

Compound gear train

In a compound gear train, two or more gears are mounted on one shaft as shown in Figure 2.8. If gear 1 is the driver gear, the gear ratio of gear 2, G_2, can be calculated in a straightforward fashion, as in Equation (2.14). Since gear 4 is mounted on the same shaft as gear 2, its angular speed is ω_2 and the relationship between its pitch radius and that of gear 3 is obtained by equating their transverse velocities at the contact point:

$$r_3\omega_3 = r_4\omega_2 \Rightarrow \frac{r_3}{r_4} = \frac{\omega_2}{\omega_3} \tag{2.16}$$

The gear ratio for gear 3 is the ratio between the angular speed of the driver gear (gear 1) and gear 3:

$$G_3 = \frac{\omega_1}{\omega_3}$$

or

$$G_3 = \frac{\omega_1}{\omega_3} \times \frac{\omega_2}{\omega_2} = \frac{\omega_1}{\omega_2} \times \frac{\omega_2}{\omega_3} \tag{2.17}$$

Substituting Equation (2.14) into Equation (2.17), gives

$$G_3 = G_2 \times \frac{\omega_2}{\omega_3} \tag{2.18}$$

From Equation (2.16), Equation (2.18) can also be written as:

$$G_3 = G_2 \times \frac{r_3}{r_4} \tag{2.19}$$

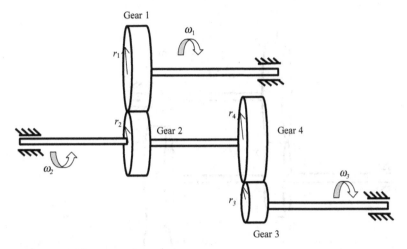

Figure 2.8: A compound gear train

Example 2.4 Wheels and gears: compound gear train

In the compound gear train shown in Figure 2.9, the driver gear, 1, has a radius of 40 mm and rotates with a constant angular velocity of 50 rev/min. The gear ratios for gear 2 and gear 3 are 1/2 and 1/6, respectively. If gear 3 has a radius of 10 mm, determine:

a) the angular speeds of gears 2 and 3;
b) the radii of gears 2 and 4.

Solution

a) Using Equations (2.14) and (2.17), the angular speeds of gears 2 and 3 are:

$$G_2 = \frac{\omega_1}{\omega_2} \Rightarrow \frac{1}{2} = \frac{50}{\omega_2} \Rightarrow \omega_2 = 100\,\text{rev/min}$$

$$G_3 = \frac{\omega_1}{\omega_3} \Rightarrow \frac{1}{6} = \frac{50}{\omega_3} \Rightarrow \omega_3 = 300\,\text{rev/min}$$

b) From Equation (2.14), the radius of gear 2 is:

$$G_2 = \frac{r_2}{r_1} \Rightarrow \frac{1}{2} = \frac{r_2}{40} \Rightarrow r_2 = 20\,\text{mm}$$

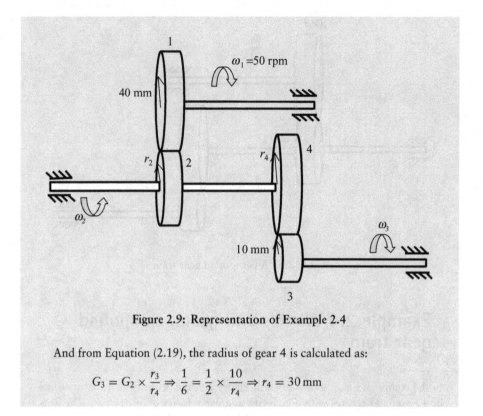

Figure 2.9: Representation of Example 2.4

And from Equation (2.19), the radius of gear 4 is calculated as:

$$G_3 = G_2 \times \frac{r_3}{r_4} \Rightarrow \frac{1}{6} = \frac{1}{2} \times \frac{10}{r_4} \Rightarrow r_4 = 30 \, \text{mm}$$

Engine gearbox

An engine gearbox is a practical example of a compound gear train. As shown in Figure 2.10, for a five-speed gearbox, 10 gears are mounted on three shafts. The following gear transmissions can be obtained:

- First gear: Engage 1 with 1′
- Second gear: Engage 2 with 2′
- Third gear: Engage 3 with 3′
- Fourth gear: Engage 4 with 4′
- Fifth gear: Engage 5 with 5′ (the wheel shaft with the engine shaft)

The angular speed of the engine shaft is ω_e, of the wheel shaft is ω_w and of the layshaft (intermediate shaft) is ω_i. Gear 6′ is engaged with driver gear 6 all the time and has a gear ratio of

$$G_6 = \frac{\omega_e}{\omega_i} = \frac{r_6'}{r_6} \tag{2.20}$$

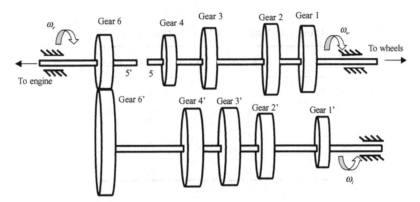

Figure 2.10: A five-speed engine gear box

For the first gear transmission, the gear ratio is obtained in a way similar to that for compound gear trains:

$$G_1 = \frac{\omega_e}{\omega_w} \quad \text{or} \quad G_1 = \frac{\omega_e}{\omega_w} \times \frac{\omega_i}{\omega_i} = \frac{\omega_e}{\omega_i} \times \frac{\omega_i}{\omega_w}$$

Thus

$$G_1 = \frac{r_1}{r_1'} \times \frac{r_6'}{r_6} \quad \text{or} \quad G_1 = G_6 \times \frac{r_1}{r_1'} \tag{2.21}$$

Similarly, for other gear ratios:

$$G_2 = G_6 \times \frac{r_2}{r_2'} \tag{2.22}$$

$$G_3 = G_6 \times \frac{r_3}{r_3'} \tag{2.23}$$

$$G_4 = G_6 \times \frac{r_4}{r_4'} \tag{2.24}$$

Example 2.5 Wheels and gears: engine gearbox

For the engine gearbox shown in Figure 2.10, the gear ratios and the radii of the gears on the wheels and engine shafts are given in Table 2.1. If the engine shaft runs with a speed of 3000 rev/min, determine:

a) the angular speed that can be produced by each gear;
b) the radius of the gears on the layshaft.

Table 2.1: Engine gearbox configuration

Gear	Gear ratio	Gear radius (r, mm)
1	3.5	30
2	2.3	25
3	1.5	20
4	1.12	17
5	1	–
6	2	16

Solution

The speed for each gear is the engine speed divided by the gear ratio:

$$\omega_{w1} = \frac{\omega_e}{G_1} = \frac{3000}{3.5} = 857.14 \, \text{rev/min}$$

$$\omega_{w2} = \frac{\omega_e}{G_2} = \frac{3000}{2.3} = 1304.35 \, \text{rev/min}$$

$$\omega_{w3} = \frac{\omega_e}{G_3} = \frac{3000}{1.5} = 2000 \, \text{rev/min}$$

$$\omega_{w4} = \frac{\omega_e}{G_4} = \frac{3000}{1.1} = 2727.27 \, \text{rev/min}$$

$$\omega_{w5} = \omega_e = 3000 \, \text{rev/min}$$

For the radii on the layshaft (approximated to the nearest millimetre), using Equations (2.20) to (2.24) gives:

$$G_1 = G_6 \times \frac{r_1}{r_1'} \Rightarrow 3.5 = 2 \times \frac{30}{r_1'} \Rightarrow r_1' = 17 \, \text{mm}$$

$$G_2 = G_6 \times \frac{r_2}{r_2'} \Rightarrow 2.3 = 2 \times \frac{25}{r_2'} \Rightarrow r_2' = 22 \, \text{mm}$$

$$G_3 = G_6 \times \frac{r_3}{r_3'} \Rightarrow 1.5 = 2 \times \frac{20}{r_3'} \Rightarrow r_3' = 27 \, \text{mm}$$

$$G_4 = G_6 \times \frac{r_4}{r_4'} \Rightarrow 1.12 = 2 \times \frac{17}{r_4'} \Rightarrow r_4' = 30 \, \text{mm}$$

$$G_6 = \frac{r_6'}{r_6} \Rightarrow 2 = \frac{r_6'}{16} \Rightarrow r_6' = 32 \, \text{mm}$$

2.4 KINEMATICS OF LINKAGES AND MECHANISMS

Kinematic analysis of linkages and mechanisms is very important in designing machine elements. A machine can be defined as a combination of rigid bodies or links that are connected to transmit forces and motion. A mechanism is a simplified model of a machine and can produce similar motion. Before we analyze the motion of linkages and mechanisms, you will find the following definitions useful:

- **A mechanism** is an assembly of rigid bodies or links, which are in motion relative to each other. The motion of one body in the assembly causes motion in the others. The rigid bodies in a mechanism are called links and the relative motion between them is described as a kinematic pair. A mechanism is usually drawn as a line diagram and represents a machine used to transmit forces. Figure 2.11 shows a slider–crank mechanism.
- **A linkage** is an assembly of rigid bodies or links that are connected with joints to form a closed chain. A linkage becomes a mechanism if two or more links move relative to a fixed frame.
- **Links** are the rigid bodies that form a mechanism. The slider–crank mechanism in Figure 2.11 consists of four links, which are numbered 1, 2, 3 and 4. Link 1 is the connecting rod, link 2 is the crank, link 3 is the fixed frame and link 4 is the piston or slider.
- **A kinematic pair** describes the relative motion between two links in contact with each other. There are different types of kinematic pair. For example, there is a sliding pair between the piston (link 4) and the fixed frame (link 3) and a turning pair between the crank (link 2) and the connecting rod (link 1) in Figure 2.11. Another type of kinematic pair, which is not shown in Figure 2.11, is a screw pair, for example, between a nut and a bolt. In Figure 2.11, there are three turning pairs between (1, 2), (2, 3) and (1, 4) and one sliding pair between (3, 4).
- **A kinematic chain** is a group of links that are connected to each other so that relative motion is possible by fixing one of them. The slider–crank mechanism in Figure 2.11 is a four-link chain in which one of the links, link 3, is fixed so that the chain produces motion by the rotation of the crank, link 2, and the subsequent motion of the connecting rod, link 1, and the harmonic translation of the slider, link 4.

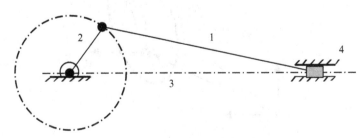

Figure 2.11: Slider–crank mechanism

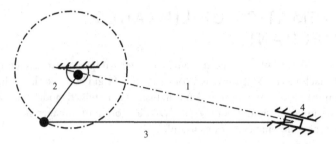

Figure 2.12: Inversion of the slider–crank mechanism

- **Inversion** enables a kinematic chain to give several different mechanisms. The mechanism in Figure 2.11 is obtained by fixing link 3. However, if we fix link 1 instead and the crank, link 2, is allowed to rotate about its other end, a different mechanism is obtained, as shown in Figure 2.12.
- **Kinematic constraint** means that a motion of a point in a specific direction is prevented by a device, e.g. a bearing. For example, in Figure 2.12, link 4 is constrained to move in the vertical direction and link 3 is constrained to move in both horizontal and vertical directions.

2.4.1 Kinematics of linkages

The motion of a rigid body can be described as a combination of translation and rotation. The relative motion of a rigid body can be demonstrated by considering two sets of Cartesian co-ordinate systems as shown in Figure 2.13. The (x, y) co-ordinate system has its origin at O, while the (x', y') co-ordinate system has its origin at A. The position vector of point B, r_B, is related to the position vector of point A, r_A, and the relative position vector (B relative to A), $r_{B/A}$, through the following relationship:

$$r_B = r_A + r_{B/A} \tag{2.25}$$

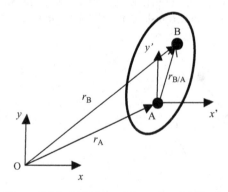

Figure 2.13: Relative motion of a rigid body

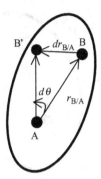

Figure 2.14: Rotation of a rigid body

where r_A, and r_B are measured from point O in the (x, y) co-ordinate system, while $r_{B/A}$ is measured from point A in the (x', y') co-ordinate system.

If point A undergoes a translation, dr_A, and point B undergoes a translation, dr_B, and a rotation $d\theta$ about A (from point B to B′ as shown in Figure 2.14), a similar equation to Equation (2.25) for the displacement is obtained as:

$$dr_B = dr_A + dr_{B/A} \tag{2.26}$$

From Figure 2.14, the relative displacement $dr_{B/A}$ is deduced as $\sin d\theta = \frac{dr_{B/A}}{r_{B/A}}$ and for small angle $d\theta$, $\sin d\theta \approx d\theta$, therefore $d\theta$ is given by $d\theta = \frac{dr_{B/A}}{r_{B/A}}$:

$$dr_{B/A} = r_{B/A}\, d\theta \tag{2.27}$$

Differentiating the displacements, Equation (2.26), with respect to time in order to derive a relationship between the velocity of A, the velocity of B and their relative velocity, gives:

$$\frac{dr_B}{dt} = \frac{dr_A}{dt} + \frac{dr_{B/A}}{dt} \tag{2.28}$$

which can be written as:

$$v_B = v_A + v_{B/A} \tag{2.29}$$

where v_A and v_B are the velocities of points A and B, respectively, measured in the (x, y) co-ordinate system, and $v_{B/A}$ is the relative velocity B/A measured in the (x', y') co-ordinate system. From Equation (2.27), the relative velocity $v_{B/A}$ is given by:

$$v_{B/A} = \frac{dr_{B/A}}{dt} = r_{B/A}\frac{d\theta}{dt} = r_{B/A}\dot{\theta} \tag{2.30}$$

Thus, Equation (2.28) can be written as:

$$v_B = v_A + r_{B/A}\dot{\theta} \tag{2.31}$$

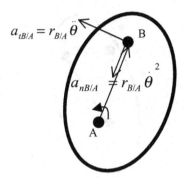

Figure 2.15: Relative acceleration components

Differentiating Equation (2.28) with respect to time, the acceleration of two points in a rigid body can be related as:

$$\frac{dv_B}{dt} = \frac{dv_A}{dt} + \frac{dv_{B/A}}{dt} \tag{2.32}$$

which can be written as:

$$a_B = a_A + a_{B/A} \tag{2.33}$$

where a_A and a_B are the accelerations of points A and B, respectively, measured in the (x, y) co-ordinate system. The relative acceleration $a_{B/A}$ has two components, normal and tangential as shown in Figure 2.15 (see Equations (2.9) and (2.10)):

$$a_{t B/A} = r_{B/A}\ddot{\theta} \tag{2.34}$$
$$a_{n B/A} = r_{B/A}\dot{\theta}^2 \tag{2.35}$$

Equations (2.31) and (2.33) are vector summations and their use is demonstrated by the examples.

Example 2.6 Fixed–frame mechanism I

The linkage shown in Figure 2.16 consists of two fixed frames perpendicular to one another and a bar AB. The bar AB is 0.3 m long and is restricted to move along the fixed frames vertically at A and horizontally at B. If the velocity of A is 1.5 m/s downwards when $\theta = 30°$, determine the velocity of B.

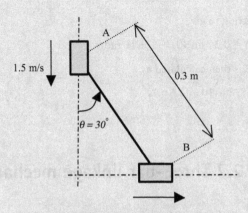

Figure 2.16: Representation of Example 2.6

Solution

From Equation (2.30), the relative velocity $v_{B/A}$ is:

$$v_{B/A} = r_{B/A}\dot{\theta} = 0.3\dot{\theta}$$

which is acting in the transverse direction to the link AB as shown in Figure 2.17. Apply Equation (2.31) in the x-direction to give:

$$v_B = 0 + 0.3\dot{\theta}\cos 30 \Rightarrow v_B = 0.2598\dot{\theta}$$

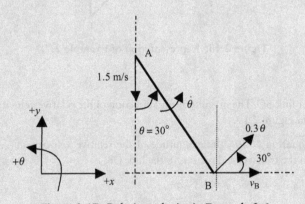

Figure 2.17: Relative velocity in Example 2.6

and in the *y*-direction to give:

$$0 = -1.5 + 0.3\dot{\theta}\sin 30 \Rightarrow \dot{\theta} = 10\,\text{rad/s}$$

The velocity of B is then calculated as:

$$v_B = 0.2598 \times 10 = 2.6\,\text{m/s}$$

Example 2.7 Three–bar linkage mechanism I

For the three-bar linkage mechanism shown in Figure 2.18, point C is constrained to move vertically, while point A is constrained in both *x* and *y* directions but is allowed to rotate freely. Determine the angular velocities of the
links AB and CB at the instant shown if C is moving with a velocity of 1.5 m/s
downwards.

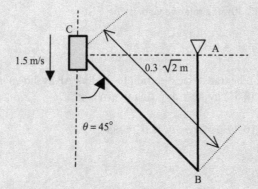

Figure 2.18: Representation of Example 2.7

Solution

a) Consider link BC. The magnitude and direction of the relative velocity $v_{B/C}$ are
   shown in Figure 2.19.

   From Equation (2.30), the magnitude of the relative velocity, $v_{B/C}$, which is
   acting in the transverse direction to the link CB, is:

   $$v_{B/C} = r_{B/C}\dot{\theta}_{C/B} = 0.3\sqrt{2}\dot{\theta}_{C/B}$$

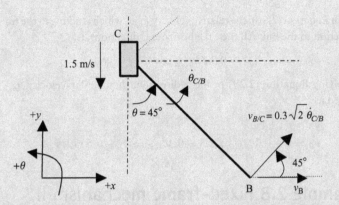

Figure 2.19: Magnitude and direction of the relative velocity of BC

Applying Equation (2.29) in the x-direction gives:

$$v_B = v_C + v_{B/C}$$
$$v_B = 0 + 0.3\sqrt{2}\dot{\theta}_{C/B}\cos 45 \Rightarrow v_B = 0.3\dot{\theta}_{C/B} \qquad\qquad\text{(E2.7)}$$

Applying Equation (2.29) in the y-direction gives:

$$0 = -1.5 + 0.3\sqrt{2}\dot{\theta}_{C/B}\sin 45 \Rightarrow \dot{\theta}_{C/B} = 5\,\text{rad/s}$$

From Equation (E2.7), the velocity of B is:

$$v_B = v_C + v_{B/C}$$
$$v_B = 0.3 \times 5 = 1.5\,\text{m/s}$$

b) Consider link AB. The magnitude and direction of the relative velocity $v_{B/A}$ are shown in Figure 2.20.

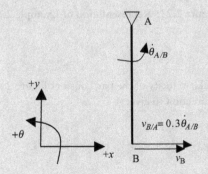

Figure 2.20: Magnitude and direction of the relative velocity of AB

From Equation (2.30), the relative velocity $v_{B/A}$, which is acting in the transverse direction to the link AB (i.e., the horizontal direction), is:

$$v_{B/A} = r_{B/A}\dot{\theta}_{A/B} = 0.3\sqrt{2}\dot{\theta}_{A/B}\cos 45 = 0.3\dot{\theta}_{A/B}$$

Applying Equation (2.29) in the x-direction, the angular velocity $\dot{\theta}_{A/B}$ is obtained as:

$$v_B = v_A + v_{B/A}$$
$$v_B = 0 + 0.3\dot{\theta}_{A/B} \Rightarrow 1.5 = 0.3\dot{\theta}_{A/B} \Rightarrow \dot{\theta}_{A/B} = 5\,\text{rad/s}$$

Example 2.8 Fixed–frame mechanism II

A linkage consists of a bar AB of 9 m length and two fixed frames. The bar AB is restricted to moving along the inclined planes as shown in Figure 2.21. The bar is horizontal when the velocity and acceleration of point A are 1.5 m/s and 3 m/s^2, respectively. Determine the acceleration of point B and the angular acceleration of the bar.

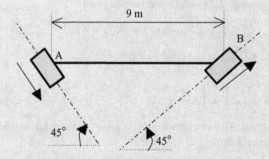

Figure 2.21: Representation of Example 2.8

Solution

To calculate the angular velocity of the bar, consider Figure 2.22 and apply Equation (2.29) in the x-direction to give:

$$v_B = v_A + v_{B/A}$$
$$v_B \cos 45 = v_A \cos 45 + 0 \Rightarrow v_B = v_A = 1.5\,\text{m/s}$$

Applying Equation (2.29) in the y-direction gives:

$$v_B \sin 45 = v_A \sin 45 + v_{B/A} \Rightarrow 1.5 \sin 45 = -1.5 \sin 45 + 9\dot{\theta} \Rightarrow \dot{\theta}$$
$$= 0.2357 \, \text{rad/s}$$

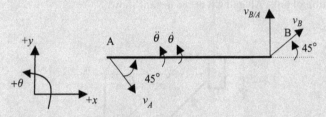

Figure 2.22: Angular velocity of AB

To calculate the angular accelerations of points A and B, consider Figure 2.23 and apply Equation (2.33)

$$a_B = a_A + a_{B/A}$$

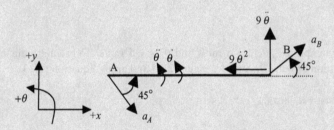

Figure 2.23: Angular accelerations of points A and B

in the x-direction:

$$a_B \cos 45 = a_A \cos 45 - 9\dot{\theta}^2 \Rightarrow a_B \cos 45 = 3 \cos 45 - 9 \times 0.2357^2$$
$$\Rightarrow a_B = 2.29 \, \text{m/s}^2$$

and in the y-direction:

$$a_B \sin 45 = a_A \sin 45 + 9\ddot{\theta} \Rightarrow 2.29 \times \sin 45 = -3 \times \sin 45 + 9\ddot{\theta} \Rightarrow \ddot{\theta}$$
$$= 0.415 \, \text{rad/s}^2$$

Example 2.9 Three-bar linkage mechanism II

In the three-bar linkage mechanism in Example 2.7, if point C moves with an acceleration 1 m/s² downwards, as shown in Figure 2.24, determine the angular acceleration of links AB and CB at the instant shown.

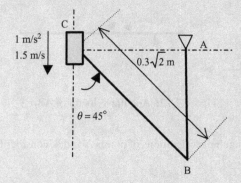

Figure 2.24: Representation of Example 2.9

Solution

From Example 2.7, $\dot\theta_{C/B} = 5$ rad/s and $\dot\theta_{A/B} = 5$ rad/s. To calculate the angular acceleration of bar AB, consider Figure 2.25 and apply Equation (2.33)

$$a_B = a_A + a_{B/A}$$

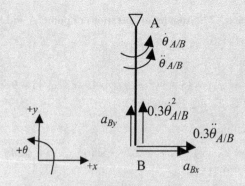

Figure 2.25: Angular acceleration of bar AB

in the x-direction, with $a_{Ax} = 0$ and $a_{B/Ax} = 0.3\ddot{\theta}_{A/B}$:

$$a_{Bx} = 0 + 0.3\ddot{\theta}_{A/B} \Rightarrow a_{Bx} = 0.3\ddot{\theta}_{A/B} \text{ m/s}^2 \qquad \text{(E2.9a)}$$

in the y-direction, with $a_{Ay} = 0$ and, $a_{B/Ay} = 0.3 \times \dot{\theta}_{A/B}^2$:

$$a_{By} = 0 + 0.3 \times \dot{\theta}_{A/B}^2 = 0.3 \times 5^2 = 7.5 \text{ m/s}^2$$

For bar CB, consider Figure 2.26, with $a_{Cx} = 0$ and $a_{B/Cx} = 0.3\sqrt{2}\ddot{\theta}_{C/B} \cos 45 - 0.3\sqrt{2}\dot{\theta}_{C/B}^2 \sin 45$. The accelerations in the x-direction are:

$$a_{Bx} = a_{Cx} + a_{B/Cx}$$
$$a_{Bx} = 0 + 0.3\sqrt{2}\ddot{\theta}_{C/B} \cos 45 - 0.3\sqrt{2}\dot{\theta}_{C/B}^2 \sin 45$$
$$= 0.3\ddot{\theta}_{C/B} - 0.3 \times 5^2 = 0.3\ddot{\theta}_{C/B} - 7.5$$

from which

$$a_{Bx} = 0.3\ddot{\theta}_{C/B} - 7.5 \qquad \text{(E2.9b)}$$

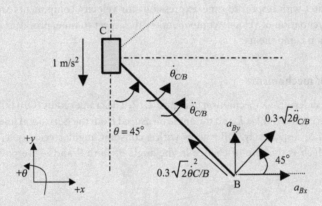

Figure 2.26: Angular acceleration of bar CB

The accelerations in the y-direction, with $a_{By} = 7.5$ m/s^2, $a_{Cy} = -1$ m/s^2 and $a_{B/Cy} = 0.3\sqrt{2}\ddot{\theta}_{C/B} \sin 45 + 0.3\sqrt{2}\dot{\theta}_{C/B}^2 \cos 45$, are:

$$a_B = a_{Cy} + a_{B/Cy}$$
$$a_{By} = a_{Cy} + 0.3\sqrt{2}\ddot{\theta}_{C/B} \sin 45 + 0.3\sqrt{2}\dot{\theta}_{C/B}^2 \cos 45$$
$$7.5 = -1 + 0.3 \times \ddot{\theta}_{C/B} + 0.3 \times 5^2$$

from which the angular acceleration of bar CB is obtained as:

$$\ddot{\theta}_{C/B} = 3.33 \text{ rad/s}^2$$

Substituting the angular acceleration of bar CB into Equation (E2.9b), the acceleration of B in the x-direction is:

$$a_{Bx} = 0.3 \times 3.33 - 7.5 = -6.5 \, \text{m/s}^2$$

From Equation (E2.9a), the angular acceleration of bar AB is:

$$\ddot{\theta}_{A/B} = \frac{a_{Bx}}{0.3} = -21.67 \, \text{rad/s}^2$$

2.4.2 Kinematics of mechanisms

The kinematics of mechanisms is concerned with techniques for finding the displacements, velocities and accelerations of the links in a mechanism. In order to study the kinematics of the mechanism, the co-ordinates of any point of interest on the link are defined using a co-ordinate system and related to the angular position of the crank. By differentiating the co-ordinates with respect to time, expressions for velocity components are obtained. Further differentiation of velocity components with respect to time, produces expressions for acceleration components.

Slider–crank mechanism

Consider the slider–crank mechanism in Figure 2.27. The crank radius (OB) is r, the length of the connecting rod (AB) is L and the angle measured from the horizontal line to OB and AB are θ and φ, respectively. OB rotates with a constant angular velocity ω, equivalent to $\dot{\theta} = \frac{d\theta}{dt}$, which is the rate of change in the angle θ. Both θ and ω are positive in the

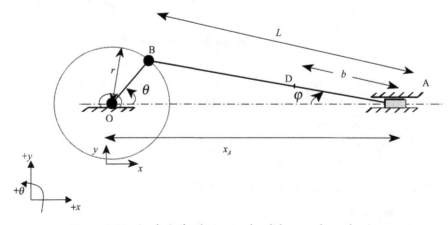

Figure 2.27: Analytical solution to the slider–crank mechanism

anti-clockwise sense, while the angle φ is measured in the clockwise sense. If the origin is taken as point O, around which the crank rotates, the co-ordinate of the slider A, which measures its displacement at a particular time, is given by:

$$x_A = r \cos \theta + L \cos \varphi \tag{2.36}$$

And θ is related to φ by:

$$r \sin \theta = L \sin \varphi \tag{2.37}$$

Substituting Equation (2.37) into (2.36) and using the relationship $\cos^2 \varphi + \sin^2 \varphi = 1$ gives:

$$x_A = r \cos \theta + L \sqrt{1 - \frac{r^2}{L^2} \sin^2 \theta} \tag{2.38}$$

Expanding the term $\sqrt{1 - \frac{r^2}{L^2} \sin^2 \theta}$ in Equation (2.38), using the binomial theorem, leads to:

$$x_A = r \cos \theta + L \left(1 - \frac{r^2}{2L^2} \sin^2 \theta - \frac{r^4}{8L^4} \sin^4 \theta - \dots \right) \tag{2.39}$$

For a slider–crank mechanism of a reciprocating engine, the ratio r/L is, in general, less than 1/3; moreover, $\sin \theta$ is less than or equal to 1. Thus the higher terms in Equation (2.39) can be neglected and the first two terms in Equation (2.39) would provide an acceptable approximation. Therefore, Equation (2.39) becomes:

$$x_A = r \cos \theta + L \left(1 - \frac{r^2}{2L^2} \sin^2 \theta \right) \tag{2.40}$$

Differentiating Equation (2.40) with respect to time, gives:

$$\dot{x}_A = -r\dot{\theta} \sin \theta + L \left(0 - \frac{r^2}{2L^2} 2\dot{\theta} \sin \theta \cos \theta \right) \tag{2.41}$$

Rearranging Equation (2.41) and using the notation ω for $\dot{\theta}$ and the relationship $\sin 2\theta = 2\sin\theta \cos\theta$, the velocity of A, $\dot{x}_A$, is found as:

$$\dot{x}_A = -r\omega \left(\sin \theta + \frac{r}{2L} \sin 2\theta \right) \tag{2.42}$$

Differentiating again with respect to time:

$$\ddot{x}_A = -r\ddot{\theta} \left(\sin \theta + \frac{r}{2L} \sin 2\theta \right) - r\dot{\theta} \left(\dot{\theta} \cos \theta + \frac{r}{2L} (2\dot{\theta}) \cos 2\theta \right) \tag{2.43}$$

If the crank rotates with a constant angular velocity, the angular acceleration, $\ddot{\theta}$, should be equal to zero. The acceleration of A, $\ddot{x}_A$, is then obtained as:

$$\ddot{x}_A = -r\omega^2 \left(\cos \theta + \frac{r}{L} \cos 2\theta \right) \tag{2.44}$$

The point D on the link AB has the co-ordinates (x_D, y_D) from point O. From Figure 2.27, x_D and y_D are:

$$x_D = x_A - b\cos\varphi \qquad (2.45)$$

$$y_D = b\sin\varphi \qquad (2.46)$$

Using the relationship $\cos^2\varphi + \sin^2\varphi = 1$, Equation (2.45) becomes:

$$x_D = x_A - b\sqrt{1 - \sin^2\varphi} \qquad (2.47)$$

Substituting Equation (2.37) into Equation (2.47), gives:

$$x_D = x_A - b\sqrt{1 - \frac{r^2}{L^2}\sin^2\theta} \qquad (2.48)$$

Approximating the square-root term in Equation (2.48) using the binomial theorem gives:

$$x_D = x_A - b\left(1 - \frac{r^2}{2L^2}\sin^2\theta\right) \qquad (2.49)$$

Similarly substituting Equation (2.37) into Equation (2.46), y_D becomes:

$$y_D = \frac{rb}{L}\sin\theta \qquad (2.50)$$

Differentiating Equations (2.49) and (2.50) with respect to time gives:

$$\dot{x}_D = \dot{x}_A + \frac{r^2\omega b}{2L^2}\sin 2\theta \qquad (2.51)$$

$$\dot{y}_D = \frac{r\omega b}{L}\cos\theta \qquad (2.52)$$

Differentiating Equations (2.51) and (2.52) with respect to time and considering constant angular velocity, the acceleration components of point D are obtained as:

$$\ddot{x}_D = \ddot{x}_A + \frac{r^2\omega^2 b}{L^2}\cos 2\theta \qquad (2.53)$$

$$\ddot{y}_D = -\frac{r\omega^2 b}{L}\sin\theta \qquad (2.54)$$

Example 2.10 Slider–crank mechanism

For the slider–crank mechanism shown in Figure 2.28, the connecting rod length (AB) is 0.5 m and the crank radius (OB) is 0.15 m. The crank rotates anti-clockwise with a constant angular velocity of 250 rev/min. For $\theta = 30°$, determine the velocity and acceleration of A and the mid-point of the connecting rod, D.

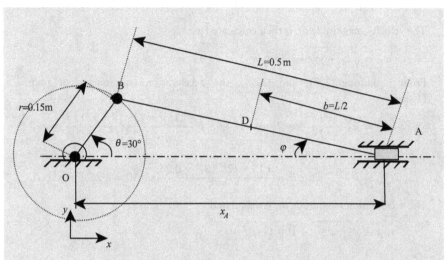

Figure 2.28: Representation of Example 2.10

Solution

The slider–crank mechanism data are summarized as follows:

$$L = 0.5\,\text{m}, \qquad r = 0.15\,\text{m}, \qquad \theta = 30° \quad \text{and}$$

$$\omega = 250\,\text{rev/min} = 250 \times \frac{2\pi}{60} = 26.18\,\text{rad/s}.$$

Using Equation (2.42), the velocity of A is obtained as:

$$\dot{x}_A = -r\omega \left(\sin\theta + \frac{r}{2L}\sin 2\theta\right) = -0.15 \times 26.18 \times \left(\sin 30 + \frac{0.15}{2 \times 0.5}\sin 60\right)$$

$$= -2.47\,\text{m/s}$$

Using Equation (2.44), the acceleration of A is obtained as:

$$\ddot{x}_A = -r\omega^2 \left(\cos\theta + \frac{r}{L}\cos 2\theta\right) = -0.15 \times (26.18)^2 \left(\cos 30 + \frac{0.15}{0.5}\cos 60\right)$$

$$= -104.46\,\text{m/s}^2$$

The distance b between the mid-point D and the slider A is: $b = L/2 = 0.25$ m.

From Equations (2.51) and (2.52), the velocity components of point D are:

$$\dot{x}_D = x_A + \frac{r^2\omega b}{2L^2}\sin 2\theta = -2.47 + \frac{(0.15)^2 \times 26.18 \times 0.25}{2 \times (0.5)^2}\sin 60 = -2.215\,\text{m/s}$$

$$\dot{y}_D = \frac{r\omega b}{L}\cos\theta = \frac{0.15 \times 26.18 \times 0.25}{0.5}\cos 30 = 1.7\,\text{m/s}$$

The velocity magnitude v_D is then calculated as:

$$v_D = \sqrt{\dot{x}_D^2 + \dot{y}_D^2} = 2.79 \, \text{m/s}$$

From Equations (2.53) and (2.54), the acceleration components of point D are:

$$\ddot{x}_D = \ddot{x}_A + \frac{r^2 \omega^2 b}{L^2} \cos 2\theta = -104.45 + \frac{(0.15)^2 \times (26.18)^2 \times 0.25}{(0.5)^2} \cos 60$$

$$= -96.739 \, \text{m/s}^2$$

$$\ddot{y}_D = -\frac{r \omega^2 b}{L} \sin \theta = -\frac{0.15 \times (26.18)^2 \times 0.25}{0.5} \sin 30 = -25.7 \, \text{m/s}^2$$

The acceleration magnitude a_D is calculated as:

$$a_D = \sqrt{\ddot{x}_D^2 + \ddot{y}_D^2} = 100.1 \, \text{m/s}^2$$

Offset slider–crank mechanism

An offset slider–crank mechanism generates a faster stroke in one direction than that generated by the standard slider–crank mechanism. A faster stroke, which means lower acceleration, produces lower forces for constant crank torque. In an offset slider–crank mechanism, the crank axis of rotation is offset from the centre-line of the slider by a distance, a, as shown in Figure 2.29. In such a case, the analytical solution derived for the slider–crank mechanism should be modified to account for the offset.

The co-ordinate x of the slider A is given by Equation (2.36). However, the angle θ is related to φ (as can be deduced from Figure 2.29) by:

$$r \sin \theta = L \sin \phi + a$$

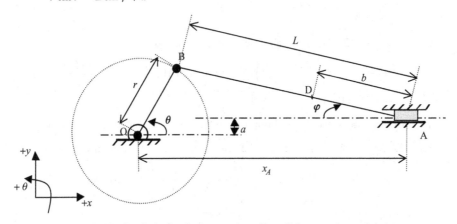

Figure 2.29: Analytical solution to the offset slider–crank mechanism

or

$$\sin \phi = \frac{r}{L} \sin \theta - \frac{a}{L} \qquad (2.55)$$

Using the relationship $\cos^2 \phi + \sin^2 \phi = 1$ (i.e. $\cos \phi = \sqrt{1 - \sin^2 \phi}$) and substituting Equation (2.55) into (2.36), gives:

$$x_A = r \cos \theta + L \sqrt{1 - \frac{r^2}{L^2} \left(\sin \theta - \frac{a}{r} \right)^2} \qquad (2.56)$$

Expanding the term $\sqrt{1 - \frac{r^2}{L^2} \left(\sin \theta - \frac{a}{r} \right)^2}$ using the binomial theorem and retaining only the first two terms, leads to:

$$x_A = r \cos \theta + L \left(1 - \frac{r^2}{2L^2} \left(\sin \theta - \frac{a}{r} \right)^2 \right) \qquad (2.57)$$

Differentiating Equation (2.57) with respect to time gives:

$$\dot{x}_A = -r\dot{\theta} \sin \theta + L \left(0 - \frac{r^2}{2L^2} 2\dot{\theta} \cos \theta \left(\sin \theta - \frac{a}{r} \right) \right) \qquad (2.58)$$

Rearranging Equation (2.58) and using the notation ω for $\dot{\theta}$ and the relationship $\sin 2\theta = 2\sin \theta \cos \theta$, the velocity of A, $\dot{x}_A$, is found to be:

$$\dot{x}_A = -r\omega \left(\sin \theta + \frac{r}{2L} \sin 2\theta - \frac{a}{L} \cos \theta \right) \qquad (2.59)$$

Differentiating Equation (2.58) again with respect to time, gives:

$$\ddot{x}_A = -r\ddot{\theta} \left(\sin \theta + \frac{r}{2L} \sin 2\theta - \frac{a}{L} \cos \theta \right)$$
$$- r\dot{\theta} \left(\dot{\theta} \cos \theta + \frac{r}{2L} (2\dot{\theta}) \cos 2\theta + \frac{a}{L} \sin \theta \right) \qquad (2.60)$$

Again, if the crank rotates with a constant angular velocity, the angular acceleration, $\ddot{\theta}$, is equal to zero, and Equation (2.60) becomes:

$$\ddot{x}_A = -r\omega^2 \left(\cos \theta + \frac{r}{L} \cos 2\theta + \frac{a}{L} \sin \theta \right) \qquad (2.61)$$

Similarly, the point D has the co-ordinates (xD, yD) from point O:

$$x_D = x_A - b \cos \varphi$$
$$y_D = b \sin \varphi + a$$

As before using $\sin \phi = \frac{r}{L} \sin \theta - \frac{a}{L}$, from Equation (2.55) and $\cos^2 \varphi + \sin^2 \varphi = 1$ and approximating the square-root using the binomial theorem gives:

$$x_D = x_A - b \left(1 - \frac{r^2}{2L^2} \left(\sin \theta - \frac{a}{L} \right)^2 \right) \qquad (2.62)$$

And writing y_D in terms of θ, leads to:

$$y_D = \frac{rb}{L}\sin\theta - \frac{ab}{L} + a \tag{2.63}$$

Differentiating (Equations (2.62) and (2.63)) with respect to time yields:

$$\dot{x}_D = \dot{x}_A + \frac{r^2\omega b}{2L^2}\sin 2\theta - \frac{r\omega ba}{L^2}\cos\theta \tag{2.64}$$

$$\dot{y}_D = \frac{r\omega b}{L}\cos\theta \tag{2.65}$$

Differentiating again with respect to time and considering constant angular velocity gives:

$$\ddot{x}_D = \ddot{x}_A + \frac{r^2\omega^2 b}{L^2}\cos 2\theta + \frac{r\omega^2 ba}{L^2}\sin\theta \tag{2.66}$$

$$\ddot{y}_D = -\frac{r\omega^2 b}{L}\sin\theta \tag{2.67}$$

Example 2.11 Offset slider–crank mechanism

An offset slider–crank mechanism, shown in Figure 2.30, has a crank radius to connecting rod ratio of 0.25. The connecting rod length (AB) is 0.5 m and the crank axis of rotation is offset from the centre-line of the slider by 0.05 m. The crank rotates anti-clockwise with a constant angular velocity 380 rev/min. Determine the velocity and acceleration of A when $\theta = 30°$ and the angle θ in the first quadrant at which the maximum velocity takes place.

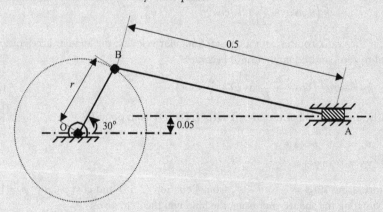

Figure 2.30: Representation of Example 2.11

Solution

The slider–crank mechanism data are summarized as follows:

$L = 0.5$ m, $r/L = 0.25$, thus $r = 0.25 \times 0.5 = 0.125$ m, $a = 0.05$ m and $\theta = 30°$.
$\omega = 380$ rev/min $= 380 \times 2\pi/60 = 39.7935$ rad/s.

From Equation (2.59), the velocity of A is:

$$\dot{x}_A = -r\omega \left(\sin\theta + \frac{r}{2L} \sin 2\theta - \frac{a}{L} \cos\theta \right) = -0.125 \times 39.7935$$

$$\times \left(\sin 30 + \frac{0.125}{2 \times 0.5} \sin 60 - \frac{0.05}{0.5} \cos 30 \right) = -2.59 \text{ m/s}$$

From Equation (2.61), the acceleration of A is:

$$\ddot{x}_A = -r\omega^2 \left(\cos\theta + \frac{r}{L} \cos 2\theta + \frac{a}{L} \sin\theta \right) = -0.125 \times (39.7935)^2$$

$$\times \left(\cos 30 + \frac{0.125}{0.5} \cos 60 + \frac{0.05}{0.5} \sin 30 \right) = -206.06 \text{ m/s}^2$$

Maximum velocity at $\ddot{x}_A = 0$, i.e.

$$\ddot{x}_A = -r\omega^2 \left(\cos\theta + \frac{r}{L} \cos 2\theta + \frac{a}{L} \sin\theta \right) = 0$$

$$\cos\theta + \frac{r}{L} \cos 2\theta + \frac{a}{L} \sin\theta = 0 \qquad\qquad (E2.11)$$

Using $\cos 2\theta = 2\cos^2\theta - 1$ and binomial theory, $\sin\theta = \sqrt{1 - \cos^2\theta} = (1 - \frac{1}{2}\cos^2\theta)$, which is valid for small $\cos\theta$, i.e. $\cos\theta \ll 1 \Rightarrow \theta > 60$.

Rewriting Equation (E2.11) gives:

$$\cos\theta + \frac{r}{L}(2\cos^2\theta - 1) + \frac{a}{L}\left(1 - \frac{1}{2}\cos^2\theta \right) = 0$$

and rearranging terms leads to:

$$\left(\frac{2r}{L} - \frac{a}{2L} \right) \cos^2\theta + \cos\theta - \left(\frac{r}{L} - \frac{a}{L} \right) = 0$$

Solving for $\cos\theta$ using the quadratic formula, gives:

$$\cos\theta = \frac{-1 \pm \sqrt{1 - 4 \times \left(\frac{2r}{L} - \frac{a}{2L} \right) \times \left(-\left(\frac{r}{L} - \frac{a}{L} \right) \right)}}{2 \times \left(\frac{2r}{L} - \frac{a}{2L} \right)}$$

For $r/L = 0.25$ and $a/L = 0.05/0.5 = 0.1$, the angle θ at which the maximum velocity takes place is obtained as:

$$\cos\theta = 0.141 \Rightarrow \theta = 81.89°$$

Four-bar linkage mechanism

A four-bar linkage mechanism is a kinematic chain that consists of four links or bars. The four links are connected to each other through turning pairs. Figure 2.31 shows a four-bar linkage mechanism, ABCD, in which linkage AD is a fixed frame. The link BC, which is not connected to the fixed frame, is called the coupler link. If the link AB rotates about A with an angular velocity $\dot{\theta}$, then AB is called the crank link and CD is called the follower link.

In order to determine the velocities and accelerations at any point in the four-bar linkage, the displacements of the point of interest are related geometrically, then differentiated with respect to time. Considering the four-bar linkage mechanism in Figure 2.32, the length of the crank link (AB) is L_1, the coupler link (BC) is L_2, the follower link (CD) is L_3 and the fixed frame (AD) is L_4. The angle measured from the horizontal axis to the crank link is θ, measured anti-clockwise; the angle measured from the horizontal axis to the follower link is φ, measured clockwise; and the angle measured from the horizontal axis to the coupler link is ψ measured anti-clockwise.

The crank link AB rotates anti-clockwise with a constant angular velocity $\dot{\theta}$. The displacement of a point E on the coupler link, measured from A is given by:

$$x_E = L_1 \cos\theta + b\cos\psi \tag{2.68}$$
$$y_E = L_1 \sin\theta + b\sin\psi \tag{2.69}$$

where b is the distance on the coupler link measured from B to E. Differentiating Equations

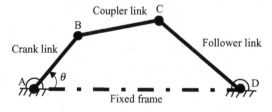

Figure 2.31: A four-bar linkage mechanism

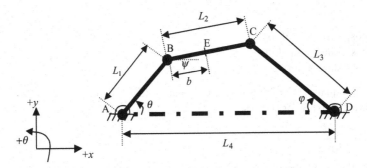

Figure 2.32: Dimensions of a four-bar linkage mechanism

(2.68) and (2.69) with respect to time, the velocity components of point E are:

$$\dot{x}_E = -L_1\dot{\theta}\sin\theta - b\dot{\psi}\sin\psi \tag{2.70}$$

$$\dot{y}_E = L_1\dot{\theta}\cos\theta + b\dot{\psi}\cos\psi \tag{2.71}$$

Differentiating Equations (2.70) and (2.71) with respect to time to obtain the acceleration components in the x and y directions and considering the constant angular velocity of the crank link, i.e. $\ddot{\theta} = 0$, gives:

$$\ddot{x}_E = -L_1\dot{\theta}^2\cos\theta - b\ddot{\psi}\sin\psi - b\dot{\psi}^2\cos\psi \tag{2.72}$$

$$\ddot{y}_E = -L_1\dot{\theta}^2\sin\theta + b\ddot{\psi}\cos\psi - b\dot{\psi}^2\sin\psi \tag{2.73}$$

The relationships between the three angles θ, ψ and ϕ can be deduced from Figure 2.32 as follows:

$$L_1\sin\theta + L_2\sin\psi = L_3\sin\varphi \tag{2.74}$$

$$L_4 = L_1\cos\theta + L_2\cos\psi + L_3\cos\varphi \tag{2.75}$$

From Equation (2.74), the angle ψ can be determined as:

$$\sin\psi = \frac{L_3\sin\varphi - L_1\sin\theta}{L_2} \tag{2.76}$$

Differentiating Equations (2.74) and (2.75) with respect to time gives:

$$L_1\dot{\theta}\cos\theta + L_2\dot{\psi}\cos\psi = L_3\dot{\varphi}\cos\varphi \tag{2.77}$$

$$0 = -L_1\dot{\theta}\sin\theta - L_2\dot{\psi}\sin\psi - L_3\dot{\varphi}\sin\varphi \tag{2.78}$$

The angular velocities $\dot{\psi}$ and $\dot{\varphi}$ are obtained by solving Equations (2.77) and (2.78). Multiplying Equation (2.77) by $\sin\varphi$ and Equation (2.78) by $\cos\varphi$, adding the two equations and solving for $\dot{\psi}$ gives:

$$\dot{\psi} = -\frac{L_1\dot{\theta}C_1}{L_2C_2} \tag{2.79}$$

where

$$C_1 = \cos\theta \sin\varphi + \sin\theta \cos\varphi$$
$$C_2 = \cos\psi \sin\varphi + \sin\psi \cos\varphi$$
$$C_3 = \cos\theta \sin\psi - \sin\theta \cos\psi$$

Multiplying Equation (2.77) by $\sin\psi$ and Equation (2.78) by $\cos\psi$, subtracting the two equations and solving for $\dot\varphi$ gives:

$$\dot\varphi = \frac{L_1\dot\theta C_3}{L_3 C_2} \tag{2.80}$$

To obtain the angular accelerations $\ddot\psi$ and $\ddot\varphi$, Equations (2.77) and (2.78) are further differentiated with respect to time as:

$$L_1\ddot\theta \cos\theta - L_1\dot\theta^2 \sin\theta + L_2\ddot\psi \cos\psi - L_2\dot\psi^2 \sin\psi$$
$$= L_3\ddot\varphi \cos\varphi - L_3\dot\varphi^2 \sin\varphi \tag{2.81}$$
$$0 = -L_1\ddot\theta \sin\theta - L_1\dot\theta^2 \cos\theta - L_2\ddot\psi \sin\psi - L_2\dot\psi^2 \cos\psi$$
$$- L_3\ddot\varphi \sin\varphi - L_3\dot\varphi^2 \cos\varphi \tag{2.82}$$

The angular accelerations $\ddot\psi$ and $\ddot\varphi$ are obtained by solving Equations (2.81) and (2.82) and considering constant angular rotation of the crank link $\dot\theta$, i.e. $\ddot\theta = 0$. Multiplying Equation (2.81) by $\sin\varphi$ and Equation (2.82) by $\cos\varphi$, adding the two equations and solving for $\ddot\psi$ gives:

$$\ddot\psi = -\left(\frac{L_1\dot\theta^2 C_4 + L_2\dot\psi^2 C_5 + L_3\dot\varphi^2}{L_2 C_2} \right) \tag{2.83}$$

where

$$C_4 = \cos\theta \cos\varphi - \sin\theta \sin\varphi$$
$$C_5 = \cos\psi \cos\varphi - \sin\psi \sin\varphi$$
$$C_6 = \cos\theta \cos\psi + \sin\theta \sin\psi$$

Similarly, multiplying Equation (2.81) by $\sin\psi$ and Equation (2.827) by $\cos\psi$, subtracting the two equations and solving for $\ddot\varphi$ gives:

$$\ddot\varphi = -\left(\frac{L_1\dot\theta^2 C_6 + L_2\dot\psi^2 + L_3\dot\varphi^2 C_5}{L_3 C_2} \right) \tag{2.84}$$

Example 2.12 Four-bar linkage mechanism

A four-bar linkage, ABCD, has a crank link length of 0.15 m, a coupler link of 0.35 m, a follower link of 0.4 m and a fixed frame of 0.5 m. The crank link rotates with a constant angular velocity of 3 rad/s. For the position shown in Figure 2.33, determine:

a) the coupler link and follower link angular velocities;
b) the coupler link and follower link angular accelerations;
c) the velocity and acceleration components of point E, located in the middle of coupler link BC, in the Cartesian co-ordinate system.

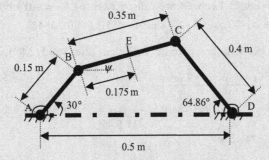

Figure 2.33: Representation of Example 2.12

Solution

The four-bar linkage data are summarized as follows:

$$L_1 = 0.15 \, \text{m}, \quad L_2 = 0.35 \, \text{m}, \quad L_3 = 0.4 \, \text{m}, \quad L_4 = 0.5 \, \text{m}, \quad \theta = 30°$$
$$\text{and} \; \varphi = 64.86°$$

From Equation (2.76), the angle ψ can be determined as:

$$\sin \psi = \frac{L_3 \sin \varphi - L_1 \sin \theta}{L_2} = \frac{0.4 \times \sin 64.86 - 0.15 \sin 30}{0.35} \Rightarrow \psi = 55.12°$$

a) The constants C_1 to C_3, from Equation (2.79), are:

$$C_1 = \cos 30 \; \sin 64.86 + \sin 30 \; \cos 64.86 = 0.9964$$
$$C_2 = \cos 55.12 \; \sin 64.86 + \sin 55.12 \; \cos 64.86 = 0.8662$$
$$C_3 = \cos 30 \; \sin 55.12 - \sin 30 \; \cos 55.12 = 0.42452$$

The angular velocities $\dot{\psi}$ and $\dot{\varphi}$, Equations (2.79) and (2.80), are obtained as:

$$\dot{\psi} = -\frac{L_1 \dot{\theta} C_1}{L_2 C_2} = -\frac{0.15 \times 3 \times 0.9964}{0.35 \times 0.8662} = -1.479 = -1.48 \, \text{rad/s}$$

and

$$\dot{\varphi} = \frac{L_1\dot{\theta}C_3}{L_3C_2} = \frac{0.15 \times 3 \times 0.42452}{0.4 \times 0.8662} = 0.5514 = 0.55\,\text{rad/s}$$

b) The constants C_4 to C_6 from Equation (2.83) are:

$$C_4 = \cos 30 \ \cos 64.86 - \sin 30 \ \sin 64.86 = -0.08472$$
$$C_5 = \cos 55.12 \cos 64.86 - \sin 55.12 \sin 64.86 = -0.4997$$
$$C_6 = \cos 30 \cos 55.12 + \sin 30 \sin 55.12 = 0.90542$$

The angular accelerations $\ddot{\psi}$ and $\ddot{\varphi}$, Equations (2.83) and (2.84), are:

$$\ddot{\psi} = -\left(\frac{L_1\dot{\theta}^2 C_4 + L_2\dot{\psi}^2 C_5 + L_3\dot{\varphi}^2}{L_2 C_2}\right)$$
$$= -\left(\frac{0.15 \times 3^2 \times (-0.08472) + 0.35 \times (-1.479)^2 \times (-0.4997) + 0.4 \times 0.5514^2}{0.35 \times 0.8662}\right) -$$
$$\ddot{\psi} = 1.238 = 1.24\,\text{rad/s}^2$$

And

$$\ddot{\varphi} = -\left(\frac{L_1\dot{\theta}^2 C_6 + L_2\dot{\psi}^2 + L_3\dot{\varphi}^2 C_5}{L_3 C_2}\right)$$
$$= -\left(\frac{0.15 \times 3^2 \times 0.90542 + 0.35 \times (-1.479)^2 + 0.4 \times 0.5514^2 \times -0.4997}{0.4 \times 0.8662}\right)$$
$$\ddot{\varphi} = -5.561 = -5.56\,\text{rad/s}^2$$

c) From Equations (2.70) and (2.71), the velocity components of point E, are:

$$\dot{x}_E = -L_1\dot{\theta}\sin\theta - b\dot{\psi}\sin\psi = -0.15 \times 3 \times \sin 30 - 0.175 \times (-1.479)$$
$$\times \sin 55.12 = -0.01267\,\text{m/s}$$
$$\dot{y}_E = L_1\dot{\theta}\cos\theta + b\dot{\psi}\cos\psi = 0.15 \times 3 \times \cos 30 + 0.175 \times (-1.479)$$
$$\times \cos 55.12 = 0.2417\,\text{m/s}$$

The total velocity magnitude is:

$$v_E = \sqrt{\dot{x}_E^2 + \dot{y}_E^2} = \sqrt{(-0.01267)^2 + 0.2417^2} = 0.24\,\text{m/s}$$

The acceleration components of point E are obtained from Equations (2.72) and (2.73) as:

$$\ddot{x}_E = -L_1\dot{\theta}^2\cos\theta - b\ddot{\psi}\sin\psi - b\dot{\psi}^2\cos\psi$$
$$= -0.15 \times 3^2 \cos 30 - 0.175 \times 1.238 \times \sin 55.12 - 0.175$$
$$\times (-1.479)^2 \cos 55.12 = -1.56577\,\text{m/s}^2$$

$$\ddot{y}_E = -L_1\dot{\theta}^2 \sin\theta + b\ddot{\psi}\cos\psi - b\dot{\psi}^2 \sin\psi$$
$$= -0.15 \times 3^2 \sin 30 + 0.175 \times 1.238 \times \cos 55.12 - 0.175$$
$$\times (-1.479)^2 \sin 55.12 = -0.8651\,\text{m/s}^2$$

The total acceleration magnitude is:

$$a_E = \sqrt{\ddot{x}_E^2 + \ddot{y}_E^2} = \sqrt{(-1.56577)^2 + (-0.8651)^2} = 1.79\,\text{m/s}^2$$

2.5 Tutorial Sheet

2.5.1 Kinematics of wheels and gears

Q2.1 A wheel of 0.15 m radius starts to rotate from rest when a weight W is attached as shown in Figure 2.34. The weight acceleration is 3 m/s^2 downwards. When the time is 2 seconds, determine:

a) the angular acceleration of the wheel;

[−20 rad/s^2]

b) the angular velocity of the wheel;

[−40 rad/s]

c) the magnitude of the velocity of a point A at the outer edge;

[6 m/s]

d) the magnitude of the acceleration of a point A at the outer edge.

[240 m/s^2]

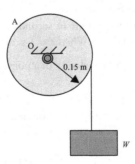

Figure 2.34: Representation of Question 2.1

Q2.2 A car travels with a constant speed of v and has tyres of radius 350 mm as shown in Figure 2.35. When the driver applies the brakes, the car has a constant deceleration of a. Consider the moment just after applying the brakes. If the wheel's angular velocity is 79.37 rad/s and its angular acceleration is −6.43 rad/s², determine:

a) the speed of the car, v, in km/hr;

[100 km/hr]

b) the deceleration of the car, a, in m/s²;

[−2.25 m/s²]

c) the radial and transverse acceleration of a point A, 250 mm from the wheel's centre.

[−1575 m/s², −1.61 m/s²]

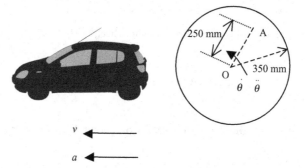

Figure 2.35: Representation of Question 2.2

Q2.3 A disk rotates with an angular velocity equivalent to $\dot{\theta} = 4t^2 + 3$, where $\dot{\theta}$ is in rad/s and t is the time in seconds. The disk has a radius of 0.5 m, as shown in Figure 2.36. When $t = 1$ s, determine:

a) the magnitude of the velocity of a point A at the outer edge of the disk;

[3.5 m/s]

d) the magnitude of the acceleration of a point A at the outer edge of the disk.

[24.82 m/s²]

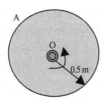

Figure 2.36: Representation of Question 2.3

Q2.4 A sphere, shown in Figure 2.37, has a radius of 0.3 m and starts to rotate about its axis from rest, i.e. at $\theta = 0$, $\dot{\theta} = 0$, with an angular acceleration equal to $\ddot{\theta} = 3\theta$, where $\ddot{\theta}$ is in rad/s^2 and θ is in radians. When $\theta = 4$ radians, determine:

a) the angular acceleration of the sphere;

[12 rad/s^2]

b) the angular velocity of the sphere;

[6.93 rad/s]

c) the magnitude of the velocity and acceleration of point A in Figure 2.37.

[1.8 m/s, 12.85 m/s^2]

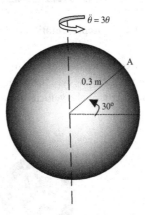

Figure 2.37: Representation of Question 2.4

Q2.5 A wheel of radius 1.2 m rolls as shown in Figure 2.38. Its centre of gravity is located at point O and at the instant shown it is in contact with the ground at point A. Its angular position is given by $\theta = 0.5t^2$, where θ is in radians and t is the time in seconds and its motion starts from rest. When $t = 2$ s, determine:

a) the velocity of point O;

[2.4 m/s]

b) the acceleration of point O;

[1.2 m/s^2]

c) the acceleration of point A;

[4.8 m/s^2]

d) the distance travelled.

[2.4 m]

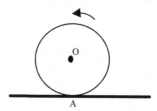

Figure 2.38: Representation of Question 2.5

Q2.6   A point B is located at a radius of 0.16 m from a tyre's centre of gravity as shown in Figure 2.39. The tyre has a radius of 0.4 m. If the velocity of point B is 4 m/s, determine the velocity of point A at the tyre's outer edge.

[10 m/s]

Figure 2.39: Representation of Question 2.6

Q2.7   If a pipe of radius 0.8 m is free to roll on the back of a truck as shown in Figure 2.40, determine the pipe's angular velocity when the truck travels with a speed of 9 m/s.

[11.25 rad/s]

Figure 2.40: Representation of Question 2.7

Q2.8   Determine the velocity and acceleration of the centre of gravity of the pipe in Figure 2.40, if it rolls with an angular velocity of 5 rad/s and angular acceleration of 2 rad/s^2.

[4 m/s, 1.6 m/s^2]

Q2.9   A gear B of radius 0.15 m rolls anticlockwise with an angular velocity of 3 rad/s between two gear racks A and C as shown in Figure 2.41. If gear rack A is fixed, determine the velocity of gear rack C.

[0.9 m/s]

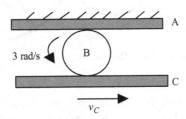

Figure 2.41: Representation of Question 2.9

Q2.10 In Question 2.9, if gear rack A moves to the left with a speed of 2 m/s as shown in Figure 2.42, determine the velocity of gear rack C.

[−1.1 m/s]

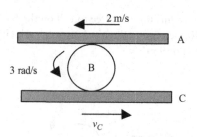

Figure 2.42: Representation of Question 2.10

Q2.11 Two gears 1 and 2 are meshed together as shown in Figure 2.43. Gear 1 has a radius of 20 mm, while gear 2 has a radius of 80 mm. Gear 1 starts motion from rest with a constant angular acceleration of 3 rad/s². At what time would the angular velocity of gear 2 reach 60 rad/s?

[80 s]

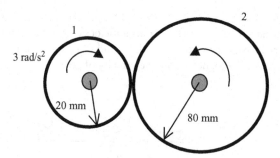

Figure 2.43: Representation of Question 2.11

Q2.12 In Figure 2.44, gear 1, of 60 mm radius, is attached to a motor and runs with an angular acceleration equal to $\ddot{\theta} = 3t^2$, where $\ddot{\theta}$ is in rad/s² and t is the time in seconds. Gear 1 has an initial velocity of 30 rad/s and meshes with gear 2, which has a radius of 180 mm. When $t = 3$ s, determine the angular velocity of gear 2.

[19 rad/s]

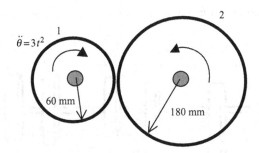

Figure 2.44: Representation of Question 2.12

Q2.13 A three-speed automotive transmission has the gear arrangements shown in Figure 2.45 for its reverse operation. If gear 1 rotates with an angular velocity of 50 rad/s, determine the angular velocity of gear 6.

[110.6 rad/s]

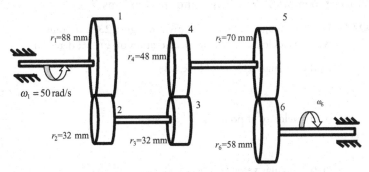

Figure 2.45: Representation of Question 2.13

Q2.14 An electric motor of a drill starts from rest and runs with a constant angular acceleration of 30 rad/s^2. When the time is 2 seconds, determine:

a) the angular velocity;

[60 rad/s]

b) the angular displacement.

[9.55 rev]

Q2.15 Determine the five gear ratios for the engine gearbox shown in Figure 2.46.

[4, 2.29, 1.33, 1.1, 1]

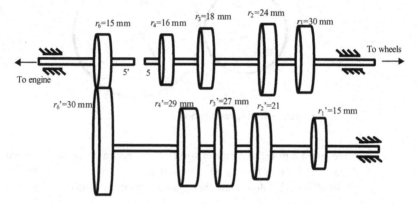

Figure 2.46: Representation of Question 2.15

2.5.2 Kinematics of linkages and mechanisms

Q2.16 For the linkage mechanism shown in Figure 2.47, if point C moves down with a velocity of 1 m/s and acceleration of 2 m/s^2, determine:

a) the angular velocity of link CB;

[2.5 rad/s]

b) the velocity of point B;

[1 m/s]

c) the angular velocity of link AB;

[2.5 rad/s]

d) the angular acceleration of link CB;

[5 rad/s^2]

e) the acceleration components of point B;

[−0.5 m/s^2, 2.5 m/s^2]

f) the angular acceleration of link AB.

[−1.25 rad/s^2]

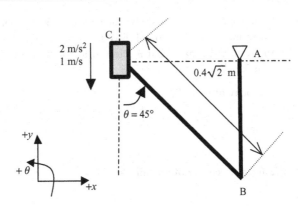

Figure 2.47: Representation of Question 2.16

Q2.17 In the linkage mechanism shown in Figure 2.48, point C moves horizontally, while the bar AB rotates with an angular velocity of 6 rad/s in the clockwise sense. Determine:

a) the absolute value of the velocity of point B;

[2.55 m/s]

b) the angular velocity of bar BC;

[3 rad/s]

c) the velocity of point C.

[2.7 m/s]

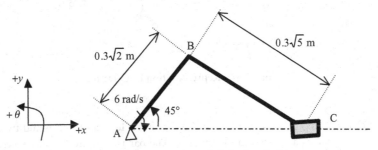

Figure 2.48: Representation of Question 2.17

Q2.18 For the linkage mechanism shown in Figure 2.49, the slider C is constrained to move in a horizontal direction, while point A is constrained in both x and y directions but is allowed to rotate freely. If the link AB rotates at 3 rad/s, determine:

a) the velocity of point B;

[0.75 m/s]

b) the angular velocity of the link BC;

[7.5 rad/s]

c) the velocity of the slider C.

[−1.06 m/s]

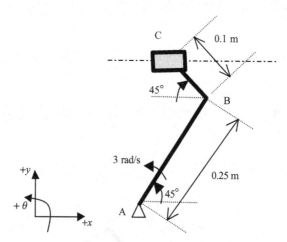

Figure 2.49: Representation of Question 2.18

Q2.19 A linkage consists of a bar AB that has a length of 0.2 m and two fixed frames as shown in Figure 2.50. The bar is restricted to move vertically at A and horizontally at B. At the instant shown, point A is moving downwards with a velocity of 0.1 m/s and an acceleration of 0.14 m/s^2. determine:

a) the angular velocity of the bar;

[1 rad/s]

b) the velocity of point B;

[0.17 m/s]

c) the angular acceleration of the bar;

[−0.33 rad/s^2]

d) the acceleration of point B.

[−0.16m/s^2]

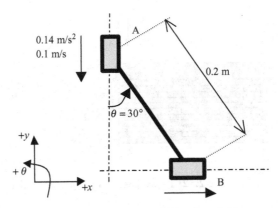

Figure 2.50: Representation of Question 2.19

Q2.20 A uniform bar of length 3 m rotates at its point A as shown in Figure 2.51. If the bar rotates with an angular velocity of 2 rad/s and an angular acceleration of 1.5 rad/s^2 at the position shown, determine the magnitude of velocity and acceleration of the centre of gravity of the rod, G.

[3 m/s, 6.4 m/s^2]

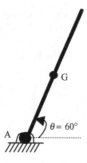

Figure 2.51: Representation of Question 2.20

Q2.21 A 9 m long ladder is supported on a wall at point A and on the ground at point B as shown in Figure 2.52. If the velocity and acceleration of point B away from the wall are 2 m/s and 1.5 m/s², respectively, determine the angular velocity and angular acceleration of the ladder.

[0.44 rad/s, 0.675 rad/s²]

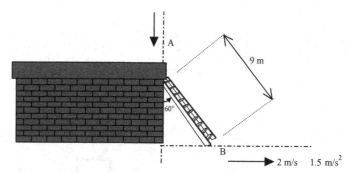

Figure 2.52: Representation of Question 2.21

Q2.22 An L-shaped link rotates about point O with an angular velocity of 2 rad/s and an angular acceleration of 1.5 rad/s², as shown in Figure 2.53. Determine:

a) the magnitude of velocity and acceleration of point A;

[0.9 m/s, 1.92 m/s²]

b) the magnitude of velocity and acceleration of point B.

[1.27 m/s, 2.72 m/s²]

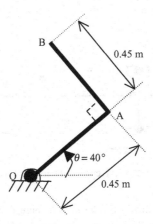

Figure 2.53: Representation of Question 2.22

Q2.23 For the slider–crank mechanism shown in Figure 2.54, the connecting rod length (AB) is 0.4 m and the crank radius (OB) is 0.1 m. The crank rotates anti-clockwise with a constant angular velocity of 3500 rev/min. For $\theta = 30°$, determine the velocity and acceleration of A and of the mid-point of the connecting rod, D.

$$[-22.29 \text{ m/s}, 25.76 \text{ m/s}, -13\,313 \text{ m/s}^2, 12\,918 \text{ m/s}^2]$$

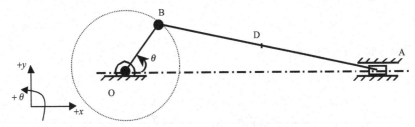

Figure 2.54: Representation of Question 2.23

Q2.24 Figure 2.55 shows an offset slider–crank mechanism, in which the crank, OB, rotates about O with constant velocity 600 rev/min anti-clockwise and drives through the rod AB a slider A that has an offset distance of 0.02 m downwards. For the position shown, calculate the velocity and acceleration of slider A.

$$[3.14 \text{ m/s}, -29.6 \text{ m/s}^2]$$

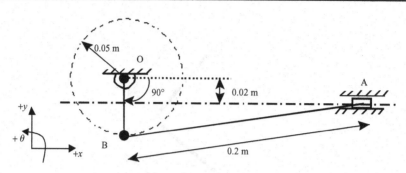

Figure 2.55: Representation of Question 2.24

Q2.25 For the slider–crank mechanism in Figure 2.54, determine the angle θ, in the first quadrant, at which the velocity of A is maximum (absolute maximum) and hence calculate the absolute maximum velocity of A.

[77°, 37.72 m/s]

Q2.26 In the slider–crank mechanism shown in Figure 2.56, the crank rotates anti-clockwise with a constant angular velocity of 3000 rev/min. The crank radius (OB) is 0.15 m and the ratio of the crank radius to the length of the connecting rod (AB) is 0.25. If the off-set distance, a, is 0.1 m downwards, determine:

a) the velocity of A;

[33.66 m/s]

b) the acceleration of A.

[12 213 m/s²]

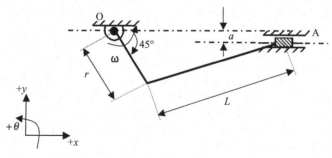

Figure 2.56: Representation of Question 2.26

Q2.27 Determine the constant angular velocity (in rev/min) of the crank shown in Figure 2.57, when the velocity of the slider A is 3 m/s to the left. What is the acceleration of A? The crank radius (OB) is 0.1 m and the ratio of the crank radius to the length of the connecting rod (AB) is 0.3.

[454.7 rev/min, −230.4 m/s²]

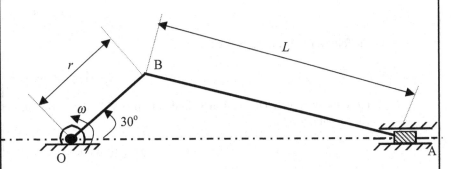

Figure 2.57: Representation of Question 2.27

Q2.28 For the slider–crank mechanism in Figure 2.58, determine the angle θ, in the first quadrant, at which the maximum velocity of the slider takes place.

[80.63°]

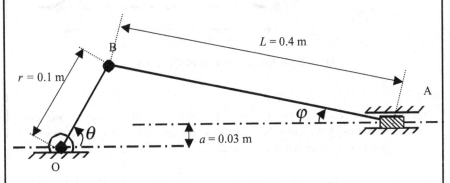

Figure 2.58: Representation of Question 2.28

Q2.29 For the four-bar linkage, shown in Figure 2.59, the crank link rotates with a constant angular velocity of 2 rad/s. For the position shown, determine:

a) the coupler link angle ψ;

[31°]

b) the coupler link angular velocity;

[−0.68 rad/s]

c) the follower link angular velocity;

[−0.27 rad/s]

d) the velocity components and magnitude of point C in the Cartesian co-ordinate system.

[−0.108, −8.21 × 10⁻³ m/s, 0.11 m/s]

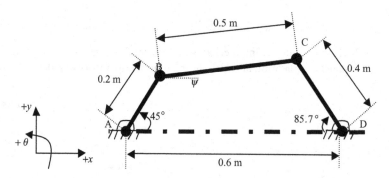

Figure 2.59: Representation of Question 2.29

Q2.30 For the four-bar linkage, shown in Figure 2.60, determine the angular velocities of the links BC and CD if the crank link AB rotates with a constant angular velocity of 3 rad/s.

[−0.53 rad/s, −1.36 rad/s]

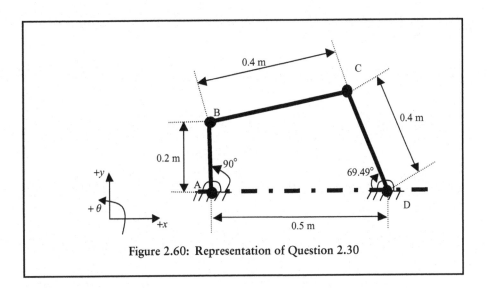

Figure 2.60: Representation of Question 2.30

CHAPTER 3

Kinetics of Particles

3.1 INTRODUCTION

Kinetics can be defined as the relationship between the forces acting on a body and the change in its motion. Newton's second law is used to solve kinetic problems that involve force, mass and motion. Although kinetics can be dealt with using three different techniques (force and acceleration, work and energy, and impulse and momentum), they are all based on Newton's second law.

The force and acceleration method is a direct application of Newton's second law, but the other two methods are derived from it through integration over displacement or time. The kinematic relationships between displacement, velocity, acceleration and time, derived in Chapter 1, are required for the application of the three methods. Each of these three techniques has advantages and disadvantages. This chapter considers the application of kinetics principles to particles. Firstly, it presents Newton's three laws of motion and Newton's law of universal gravitation. Then the force and acceleration method is explained and applied to practical problems, followed in turn by the work and energy method and the impulse and momentum method.

Before studying kinetics, the following definitions are useful.

- **Kinetics** is the study of the relationships between displacement, velocity, acceleration of a body, its mass and the forces acting on it.
- **Mass** can be defined as a property of a body that causes it to have weight in a gravitational field. It also results in a mutual attraction with other bodies that have mass. Another definition of mass is that it is a measure of resistance to change in motion.
- **Inertia** is the tendency of a body to resist a change in its motion. In Newton's second law ($F = ma$), the term 'ma' is called inertia force and resists the external force F that causes the change in motion. Another definition of inertia is that it is the property of a body that makes it resist changes in velocity.
- **Force** is defined as the nature of the mutual attraction between two bodies that possess mass. A body does not start or change its motion without the action of an external force. The motion depends on the magnitude and direction of the applied force. As force is a vector quantity, a vector summation is required to add a group of forces, or force polygon, as shown Figure 3.1.

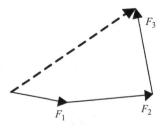

Figure 3.1: Force polygon

3.2 NEWTON'S LAWS

3.2.1 Laws of motion

Isaac Newton formulated three laws of motion in the seventeenth century. Although based upon observation and with no theoretical background, these three laws and Newton's law of universal gravitation (see Section 3.2.2) provide the fundamental principles of Newtonian mechanics, on which modern engineering dynamics and statics are based.

The fundamental laws of motions are summarized below:

- First Law: Every body remains in its state of rest or in a uniform motion unless it is enforced to change that state by the action of external forces.
- Second Law: Force is equal to the rate of change in momentum (momentum $= mv$, where m is the mass and v is the velocity). For constant mass, force equals mass times acceleration ($F = ma$).
- Third Law: For every action, there is a reaction that is equal and opposite to the action.

The first and third laws are used in static analysis, when the studied bodies are at rest or moving with constant velocity. When the bodies undergo changes in velocity (they accelerate or decelerate), then the second law should be applied.

By Newton's second law, the force is proportional or equal to the time rate of change of momentum of the mass. The momentum of a mass, m, that moves with a velocity, v, is mv. This can be written as:

$$F \propto \frac{d(m\,v)}{dt} = k\frac{d(m\,v)}{dt} \tag{3.1}$$

Differentiating Equation (3.1) with respect to time, gives:

$$F = k\,m\frac{dv}{dt} + k\,v\frac{dm}{dt} \tag{3.2}$$

where k is a constant that accounts for the inconsistencies in units. In many engineering applications, bodies and systems have a constant mass and the term dm/dt vanishes, yielding the common form of Newton's second law:

$$F = k\,m\frac{dv}{dt} = k\,m\,a \tag{3.3}$$

where F is the total external force, m is the mass of the body and a is the acceleration of the body. In the SI system, k is 1 and Newton's second law is reduced to its popular form:

$$F = ma \tag{3.4}$$

In Equation (3.4), F represents the forces acting on the body and ma represents the inertia force due to the change in motion. If the inertia force is zero, i.e. there is no acceleration, as in the case of a body at rest or moving with a constant velocity, Newton's second law is reduced to Newton's first law.

3.2.2 Law of universal gravitation

Every body in the universe attracts every other body with a force proportional to the product of their masses and inversely proportional to the square of the distance between them. This force is directed along the line of centres of the two bodies.

The force described above is a mutual attraction between two bodies and is known as gravitational force. Although the gravitational force is commonly used to describe a body's attraction to the Earth, it can simply describe the attraction force between two bodies. Newton's law of universal gravitation can be written in its general form as:

$$F \propto \frac{m_1 m_2}{r^2} \tag{3.5}$$

or

$$F = G \frac{m_1 m_2}{r^2} \tag{3.6}$$

where G is the universal constant of gravitation, given by $6.673 \times 10^{-11} \frac{m^3}{kg.s^2}$.

The gravitational force F in Equation (3.6) can be calculated for any two bodies that possess mass. When considering Earth, which is very heavy compared to any body on its surface, the gravitational force becomes significantly large. In such a case Newton's law of gravitation, becomes:

$$F = \left[\frac{GM}{R^2} \right] \times m \tag{3.7}$$

where M is the mass of the Earth and is approximately equal to 5.9742×10^{24} kg, R is the mean radius of the Earth and m is the mass of the body. Equation (3.7) is equivalent to $F = mg$, where g is the acceleration of gravity and is equal to

$$g = \frac{GM}{R^2} \tag{3.8}$$

As the Earth's radius varies from 6356.750 km to 6378.135 km, the acceleration of gravity varies from 9.8 m/s^2 to 9.868 m/s^2. At sea level, a value of $g = 9.81$ m/s^2 is acceptable. The gravitational force between a body and the surface of the Earth is known as the weight of the body and it acts towards the centre of the Earth.

Example 3.1 Newton's law of universal gravitation

The average distance between the Earth and the Moon is about 248 600 miles, while the distance between the Earth and the Sun is estimated as 93.2 million miles. The mass of the Sun is 2×10^{30} kg, the mass of the Earth is 5.9742×10^{24} kg and the mass of the Moon is approximately 7.35×10^{22} kg. The Moon's mean radius is 1738 km. Use Newton's law of universal gravitation to determine:

a) the Moon's gravitational acceleration;
b) the force of attraction between the Earth and the Moon;
c) the force of attraction between the Earth and the Sun;
d) Compare the values obtained in parts (b) and (c).

(Use 1 mile = 1.609 km)

Solution

a) $R_{EM} = 248\ 600$ miles $= 399\ 997$ km, $R_{ES} = 93.2$ million miles $= 149.958$ million km, $M_S = 2 \times 10^{30}$ kg, $M_E = 5.9742 \times 10^{24}$ kg, $M_M = 7.35 \times 10^{22}$ kg and $R_M = 1738$ km.

From Equation (3.8), the Moon's gravitational acceleration is:

$$g_M = \left[\frac{GM_M}{R_M^2} \right] = \frac{6.673 \times 10^{-11} \times 7.35 \times 10^{22}}{1\ 738\ 000^2} = 1.62\,\text{m/s}^2$$

b) From Equation (3.6), the force of attraction between the Earth and the Moon is calculated as:

$$F_{EM} = G\frac{M_E\,M_M}{R_{EM}^2} = 6.673 \times 10^{-11} \times \frac{5.9742 \times 10^{24} \times 7.35 \times 10^{22}}{399\ 997^2}$$
$$= 1.83 \times 10^{26}\,\text{N}$$

c) From Equation (3.6), the force of attraction between the Earth and the Sun is calculated as:

$$F_{Es} = G\frac{M_E\,M_s}{R_{Es}^2} = 6.673 \times 10^{-11} \times \frac{5.9742 \times 10^{24} \times 2 \times 10^{30}}{(149.958 \times 10^9)^2}$$
$$= 3.546 \times 10^{22}\,\text{N}$$

d) The ratio between the force of attraction between the Earth and the Moon and that between the Earth and the Sun is:

$$\frac{F_{ES}}{F_{EM}} = \frac{3.546 \times 10^{22}}{1.83 \times 10^{20}} = 193.7$$

3.2.3 Newton's second law

For a particle of mass m moving in a rectilinear motion with acceleration a, Newton's second law, Equation (3.4), is applicable. If the particle moves in a curvilinear motion, using the Cartesian co-ordinate system, Newton's second law in the x and y directions is written as:

$$\sum F_x = m\ddot{x}, \qquad \sum F_y = m\ddot{y} \tag{3.9}$$

where $\ddot{x}$ and $\ddot{y}$ are the acceleration components in the x and y directions, respectively. From Equation (3.7), it can be seen that if a particle is located close to the Earth's surface, the gravitational force F is given by:

$$F = mg \qquad (3.10)$$

Thus, in the case of gravitational motion $mg = ma$ and $a = g$, which indicates that a body in a free-fall condition experiences a constant acceleration of g.

3.2.4 Friction

Friction forces are tangential forces, which are produced when two surfaces are in contact and moving relative to one another. Friction forces act in the opposite direction to the motion and resist it; however, they are of limited magnitude and they cannot prevent motion, in general. Motion starts when the applied forces are larger than the friction forces.

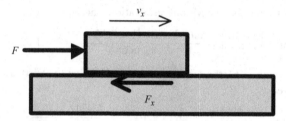

Figure 3.2: Friction force

As shown in Figure 3.2, when two surfaces are in contact with one another and a force F is applied to one of them, a friction force F_x is generated to resist motion. This force is due to the intense pressure at the contacting surfaces that results in cold welds at the contact points. By gradually increasing the applied force F, till point 1 in Figure 3.3, the cold welds prevent motion through an equal and opposite friction force F_x. At point 1, the

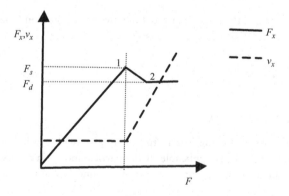

Figure 3.3: Static and dynamic friction

shear strength of the cold welds is reached and the static friction force F_s is attained. At this point, the value of the friction force F_x has reached a maximum. By further increasing F, the welds fail in a drastic manner, a sudden reduction in F_x takes place, motion starts and the velocity v_x increases. This reduction in friction force continues until an almost constant dynamic friction force F_d is attained (point 2 in Figure 3.3). As the static friction force F_s is larger than the dynamic friction force F_d, it follows that the force required to start motion is larger than that required to maintain it.

The static and dynamic friction forces can be approximated using the following expressions:

$$F_s \propto N, \qquad F_d \propto N \tag{3.11}$$

and

$$F_s = \mu_s N, \qquad F_d = \mu_d N \tag{3.12}$$

where N is the normal reaction between the two surfaces, μ_s is the coefficient of the static friction and μ_d is the coefficient of the dynamic friction. Equations (3.11) and (3.12) have been experimentally proven and universally accepted.

For dynamic applications, motion and the prediction of onset of motion are of primary concern. The dynamic coefficient of friction is thus the most appropriate to use and the subscripts in Equation (3.12) can be dropped (i.e. $F = \mu N$). In general, the friction force is larger than μN and is only equal to μN when motion starts or during motion.

3.3 FORCE AND ACCELERATION

The following steps are involved in order to apply Newton's second law to a particle:

Step 1. Define and identify the particles or bodies whose motion should be described. In Figure 3.4(a), the two masses m_1 and m_2 are identified.

Step 2. Isolate the identified particles from all other particles in the system. In Figure 3.4(b), the two bodies are surrounded by an imaginary envelope. The bodies lying outside this envelope are replaced by forces in Step 4.

Step 3. Define a co-ordinate system and sign conventions as shown in Figure 3.4(c).

Step 4. Draw a free-body diagram for each particle, Figure 3.4(d), by identifying and drawing all the forces acting on the bodies and contributing to their motion. This is a very important first step in analyzing the kinetics of a body.

Step 5. Apply Newton's second law, Equation (3.9), taking into consideration the free-body diagrams and identifying all forces contributing to the motion, as shown in Figure 3.4(e).

Step 6. Define any additional equations using frictional relationships, Equation (3.12), or kinematic equations, i.e. relationships between displacement, velocity and acceleration (see Chapter 1), as shown in Figure 3.4(f).

Step 7. Solve the system of equations.

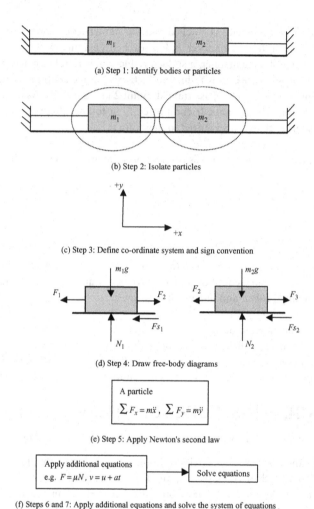

(a) Step 1: Identify bodies or particles

(b) Step 2: Isolate particles

(c) Step 3: Define co-ordinate system and sign convention

(d) Step 4: Draw free-body diagrams

A particle

$$\sum F_x = m\ddot{x}, \quad \sum F_y = m\ddot{y}$$

(e) Step 5: Apply Newton's second law

Apply additional equations
e.g. $F = \mu N$, $v = u + at$ → Solve equations

(f) Steps 6 and 7: Apply additional equations and solve the system of equations

Figure 3.4: Applying Newton's second law to particles

Example 3.2 Force and acceleration: mass on an inclined plane I

A body of mass 10 kg slides on a plane inclined at an angle $\theta = 30°$ to the horizontal and connected by a light, inextensible string to a suspended body of mass 5 kg as shown in Figure 3.5. The coefficient of friction μ between the mass and the plane is 0.3. Determine the acceleration of the system and the tension in the string. Neglect friction in the pulley.

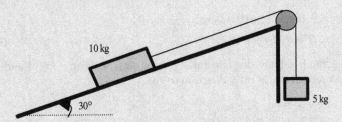

Figure 3.5: Representation of Example 3.2

Solution

Consider the free-body diagram for the 10 kg mass in Figure 3.6(a) and apply Newton's second law in the x-direction to give:

$$T - 10g \sin 30 - \mu N = 10a \qquad \text{(E3.2a)}$$

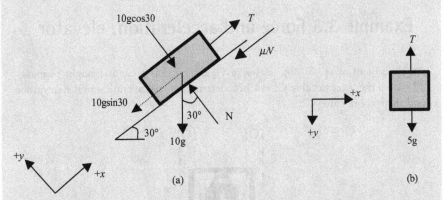

Figure 3.6: Free-body diagrams for a) the 10 kg mass and b) the 5 kg mass in Example 3.2

where a is the acceleration of the system.

Applying Newton's second law in the y-direction (Figure 3.6(a)) gives:

$$N - 10g \cos 30 = 0 \Rightarrow N = 10g \cos 30 \qquad \text{(E3.2b)}$$

Combining Equations (E3.2a) and (E3.2b) gives:

$$T - 10g\sin 30 - 10\mu g\cos 30 = 10a$$

$$T - 10g \times \frac{1}{2} - 10 \times 0.3g \times \frac{\sqrt{3}}{2} = 10a$$

$$\Rightarrow T = 7.598g + 10a \qquad\qquad\text{(E3.2c)}$$

Consider the free-body diagram for the 5 kg mass in Figure 3.6(b) and apply Newton's second law in the y-direction to give:

$$5g - T = 5a$$

$$\Rightarrow T = 5g - 5a \qquad\qquad\text{(E3.2d)}$$

Combining Equations (E3.2c) and (E3.2d) and solving for a, leads to:

$$7.598g + 10a = 5g - 5a$$

$$\Rightarrow 15a = -2.598g \Rightarrow a = -0.1732g = -1.7\,\text{m/s}^2$$

For the tension in the string, using Equation (E3.2d), gives:

$$\Rightarrow T = 5g - 5 \times -1.7 = 57.55\,\text{N}$$

Example 3.3 Force and acceleration: elevator

An elevator of mass 450 kg, shown in Figure 3.7, has a counterweight of mass 120 kg. If the force in cable C1 is 1 kN, determine the acceleration of the elevator and the force in cable C2.

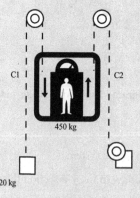

Figure 3.7: Representation of Example 3.3

Solution

Consider the free-body diagrams for each mass as shown in Figure 3.8. Applying Newton's second law for the 120 kg mass gives:

$$120g - F_1 = 120a$$

from which the acceleration is calculated as:

$$a = \frac{120 \times g - F_1}{120} = \frac{120 \times 9.81 - 1000}{120} = 1.4766 = 1.48 \, \text{m/s}^2$$

For the 450 kg mass, applying Newton's second law gives:

$$F_1 + F_2 - 450g = 450a \Rightarrow 1000 + F_2 = 450(g + 1.4766)$$

From which the force in cable C2 is calculated as:

$$F_2 = 3078.97 = 3079 \, \text{N}$$

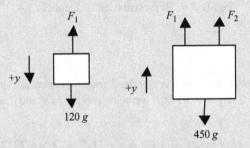

Figure 3.8: Free-body diagrams for a) the counterweight and b) the elevator in Example 3.3

Example 3.4 Force and acceleration: water–park ride

A sledge on a water-park ride has a mass of 815 kg and slides from rest into the pool as shown in Figure 3.9. The coefficient of friction between the sledge and the incline plane is μ. If the speed of the sledge when it reaches the pool is 20 m/s, determine the coefficient of friction μ.

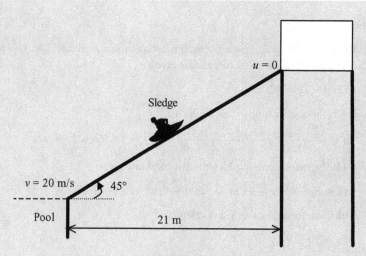

Figure 3.9: Representation of Example 3.4

Solution

The acceleration of the sledge is obtained from kinematics, Equation (1.6), as (using initial velocity $u = 0$, final velocity $v = 20$ m/s and travelled distance $s = 21\sqrt{2}$ m):

$$v^2 = u^2 + 2as \Rightarrow 20^2 = 0 + 2 \times a \times 21\sqrt{2} \Rightarrow a = 6.7343\,\text{m/s}^2$$

Consider the free-body diagram for the sledge in Figure 3.10. The normal force, N, is calculated by applying Newton's second law in the y-direction as:

$$N - 815g\cos 45 = 0 \Rightarrow N = 815g\cos 45 = 815 \times 9.81 \times \cos 45 = 5653.4248\,\text{N}$$

From Equation (3.12), the friction force, F, is calculated as:

$$F = \mu \times 815g\cos 45 = \mu \times 815 \times 9.81 \times \cos 45 = 5653.4248\mu \quad \text{(E3.4a)}$$

And applying Newton's second law in the x-direction gives:

$$815g\sin 45 - F = 815 \times 6.7343 \Rightarrow F = 164.9293\,\text{N} \qquad \text{(E3.4b)}$$

From Equations (E3.4a) and (E3.4b), the coefficient of friction is obtained as:

$$\mu = \frac{164.9293}{5653.4248} = 0.029$$

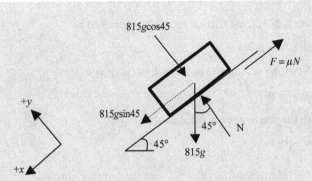

Figure 3.10: Free-body diagram for Example 3.4

Example 3.5 Force and acceleration: conveyor belt

A conveyor belt carries a package of mass m and moves with an acceleration a, as shown in Figure 3.11. The coefficient of friction between the conveyor and the package is μ. If no external force is applied, show that the acceleration of the system is independent of the mass m and equals $-\mu g$.

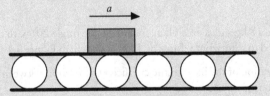

Figure 3.11: Representation of Example 3.5

Solution

Consider the free-body diagram of the package in Figure 3.12, the normal force, N, is calculated by applying Newton's second law in the y-direction as:

$$N - mg = 0 \Rightarrow N = mg$$

From Equation (3.12), the friction force, F, is obtained as:

$$F = \mu N = \mu \times mg$$

Applying Newton's second law in the x-direction gives:

$$-\mu mg = ma$$

From which the acceleration is calculated as:

$$a = -\mu g$$

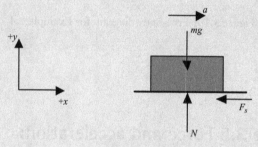

Figure 3.12: Free-body diagram for Example 3.5

Example 3.6 Force and acceleration: mass on an inclined plane II

A block of mass 8 kg slides on a plane inclined at an angle 20° with the horizontal and is subjected to a horizontal force of 40 N as shown in Figure 3.13. Determine:

a) the acceleration of the block if the coefficient of friction between the body and the surface is $\mu = 0.1$.

b) the minimum coefficient of friction μ so that the block remains at rest

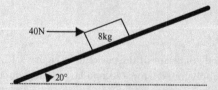

Figure 3.13: Representation of Example 3.6

Solution

The free-body diagram of the block is shown in Figure 3.14.

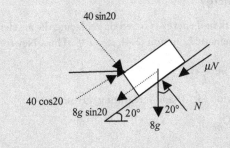

Figure 3.14: Free-body diagram for Example 3.6

a) Applying Newton's second law in the y-direction gives:

$$N - 8g\cos 20 - 40 \times \sin 20 = 0 \Rightarrow N = 87.42\,\text{N} \qquad (\text{E3.6a})$$

And in the x-direction, gives:

$$40 \times \cos 20 - 8g\sin 20 - \mu N = 8 \times a \qquad (\text{E3.6b})$$

Substituting Equation (E3.6a) into (E3.6b), the acceleration is calculated as:

$$a = \frac{1}{8}(40 \times \cos 20 - 8g\sin 20 - 0.1 \times 87.42) = 0.25\,\text{m/s}^2$$

b) When the block is at rest, the acceleration is zero, i.e. $a = 0$, and Equation (E3.6b) becomes:

$$40 \times \cos 20 - 8g\sin 20 - \mu N = 8 \times a = 0$$

from which the minimum coefficient of friction is calculated as:

$$\mu = \frac{1}{N}(40 \times \cos 20 - 8g\sin 20) = 0.123$$

3.4 WORK AND ENERGY

Another technique for solving kinetic problems is based on the method of work and energy. This method may lead to a simpler solution if acceleration is not involved in the calculation or if forces are expressed in terms of displacements. The work and energy method is based on Newton's second law.

3.4.1 Kinetic energy

The method of work and energy is based on replacing the acceleration a in Newton's second law by the kinematic Equation (1.4), $a = v\frac{dv}{ds}$. Thus, Newton's second law becomes:

$$F = ma = mv\frac{dv}{ds} \tag{3.13}$$

Rearranging Equation (3.13) and integrating, gives:

$$\int F ds = \int mv\, dv \tag{3.14}$$

In case of constant mass, Equation (3.14) becomes:

$$\int F ds = m \int v\, dv \tag{3.15}$$

If the body moves from an initial position $s = s_1$ with initial velocity $v = v_1$ to a final position $s = s_2$ with final velocity $v = v_2$, the integration limits in Equation (3.15) are:

$$\int\limits_{s_1}^{s_2} F ds = m \int\limits_{v_1}^{v_2} v\, dv \tag{3.16}$$

Integrating the right-hand side of Equation (3.16) yields:

$$\int\limits_{s_1}^{s_2} F ds = \frac{1}{2}m(v_2)^2 - \frac{1}{2}m(v_1)^2 \tag{3.17}$$

where $\frac{1}{2}m(v_2)^2 - \frac{1}{2}m(v_1)^2$ is the change in kinetic energy (KE) of the mass m.

The left-hand side in Equation (3.17), the integration term $\int_{s_1}^{s_2} F ds$, is the work done by the force F, which depends upon how the force F is related to the translation ds. There are three main types of work or energy that can contribute to this term:

- Work done by external forces, acting on a body in motion
- Potential energy or the work done by a dropped weight
- Strain energy or the work done by forces acting on an elastic body, e.g. a mechanical spring

Work done by external forces

Consider a body moving from an initial position, at distance s_1 from an origin O, to a final position, at distance s_2 from the same origin, as shown in Figure 3.15. If an external force F, which makes an angle θ with the motion path of the body, is applied during the

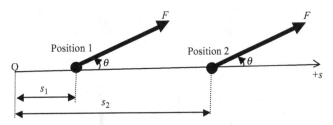

Figure 3.15: Work of a constant force in rectilinear motion

translation from position 1 to position 2, the work W_e done by the external force F is given by:

$$W_e = \int_{s_1}^{s_2} F \cos\theta \, ds \qquad (3.18)$$

Integrating Equation (3.18), considering F as a constant, yields:

$$W_e = F(s_2 - s_1)\cos\theta \qquad (3.19)$$

From Equation (3.19), it can be seen that the work of a constant force which moves a body from position 1 to position 2 is equal to the product of the force component in the motion path direction ($F\cos\theta$) and the displacement (s). Since forces and displacements are vectors, the product '$F\cos\theta$ times s' is regarded as the dot product of the vectors F and s. As the result of a dot product is a scalar, it follows that the work and energy are scalar quantities.

Potential energy

Consider a body of mass m moving from an initial position at a vertical height h_1 above a reference datum as shown in Figure 3.16, to a final position at a height h_2 above the same datum. If the body moves along a vertical path under the action of its weight mg, an expression similar to Equation (3.19) can be derived if the force F is replaced by mg.

In a similar way to Equation (3.18), the work of the weight mg, or potential energy (PE), is given by:

$$PE = \int_{s_1}^{s_2} mg \cos\theta \, ds \qquad (3.20)$$

where, as before, s is measured along the path. From Figures 3.16 and 3.15, the relationship between the vertical distance dh and the distance along the path ds is given by:

$$dh = ds \quad \text{and} \quad \theta = 180 \Rightarrow \cos\theta = -1$$

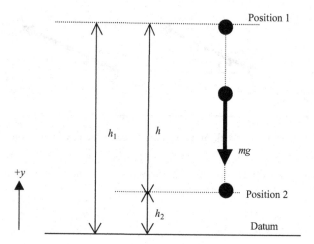

Figure 3.16: Work of the weight of a mass

Thus replacing $\cos\theta\,ds$ in Equation (3.18) by $-dh$ and changing the integration limits, Equation (3.20) becomes:

$$PE = -\int_{h_1}^{h_2} mg\,dh \qquad (3.21)$$

If mg is constant, then the integration of Equation (3.21) gives:

$$PE = -mg\int_{h_1}^{h_2} dh = -mg(h_2 - h_1) = mgh \qquad (3.22)$$

It can be seen from Equation (3.22) that the potential energy of a body of mass m is equal to mgh, which is the product of the weight (mg) and the vertical displacement (h). When the body drops from an initial vertical height h_1 above a datum to a final vertical height h_2 above the same datum, the work done is equal to the difference between its initial potential energy (mgh_1) and its final potential energy (mgh_2).

Strain energy

Strain energy is the energy possessed by an elastic body, under the action of external forces, due to its deformation. An example of an elastic body that can store strain energy is a spring. A spring generates a force that resists the deformation. In many engineering applications, a spring is used to store mechanical energy. For a linear spring, the force generated is linearly proportional to its deformation. If a spring is deformed by an amount x under the action of an external force F, the relationship between F and x can be expressed

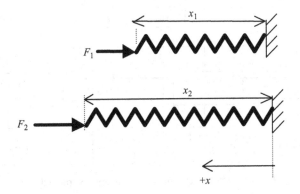

Figure 3.17: Work of the force exerted by a spring

as $F = kx$, where k is a constant called the spring stiffness. As shown in Figure 3.17, if a spring is deformed from an initial position 1 to a final position 2, the generated forces F_1 and F_2 act in opposite direction to the deformations x_1 and x_2, respectively.

Thus, from Equation (3.18) and Figure 3.17, the work of the spring due to a force F, or strain energy (SE), is given by:

$$SE = \int_{x_1}^{x_2} F \cos\theta \, ds \qquad (3.23)$$

Comparing Figure 3.15 to Figure 3.17, ($\theta = 180°, \cos\theta = -1$), thus Equation (3.23) becomes:

$$SE = -\int_{x_1}^{x_2} F \, dx \qquad (3.24)$$

If a linear spring is assumed, using the linear relationship between force and deformation, i.e. $F = kx$, and performing the integration yields:

$$SE = -\int_{x_1}^{x_2} kx \, dx = \frac{1}{2}kx_1^2 - \frac{1}{2}kx_2^2 \qquad (3.25)$$

It can be seen from Equation (3.25) that the strain energy of the force exerted by a linear spring can be expressed as half the spring stiffness times the square of the deformation. If the spring deforms from an initial extension x_1 to a final extension x_2, the strain energy is equal to the difference between the initial strain energy, $\frac{1}{2}kx_1^2$ and the final strain energy, $\frac{1}{2}kx_2^2$. If x_2 is larger than x_1, the spring stretches and the strain energy is negative, while if x_1 is larger than x_2, then the spring contracts and the strain energy is positive.

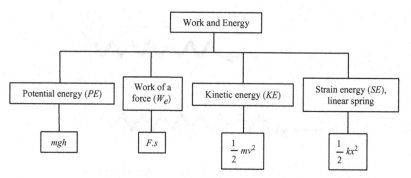

Figure 3.18: Work and energy terms for particles

3.4.2 Conservation of energy for particles

The conservation of energy law states that energy is always conserved, i.e. it cannot be created or destroyed. However, energy may be converted into other forms or non-recoverable forms, for instance friction may be converted into heat. If no such conversion takes place, the system is mechanically conservative and the principle of conservation of energy is applicable. For a conservative system, the work done by the external forces acting on the body is equal to the change in the total energy of the body at an initial position 1, initial energy, to a final position 2, final energy. This can simply be written as:

$$W_e = (PE_2 + KE_2 + SE_2) - (PE_1 + KE_1 + SE_1)$$

or

$$PE_1 + KE_1 + SE_1 + W_e = PE_2 + KE_2 + SE_2 \tag{3.26}$$

where PE_1, KE_1 and SE_1 are the potential energy, kinetic energy and strain energy of the system, respectively, at position 1; PE_2, KE_2 and SE_2 are the potential energy, kinetic energy and strain energy of the system, respectively, at position 2; and W_e is the work done by the external forces. A summary of the work and energy terms of relating to a particle is given in Figure 3.18.

Example 3.7 Work and energy: deflection of a spring

A car suspension spring has a stiffness of 16 000 N/m and is in the un-deformed state of position 1 in Figure 3.19. If a mass of 50 kg is suddenly released so that the spring contracts to position 2, determine the deflection of the spring using the energy conservation method.

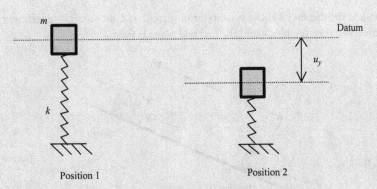

Figure 3.19: Representation of Example 3.7

Solution

Position 1: $PE_1 = 0$ (datum is taken at position 1), $KE_1 = 0$ (stationary position), $SE_1 = 0$ (spring is in its un-deformed state).

Position 2: $PE_2 = mgu_y = -50gu_y$, $KE_2 = \frac{1}{2}mv^2 = 0$ (stationary position),

$$SE_2 = \frac{1}{2}ku_y^2 = \frac{1}{2} \times 16000u_y^2 = 8000u_y^2$$

Work done by external forces is $W_e = 0$ (there are no external forces). Applying the conservation of energy principle, Equation (3.26), gives:

$$0 + 0 = -50gu_y + 8000u_y^2$$

From which the dynamic deflection of the spring is calculated as:

$$\Rightarrow u_y = \frac{50 \times 9.81}{8000} = 0.0613\,\text{m} = 6.13\,\text{cm}$$

It should be noted that the dynamic deflection is twice the static deflection ($mg/k = \frac{50 \times 9.81}{16\,000} = 0.03065$ m $= 3.065$ cm).

Example 3.8 Work and energy: mass on an inclined plane

A block of mass 5 kg is placed on an inclined plane $\theta = 30°$ as shown in Figure 3.20. If a force of 30 N is applied to move the block from position 1 to position 2, calculate the distance s if the velocity at position 2 is 1.88 m/s. The coefficient of friction between the block and the inclined plane is 0.05.

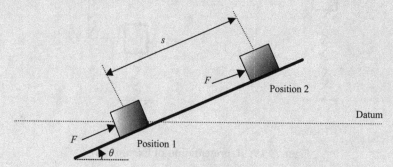

Figure 3.20: Representation of Example 3.8

Solution

Position 1: $PE_1 = 0$ (datum is taken at position 1), $KE_1 = 0$ (motion starts), $SE_1 = 0$ (no elastic bodies)

Position 2: $PE_2 = mgs \sin \theta = 5 \times 9.81 \times s \times \sin 30 = 24.545s$, $KE_2 = \frac{1}{2}mv^2$
Using $v = 1.88$ m/s gives $KE_2 = \frac{1}{2} \times 5 \times 1.88^2 = 8.836$ N.m and $SE_2 = 0$ (no elastic bodies)
Work done by force $F = F.s = 30s$
Work done by the friction force (friction force is calculated in a way similar to that presented in Example 3.4, Equation (3.4a)):
$F.s = -\mu mg \cos 30 \times s = -0.05 \times 5 \times 9.81 \times \cos 30 \times s = -2.1239s$ (negative because the friction force acts opposite to the motion)

Total work of external forces $= 30s - 2.1239s = 27.876s$

Applying the conservation of energy principle, Equation (3.26), gives:

$$0 + 27.876s = 24.545s + 8.836$$
$$s = 2.65 \, \text{m}$$

Example 3.9 Work and energy: braking car

A driver applies the brakes to his car when travelling with a speed of v_1 so that the car skids by a distance s_1, as shown in Figure 3.21. Show that if he was travelling with a speed of v_2, the skidding distance, s_2, would be:

$$s_2 = s_1 \times \frac{v_2^2}{v_1^2}$$

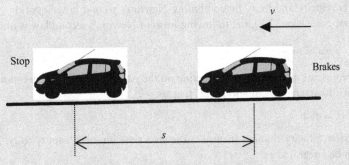

Figure 3.21: Representation of Example 3.9

Solution

Case 1: When the car is travelling with a speed of v_1, the energies are calculated as:

Position 1 (just before applying the brakes): $PE_1 = 0$, $KE_1 = \frac{1}{2}mv_1^2$, $SE_1 = 0$

Position 2 (at stop): $PE_2 = 0$ $KE_2 = 0$, $SE_2 = 0$

Work of external forces (friction) is $-Fs_1$

Applying the conservation of energy principle, Equation (3.26), gives:

$$-Fs_1 + 0 = \frac{1}{2}mv_1^2 \Rightarrow -Fs_1 = \frac{1}{2}mv_1^2 \qquad \text{(E3.9a)}$$

Case 2: When the car is travelling with a speed of v_2, the energies are calculated as:

Position 1 (just before applying the brakes): $PE_1 = 0$, $KE_1 = \frac{1}{2}mv_2^2$, $SE_1 = 0$

Position 2 (at stop): $PE_2 = 0$ $KE_2 = 0$, $SE_2 = 0$

Work of external forces (friction) is $-Fs_2$

Applying the conservation of energy principle, Equation (3.26), gives:

$$-Fs_2 + 0 = \frac{1}{2}mv_2^2 \Rightarrow -Fs_2 = \frac{1}{2}mv_2^2 \qquad \text{(E3.9b)}$$

Dividing Equation (E3.9b) by Equation (E3.9a) gives:

$$\frac{s_2}{s_1} = \frac{v_2^2}{v_1^2} \Rightarrow s_2 = s_1 \times \frac{v_2^2}{v_1^2}$$

3.5 IMPULSE AND MOMENTUM

3.5.1 Linear impulse and momentum

The method of impulse and momentum is based on direct integration of Newton's second law with respect to time, which leads to simple equations that relate forces acting on a body to its velocity and time. By combining Newton's second law, Equation (3.4), and the kinematic equation (1.2), the following form of Newton's second law is obtained:

$$F = ma = m\frac{dv}{dt} \tag{3.27}$$

where F represents all external forces acting on the particle, a is the acceleration and v is the velocity. Rearranging Equation (3.27), gives:

$$F\,dt = m\,dv \tag{3.28}$$

If the particle velocity changes from v_1 to v_2 in a time internal from t_1 to t_2, Equation (3.28) can be integrated as:

$$\int_{t_1}^{t_2} F\,dt = \int_{v_1}^{v_2} m\,dv \tag{3.29}$$

In general, the force F can be a function of time and the mass m is constant. Performing the integration of the right-hand side of Equation (3.29), leads to:

$$\int_{t_1}^{t_2} F\,dt = mv_2 - mv_1 \tag{3.30}$$

Equation (3.30) is known as the principle of linear impulse and momentum. It provides a simple way to estimate the final velocity v_2 if the initial velocity v_1 and the forces acting on the body are known. The right-hand side in Equation (3.30) $(mv_2 - mv_1)$ represents the difference in linear momentum (mv) between an initial velocity v_1 and a final velocity v_2 of a particle of mass m. The left-hand side in Equation (3.30) $(\int_{t_1}^{t_2} F\,dt)$ represents the linear impulse, which measures the effect of external forces F during a time interval from t_1 to t_2. Equation (3.30) is often rearranged in the following common form:

$$mv_1 + \int_{t_1}^{t_2} F\,dt = mv_2 \tag{3.31}$$

Equation (3.31) states that the initial linear momentum of a body at time t_1 plus the linear impulse due to all external forces acting on the body in the time interval t_1 to t_2 equals the final linear momentum at time t_2.

If the linear impulses are zero, i.e. there are no external forces, Equation (3.31) is reduced to:

$$mv_1 = mv_2 \tag{3.32}$$

For a system of particles, a summation for all particles is applied to both sides in Equation (3.32), i.e.

$$\sum mv_1 = \sum mv_2 \qquad (3.33)$$

which represents the principle of the conservation of linear momentum.

The usefulness of Equations (3.31) and (3.33) is illustrated by the following examples.

Example 3.10 Impulse and momentum: mass under force

A force of 100 N making an angle of 30° is applied for 5 s to a mass of 50 kg as shown in Figure 3.22. If the mass is initially at rest, determine the final velocity of the mass (when $t = 5$ s) and the normal reaction.

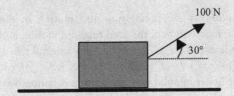

Figure 3.22: Representation of Example 3.10

Solution

Resolve the force as shown in Figure 3.23.

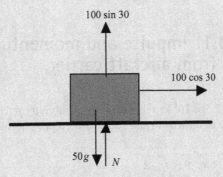

Figure 3.23: Resolving the force in Example 3.10

In the x direction, applying the linear impulse and momentum equation, Equation (3.31) gives:

$$mv_{1x} + F \times \cos 30 \times t = mv_{2x}$$

The linear momentum at the initial time (initial velocity is zero) is $mv_{1x} = 0$; the linear momentum at the final time is $mv_{2x} = 50 \times v_{2x}$ N.s; and the linear impulse during the time interval $t = 0$ to 5 s due to a force F is $F \times \cos 30 \times t = 100 \times \cos 30 \times 5$ N.s. The linear impulse and momentum equation thus becomes:

$$0 + 100 \times \cos 30 \times 5 = 50 \times v_{2x}$$

from which the final velocity is calculated as:

$$v_{2x} = 8.66 \, \text{m/s}$$

In the y direction, applying the linear impulse and momentum equation, Equation (3.31), gives:

$$mv_{1y} + N \times t - mg \times t + F \times \sin 30 \times t = mv_{2y}$$

The linear momentum at the initial time (no motion in the y direction) is $mv_{1y} = 0$; the linear momentum at the final time (no motion in the y direction) is $mv_{2x} = 0$; and the linear impulse during the time interval $t = 0$ to 5 s due to F, N and mg is $N \times t - mg \times t + F \times \sin 30 \times t = N \times 5 - 50 \times 9.81 \times 5 + 100 \times \sin 30 \times 5$ N.s. The linear impulse and momentum equation becomes:

$$0 + N \times 5 - 50 \times 9.81 \times 5 + 100 \times \sin 30 \times 5 = 0$$

From which the normal force is calculated as:

$$N = 540.5 \, \text{N}$$

Example 3.11 Impulse and momentum: plane launching from aircraft carrier

A jet plane of mass 7.5×10^3 kg takes off from an aircraft carrier that travels with a speed of 50 km/hr as shown in Figure 3.24. The engine's horizontal thrust varies as follows:

$$F = 2 \, \text{kN} \quad \text{for} \quad 0 \leq t \leq 2$$
$$F = 10 \, \text{kN} \quad \text{for} \quad 2 \leq t \leq 5$$

where t is time in seconds. Determine the plane's horizontal speed after 5 s.

Figure 3.24: Representation of Example 3.11

Solution

Applying the linear impulse and momentum equation, Equation (3.31), in the x direction gives:

$$mv_1 + \int_{t_1}^{t_2} F\,dt = mv_2$$

The initial linear momentum is $mv_1 = 7.5 \times 10^3 \times 50 \times \frac{1000}{3600}$ N.s; the final linear momentum $mv_2 = 7.5 \times 10^3 \times v_2$ N.s; the linear impulse during the time interval $0 \leq t \leq 2$ is $\int_{t_1}^{t_2} F\,dt = \int_0^2 2 \times 10^3\,dt$ N.s and the linear impulse during the time interval $2 \leq t \leq 5$ is $\int_{t_1}^{t_2} F\,dt = \int_0^2 10 \times 10^3\,dt$ N.s. The equation of linear impulse and momentum becomes:

$$7.5 \times 10^3 \times 50 \times \frac{1000}{3600} + \int_0^2 2 \times 10^3\,dt + \int_2^5 10 \times 10^3\,dt = 7.5 \times 10^3 \times v_2$$

From which the final velocity is obtained as:

$$v_2 = 18.422\,\text{m/s} = 18.422 \times \frac{3600}{1000} = 66.32\,\text{km/hr}$$

Example 3.12 Impulse and momentum: car crash

A car, A, of mass 2×10^3 kg travelling at speed 1.2 m/s crashes head on with a car, B, travelling at speed 1.8 m/s in the opposite direction, as shown in Figure 3.25. If B has a mass of 1×10^3 kg, determine the common speed of the cars just after collision assuming that the wheels are free to roll during collision. If the coupling takes place in 1 s, determine the average force between the two cars.

1.2 m/s

1.8 m/s

A

B

Figure 3.25: Representation of Example 3.12

Solution

Apply the conservation of linear momentum, Equation (3.33), and consider the cars just before the crash $m_A \times v_A + m_B \times v_B$ and just after the crash $(m_A + m_B) \times v_{AB}$ to give:

$$m_A \times v_A + m_B \times v_B = (m_A + m_B) \times v_{AB}$$

where the linear momentum of car A before the crash is $m_A \times v_A = 2 \times 10^3 \times 1.2$ N.s, the linear momentum of car B before the crash is $m_B \times v_B = -1 \times 10^3 \times 1.8$ N.s (negative because car B is moving in the opposite direction to car A) and the linear momentum of cars A and B after the crash is $(m_A + m_B) \times v_{AB} = (2 + 1) \times 10^3 \times v_{AB}$ N.s. The conservation of linear momentum equation becomes:

$$2 \times 10^3 \times 1.2 - 1 \times 10^3 \times 1.8 = (2 + 1) \times 10^3 \times v_{AB}$$

from which the velocity of cars A and B after the crash is calculated as:

$$v_{AB} = 0.2 \, \text{m/s}$$

To calculate the force between the two cars, applying the linear impulse and momentum equation, Equation (3.31), to either car A or B gives (for car A):

$$m_A v_{A1} - \int_0^1 F \, dt = m_A v_{A2}$$

The impulse is negative as the force is acting in the opposite direction to the velocity of car A at the moment of crash as shown in Figure 3.26.

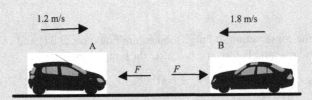

Figure 3.26: Forces in Example 3.12

The linear momentum just before the crash $m_A v_{A1} = 2 \times 10^3 \times 1.2$ N.s, the linear momentum just after the crash $m_A v_{A2} = 2 \times 10^3 \times 0.2$ N.s and the linear impulse during coupling $\int_0^1 F\, dt = F \times 1$ N.s. Thus, the linear impulse and momentum equation becomes:

$$2 \times 10^3 \times 1.2 - F \times 1 = 2 \times 10^3 \times 0.2$$

from which the average force between the two cars is calculated as:

$$F = 2000\,\text{N} = 2\,\text{kN}$$

Example 3.13 Impulse and momentum: man jumping

A man A of mass 80 kg runs with a horizontal velocity of 1.5 m/s and jumps onto a boat B of mass 100 kg as shown in Figure 3.27. If the boat is at rest when the man makes the jump, determine the boat and the man's speed just after the jump.

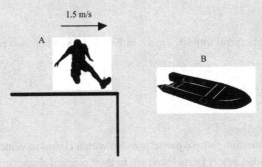

Figure 3.27: Representation of Example 3.13

Solution

Applying the conservation of linear momentum Equation (3.31) and considering the man and the boat before and just after the jump gives:

$$m_A \times v_A + m_B \times v_B = (m_A + m_B) \times v_{AB}$$

where the linear momentum of the man just before the jump is $m_A \times v_A = 80 \times 1.5$ N.s, the linear momentum of the boat just before the jump is $m_B \times v_B = 100 \times 0$ (the boat is at rest) and the linear momentum of the man and the boat just after the jump is $(m_A + m_B) \times v_{AB} = (80 + 100) \times v_{AB}$ N.s. The conservation of linear momentum equation becomes:

$$80 \times 1.5 + 100 \times 0 = (80 + 100) \times v_{AB}$$

from which the velocity of the man and the boat just after the jump is calculated as:

$$v_{AB} = 0.67 \, \text{m/s}$$

3.5.2 Impact

When two bodies collide with one another in a very short time interval, a very large impulsive force takes place at the contact point. This phenomenon is called impact. Common examples of impact are the striking of a hammer on a nail and the striking of a golf club on a ball.

Impact can be categorized into two main types, central impact and oblique impact. In central impact, the line passing through the centre of gravity of the two colliding bodies is in the direction of motion. In oblique impact, it makes an angle with the direction of motion.

Consider the case of central impact, shown in Figure 3.28, when two particles, A and B, collide.

Impact takes place in five phases:

1. Before impact
2. Deformation impulse
3. Maximum deformation, where particles move with a common velocity v
4. Restitution impulse, where particles return to their original shape (in the case of elastic impact) or remain permanently deformed (in the case of plastic impact)
5. After impact.

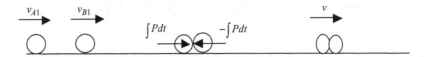

Phase 1: Before impact   Phase 2: Deformation impulse   Phase 3: Maximum deformation

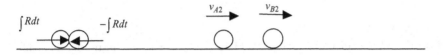

Phase 4: Restitution impulse   Phase 5: After impact

Figure 3.28: Impact analysis

Particle A has a velocity of v_{A1} before impact and v_{A2} after impact, while particle B has a velocity v_{B1} before impact and v_{B2} after impact. Applying the conservation of linear momentum Equation (3.33) to Phases 1 and 5 gives:

$$m_A v_{A1} + m_B v_{B1} = m_A v_{A2} + m_B v_{B2} \qquad (3.34)$$

Equation (3.34) has two unknowns: the velocity of A after impact, v_{A2}, and the velocity of B after impact, v_{B2}. In order to solve for these two unknowns, an additional equation is required. This can be obtained by applying the linear impulse and momentum equation to each particle. Thus, applying Equation (3.31) to particle A in Phases 1 to 3 gives

$$m_A v_{A1} - \int P dt = m_A v \qquad (3.35)$$

where v is the common velocity of particles A and B in Phase 3. Rearrange Equation (3.35), so that:

$$\int P dt = m_A v_{A1} - m_A v \qquad (3.36)$$

Similarly, applying Equation (3.31) to particle A in Phases 3 to 5 gives:

$$m_A v - \int R dt = m_A v_{A2} \qquad (3.37)$$

Rearranging Equation (3.37) gives:

$$\int R dt = m_A v - m_A v_{A2} \qquad (3.38)$$

Dividing Equation (3.38) by Equation (3.36), the coefficient of restitution e is obtained as:

$$e = \frac{\int R\,dt}{\int P\,dt} = \frac{m_A v - m_A v_{A2}}{m_A v_{A1} - m_A v} = \frac{v - v_{A2}}{v_{A1} - v} \tag{3.39}$$

Applying the same procedures to particle B, a similar equation for the coefficient of restitution is obtained as:

$$e = \frac{v_{B2} - v}{v - v_{B1}} \tag{3.40}$$

Eliminating v from equations (3.39) and (3.40), the coefficient of restitution is finally obtained as:

$$e = \frac{v_{B2} - v_{A2}}{v_{A1} - v_{B1}} \tag{3.41}$$

which depends only on initial and final velocities.

Oblique impact can be analyzed in a similar way, however, in such a case the directions of the velocities are also unknown and impulse and momentum should be analyzed in both x and y directions.

Example 3.14 Impact I

A ball, A, of mass 3 kg collides with a ball, B, of mass 1.5 kg as shown in Figure 3.29. Just before collision, ball A has a velocity of 2.5 m/s to the right, while ball B has a velocity of 1 m/s to the left. If the coefficient of restitution for the balls is 0.7, determine the balls' velocities just after collision.

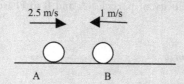

Figure 3.29: Representation of Example 3.14

Solution

Applying the conservation of linear momentum equation, Equation (3.34), gives:

$$m_A v_{A1} + m_B v_{B1} = m_A v_{A2} + m_B v_{B2}$$

where the linear momentum of ball A just before collision is $m_A v_{A1} = 3 \times 2.5$ N.s, the linear momentum of ball B just before collision is $m_B v_{B1} = -1.5 \times 1$ N.s (negative because ball B is moving in the opposite direction to ball A), the linear momentum of ball A just after collision is $m_A v_{A2} = 3 \times v_{A2}$ N.s and the linear momentum of ball B just after collision is $m_B v_{B2} = 1.5 \times v_{B2}$ N.s. The conservation of linear momentum equation becomes:

$$3 \times 2.5 - 1.5 \times 1 = 3 \times v_{A2} + 1.5 \times v_{B2}$$

from which

$$3v_{A2} + 1.5v_{B2} = 6 \qquad \text{(E3.14a)}$$

And from Equation (3.41), the coefficient of restitution is:

$$e = \frac{v_{B2} - v_{A2}}{v_{A1} - v_{B1}} \Rightarrow 0.7 = \frac{v_{B2} - v_{A2}}{2.5 - (-1)}$$

from which

$$v_{B2} - v_{A2} = 2.45 \qquad \text{(E3.14b)}$$

Solving Equations (E3.14a) and (E3.14b), the velocities of the balls after collision are obtained as:

$$v_{B2} = 2.97 \, \text{m/s}$$
$$v_{A2} = 0.516 \, \text{m/s}$$

Example 3.15 Impact II

A ball, A, of mass 1 kg strikes a block, B, of mass 9 kg rested on a horizontal plane as shown in Figure 3.30. The coefficient of restitution between the ball and the block is 0.65 and the ball has a velocity of 8 m/s just before striking the block. If the coefficient of friction between the block and the plane is 0.3, determine the time before the block stops sliding and the distance it slides.

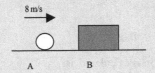

8 m/s

A   B

Figure 3.30: Representation of Example 3.15

Solution

Applying the conservation of linear momentum equation, Equation (3.34), gives:

$$m_A v_{A1} + m_B v_{B1} = m_A v_{A2} + m_B v_{B2}$$

where the linear momentum of ball A just before collision is $m_A v_{A1} = 1 \times 8$ N.s, the linear momentum of block B just before collision is $m_B v_{B1} = 9 \times 0 = 0$ (block B is at rest just before collision), the linear momentum of ball A just after collision is $m_A v_{A2} = 1 \times v_{A2}$ N.s and the linear momentum of block B just after collision is $m_B v_{B2} = 9 \times v_{B2}$ N.s. The conservation of linear momentum equation becomes:

$$1 \times 8 + 9 \times 0 = 1 \times v_{A2} + 9 \times v_{B2}$$

from which

$$v_{A2} + 9v_{B2} = 8 \qquad\qquad\qquad\qquad (E3.15a)$$

From Equation (3.41), the coefficient of restitution is given by:

$$e = \frac{v_{B2} - v_{A2}}{v_{A1} - v_{B1}} \Rightarrow 0.65 = \frac{v_{B2} - v_{A2}}{8 - (0)}$$

from which

$$v_{B2} - v_{A2} = 5.2 \qquad\qquad\qquad\qquad (E3.15b)$$

Solving Equations (E3.15a) and (E3.15b), the final velocities of the ball A and the block B after collision are obtained as:

$$v_{B2} = 1.32 \, \text{m/s}$$
$$v_{A2} = -3.88 \, \text{m/s}$$

To calculate the acceleration of block B, applying Newton's second law to the free-body diagram in Figure 3.31 in the y direction gives:

$$N_B = m_B g$$

In the x direction, it gives:

$$-\mu N_B = m_B a \Rightarrow -\mu m_B g = m_B a$$

from which the acceleration is calculated as:

$$a = -\mu g = -0.3 \times 9.81 = -2.943 \, \text{m/s}^2$$

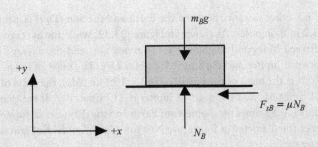

Figure 3.31: Free-body diagram for Example 3.15

Kinematic Equations (1.5) and (1.6) for rectilinear motion with constant acceleration, enable us to calculate the time before the block stops sliding after the strike and the distance it slides:

$$v = u + at \Rightarrow 0 = 1.32 - 2.943t \Rightarrow t = 0.45\,\text{s}$$
$$v^2 = u^2 + 2as \Rightarrow 0 = 1.32^2 - 2 \times 2.943 \times s \Rightarrow s = 0.296\,\text{m}$$

3.6 Tutorial Sheet

3.6.1 Force and acceleration

Q3.1 If the radius of the Earth at sea level is 6.374×10^6 m, its acceleration of gravity is 9.81 m/s^2 at sea level and the acceleration due to gravity on the surface of the Moon is 1.62 m/s^2, determine:

a) the mass of the Earth in kg;

[5.973×10^{24} kg]

b) the weight of a person of 50 kg mass at sea level;

[490.5 N]

c) the weight of a person of 50 kg mass on the Moon.

[81 N]

(Use the universal constant of gravitation, $G = 6.673 \times 10^{-11} \frac{\text{m}^3}{\text{kg.s}^2}$.)

Q3.2   The average distance between the Earth and the Sun (D_{ES}) is estimated as 93.2 million miles. As shown in Figure Q3.32, while the average distance between Mars and the Sun is 1.524 times D_{ES} and the average distance between Jupiter and the Sun is 5.2 times D_{ES}. The mass of the Sun is 2×10^{30} kg, the mass of the Earth 5.9742×10^{24} kg (M_E), the mass of Mars is 0.1 times M_E and the mass of Jupiter is 317 times M_E. If the diameter of Mars is 0.53 times the diameter of Earth and the diameter of Jupiter is 11.2 times the diameter of Earth, use Newton's law of universal gravitation to determine:

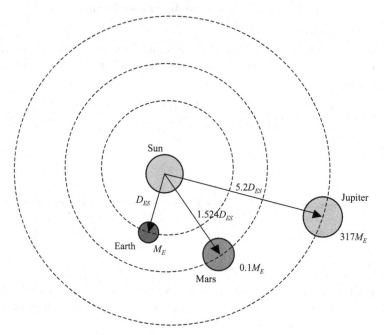

Figure 3.32: Representation of Question 3.2

a) the gravitational acceleration of Mars;

[3.49 m/s^2]

b) the gravitational acceleration of Jupiter;

[24.79 m/s^2]

c) the force of attraction between the Earth and the Sun;

[3.546×10^{22} N]

   d) the force of attraction between Mars and the Sun;

$$[1.527 \times 10^{21}\,\text{N}]$$

   e) the force of attraction between Jupiter and the Sun.

$$[4.157 \times 10^{23}\,\text{N}]$$

   (Use 1 mile = 1.609 km, $g = 9.81$ m/s^2 and the universal constant of gravitation, $G = 6.673 \times 10^{-11}\,\frac{\text{m}^3}{\text{kg.s}^2}$.)

Q3.3 If the mass of the Earth is approximately 5.97×10^{27} gm and its mean radius is 6.37×10^6 m, use Newton's law of universal gravitation to calculate the force of attraction between the Earth and a solid steel sphere of radius 2 m and mass 4×10^3 kg. Use the universal constant of gravitation, $G = 6.673 \times 10^{-11}\,\frac{\text{m}^3}{\text{kg.s}^2}$.

$$[39.27\,\text{kN}]$$

Q3.4 Use Newton's law of universal gravitation to prove that the acceleration of gravity is equal to 9.81 m/s^2. Assume that the mass of the Earth is 5.9742×10^{24} kg and the radius of Earth at sea level is 6.374×10^6 m.

Q3.5 Find the tension in a cable that carries an elevator cage of mass 1200 kg, when the elevator is:

   a) at rest, as in Figure 3.33(a);

$$[11.77\,\text{kN}]$$

   b) descending with a constant velocity, as in Figure 3.33(b);

$$[11.77\,\text{kN}]$$

   c) ascending with an acceleration of 1 m/s^2, as in Figure 3.33(c);

$$[12.97\,\text{kN}]$$

   d) descending with an acceleration of 1 m/s^2, as in Figure 3.33(d);

$$[10.69\,\text{kN}]$$

   e) ascending with a deceleration of 3 m/s^2, as in Figure 3.33(e);

$$[8.17\,\text{kN}]$$

   f) descending with a deceleration of 3 m/s^2, as in Figure 3.33(f).

$$[15.37\,\text{kN}]$$

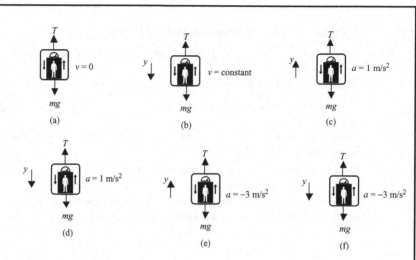

Figure 3.33: Representation of Question 3.5

Q3.6   The crane shown in Figure 3.34 lifts a package that has a mass of 1×10^3 kg with an initial acceleration of 2 m/s^2. The supporting cables make an angle θ with the horizontal. If $\theta = 45°$, determine the force in each of the two supporting cables.

[8.35 kN]

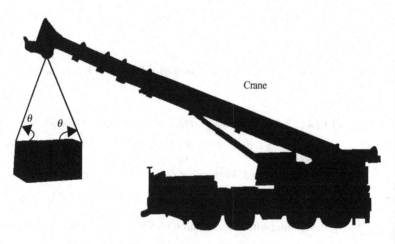

Figure 3.34: Representation of Question 3.6

Q3.7 A body of mass 8 kg slides on a horizontal plane and is connected by a light, inextensible string to a suspended body of mass 3 kg as shown in Figure 3.35. The coefficient of friction μ between the mass and the plane is 0.3. If the pulley is frictionless, determine:

a) the acceleration of the system;

$$[0.535 \text{ m/s}^2]$$

b) the tension in the string.

$$[27.83 \text{ N}]$$

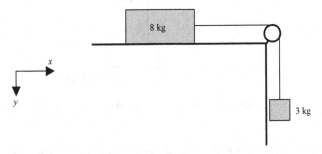

Figure 3.35: Representation of Question 3.7

Q3.8 An automobile pulls two cars, each of mass 250 kg, as shown in Figure 3.36. The automobile has a mass of 1×10^3 kg and pulls the cars with a force of 500 N. Determine:

a) the acceleration of the automobile if both cars are pulled;

$$[0.333 \text{ m/s}^2]$$

b) the acceleration of the automobile if only one car is pulled.

$$[0.4 \text{ m/s}^2]$$

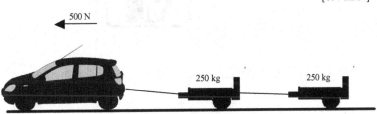

Figure 3.36: Representation of Question 3.8

Q3.9   Two masses of 20 kg and 40 kg are suspended as shown in Figure 3.37 from a light, inextensible string that passes over a frictionless pulley. If the two masses are suddenly released, determine the acceleration of the system and the tension in the string.

[3.27 m/s², 261.6 N]

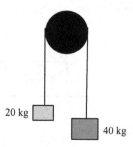

Figure 3.37: Representation of Question 3.9

Q3.10 An elevator of mass 450 kg, shown in Figure 3.38, has a counterweight of mass 120 kg. When the elevator rises 32 m, its speed reaches 8 m/s. If the pulleys and cables have negligible masses, determine:

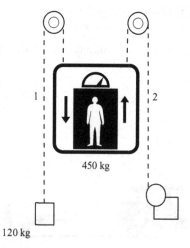

Figure 3.38: Representation of Question 3.10

a) the acceleration of the elevator;

[1 m/s^2]

b) the force in cable 1;

[1.06 kN]

c) the force in cable 2.

[3.81 kN]

Q3.11 A sledge on a water-park ride has a mass of 815 kg and slides from rest into the pool as shown in Figure Q3.39. If the coefficient of friction between the sledge and the inclined plane is 0.05, determine the speed of the sledge when it reaches the pool.

[19.78 m/s]

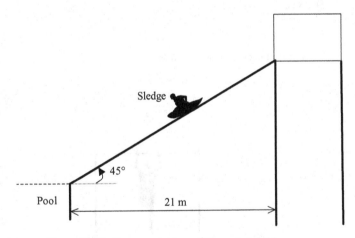

Figure 3.39: Representation of Question 3.11

Q3.12 An engine of mass 3.6×10^3 kg is suspended from a spreader beam of mass 0.5×10^3 kg, which is attached to a crane through two chains as shown in Figure Q3.12. If the crane lifts the engine with an initial acceleration of 3.6 m/s^2, determine the force in each chain.

[31.74 kN]

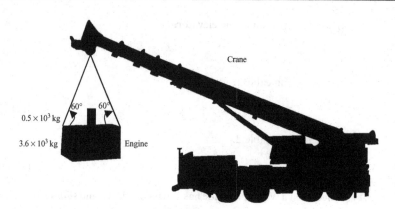

Figure 3.40: Representation of Question 3.12

Q3.13 A boy of mass 50 kg hangs from a bar as shown in Figure 3.41. Determine the force in each of his arms if the bar moves upwards with:

a) constant velocity;

[245.25 N]

b) an initial acceleration of 5 m/s^2.

[370.25 N]

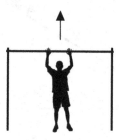

Figure 3.41: Representation of Question 3.13

Q3.14 A conveyor belt carries a package of mass m and moves with a speed of 3 m/s, as shown in Figure 3.42. The coefficient of friction between the conveyor belt and the package is 0.15. Determine the shortest time in which the belt could stop without causing the package to slide.

[2.04 s]

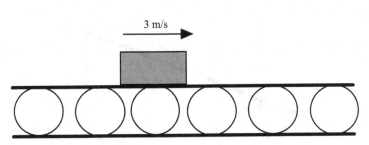

Figure 3.42: Representation of Question 3.14

Q3.15 A body of mass 10 kg lies at rest on an inclined ramp as shown in Figure 3.43. Determine the coefficient of friction between the body and the ramp so that the body is just about to slip.

[0.268]

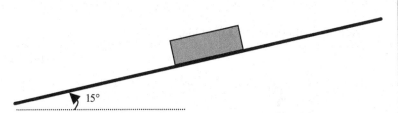

15°

Figure 3.43: Representation of Question 3.15

Q3.16 A block of mass 10 kg slides on a plane inclined at an angle θ to the horizontal and is connected by a light, inextensible string to a suspended body of mass 5 kg as shown in Figure 3.44. If the coefficient of friction between the block and the plane is 0.3 and the pulley is frictionless, determine:

a) the acceleration of the system and the tension in the string if $\theta = 30°$;

[−1.7 m/s²]

b) the angle θ at which the system would be in equilibrium.

[12°]

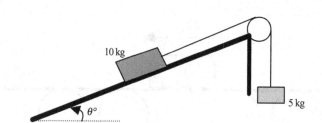

Figure 3.44: Representation of Question 3.16

3.6.2 Work and energy

Q3.17 A suspension spring has a stiffness k and is in the un-deformed state in position 1 as shown in Figure 3.45. When a mass of 15 kg is suddenly released so that the spring deforms to position 2, the dynamic deflection of the system is measured as 0.02 m. Determine:

a) the potential energy at position 2 (take datum as shown);

[−2.94 N.m]

b) the stiffness of the suspension.

[14.72 kN/m]

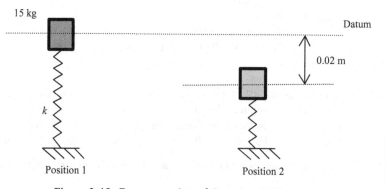

Figure 3.45: Representation of Question 3.17

Q3.18 A block of mass 3 kg is placed on an inclined plane $\theta = 30°$ as shown in Figure 3.46. If a force F is applied to move the block from position 1 to position 2 so that it travels a distance $s = 1.5$ m and its velocity at position 2 is 2.35 m/s, determine:

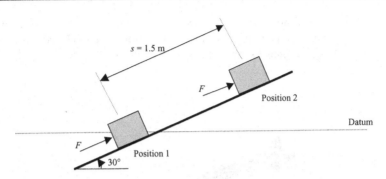

Figure 3.46: Representation of Question 3.18

a) the potential energy at position 2 (take datum as shown);

[22.07 N.m]

b) the kinetic energy at position 2;

[8.28 N.m]

c) the force F (neglect friction between the block and the plane).

[20.23 N]

Q3.19 A car of mass 1.8×10^3 kg travels down an inclined road as shown in Figure 3.47. The coefficient of friction between the car tyres and the road is 0.5. When the car's speed is 5 m/s, the driver applies the brakes so that the wheels lock and the tyres skid on the road. Determine the distance that the car skids.

[4.53 m]

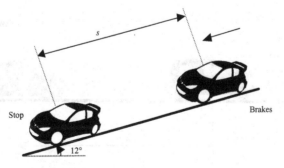

Figure 3.47: Representation of Question 3.19

Q3.20 A crane, shown in Figure 3.48, lifts a steel girder of 2.4×10^3 kg with a constant force that is equal to 32 kN. When the steel girder is lifted by 2 m, determine its speed.

[3.75 m/s]

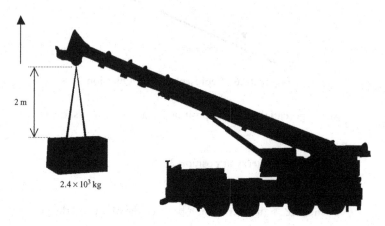

Figure 3.48: Representation of Question 3.20

Q3.21 A driver applies the brakes to his car when travelling with a speed of 11 m/s so that the car skids by 2 m, as shown in Figure 3.49. Determine the skidding distance if he was travelling with a speed of 22 m/s.

[8 m]

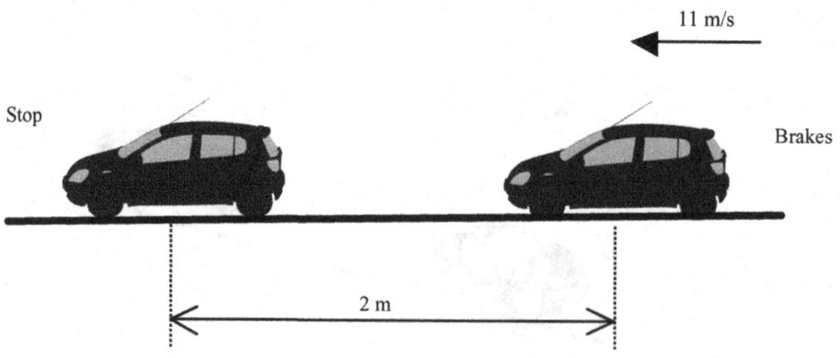

Figure 3.49: Representation of Question 3.21

Q3.22 A 10 kg crate is placed on a conveyor belt that is connected to a ramp making an angle of 30° with the horizontal and having a length of 4 m, as shown in Figure 3.50. The coefficient of friction between the crate and the ramp is 0.3. If the crate has a velocity of 2 m/s at the top of the ramp, determine the velocity of the crate at the bottom end of the ramp.

[4.78 m/s]

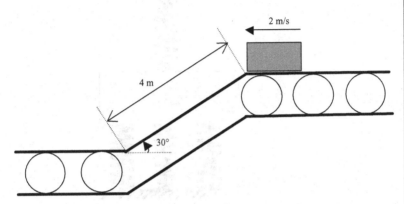

Figure 3.50: Representation of Question 3.22

Q3.23 A cyclist starts his motion on a ramp with an initial velocity of 3.71 m/s as shown in Figure 3.51. If friction is neglected and the cyclist stops pedalling on the ramp, determine the height that the cyclist reaches before coming to a stop.

[0.7 m]

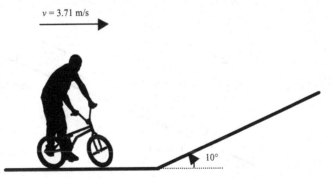

Figure 3.51: Representation of Question 3.23

Q3.24 A block of mass 25 kg is attached, as shown in Figure 3.52, to four springs, each of which has a stiffness of 2.5 kN/m. If the block is released from rest, determine the downward vertical displacement of the block.

[49 mm]

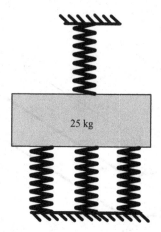

Figure 3.52: Representation of Question 3.24

Q3.25 A square block of mass m is placed on an inclined plane as shown in Figure 3.53. If a force F is applied to move the block from position 1 to position 2; show that the distance s is equal to $s = \frac{mv^2}{2(F - mg\sin\vartheta)}$, where v is the velocity at position 2. Neglect friction between the block and the plane.

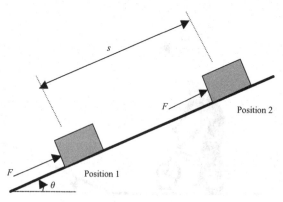

Figure 3.53: Representation of Question 3.25

3.6.3 Impulse and momentum

Q3.26 A force of 120 N making an angle of 30° is applied for t seconds to a mass of 60 kg as shown in Figure 3.54. If the mass is initially at rest and its final velocity is 10 m/s, determine the time t during which the force was applied.

[5.77 s]

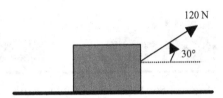

Figure 3.54: Representation of Question 3.26

Q3.27 A jet plane of mass 240×10^3 kg travels with an initial horizontal velocity of 120 m/s as shown in Figure 3.55. The engine's horizontal thrust varies as $F = 198 + 2t^2$, where F is in kN and t is the time in seconds. Determine the plane's horizontal velocity after 12 s.

[134.7 m/s]

Figure 3.55: Representation of Question 3.27

Q3.28 A car A of mass 2×10^3 kg travelling at 1.2 m/s crashes head-on and couples with a car B travelling at 2 m/s in the opposite direction, as shown in Figure 3.56. If car B has a mass of 1.2×10^3 kg, determine:

a) the common speed of the cars just after collision assuming that the wheels are free to roll during collision;

[0]

b) the average force between the two cars, if the coupling takes place in 1 s.

[2.4 kN]

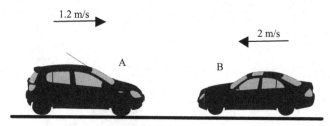

Figure 3.56: Representation of Question 3.28

Q3.29 A man A of mass 80 kg runs with a horizontal velocity of 1.5 m/s and jumps onto a boat B of mass *m* as shown in Figure 3.57. If the boat is at rest when the man makes the jump and the speed of the boat and the man just after the jump is 0.8 m/s, determine the mass of the boat.

[70 kg]

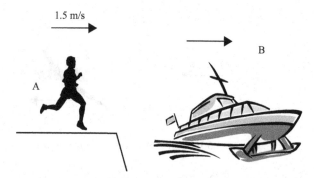

Figure 3.57: Representation of Question 3.29

Q3.30 A railway wagon A of mass 12×10^3 kg coasting at 1.2 m/s on a horizontal track crashes head-on and couples with a wagon B of mass 10×10^3 kg coasting at 1 m/s in the opposite direction, as shown in Figure 3.58. Determine:

a) the common speed of the wagons just after collision assuming that the wheels are free to roll during collision;

[0.2 m/s]

b) the average force between the two wagons, if the coupling takes place in 0.9 s.

[13.33 kN]

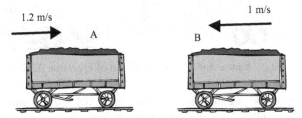

Figure 3.58: Representation of Question 3.30

Q3.31 A tugboat A of mass 340×10^3 kg pulls a barge B of mass 45×10^3 kg on the end of a rope, as shown in Figure 3.59. The barge is initially at rest when the tugboat starts pulling at 2 m/s. If the frictional effect of the water is neglected, determine the velocity of the tugboat and barge when they start moving together.

[1.77 m/s]

Figure 3.59: Representation of Question 3.31

Q3.32 A bus of mass 7×10^3 kg is travelling at 2.4 m/s when it crashes head-on and couples with a car of mass 1.2×10^3 kg travelling at 1.5 m/s, as shown in Figure 3.60. Determine the common speed of the bus and the car just after collision assuming that the wheels are free to roll during collision.

[1.83 m/s]

Figure 3.60: Representation of Question 3.32

Q3.33 Two disks A and B, as shown in Figure 3.61, each of which has mass of 2 kg, collide with one another. Just before collision, disk A has a velocity of 2 m/s to the right, while disk B has a velocity of 1 m/s to the left. If the coefficient of restitution of the disks is 0.75, determine the velocities of disks A and B just after collision.

[1.625 m/s, −0.625 m/s]

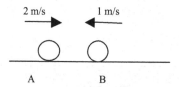

Figure 3.61: Representation of Question 3.33

Q3.34 A ball A of mass 1 kg strikes a block B of mass 9 kg resting on a horizontal plane as shown in Figure 3.62. The ball has a velocity of 10 m/s to the right just before striking the block and 5 m/s to the left just after the strike. If the coefficient of friction between the block and the plane is 0.15, determine:

a) the velocity of the block just after the strike;

[1.67 m/s]

b) the coefficient of restitution between the ball and the block;

[0.67]

c) the acceleration of the block;

[−1.47 m/s^2]

d) the time before the block stops sliding after the strike;

[1.13 s]

e) the distance that the block slides after the strike.

[0.944 m]

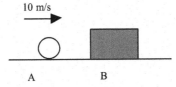

Figure 3.62: Representation of Question 3.34

CHAPTER 4
Kinetics of Rigid Bodies

4.1 INTRODUCTION

The study of the kinetics of rigid bodies can be regarded as an extension to the study of the kinetics of particles. In addition to the force, mass and acceleration relationships developed for particles in Chapter 3 (which are also applicable to rigid bodies), moment, mass moment of inertia and angular acceleration relationships are required to solve the kinetic problems of rigid bodies.

This chapter introduces the property of mass moment of inertia, which is essential for solving kinetic rotational problems. For two-dimensional kinetic problems, i.e. planar kinetics, three equations of motion are needed, two force equations and one moment equation. For translational motion, the equations derived in Chapter 3 for particles are applicable to the motion of the centres of gravity of rigid bodies. The kinematic equations developed in Chapter 2 for both translation and rotation of rigid bodies, are also required for the kinetic analysis of rigid bodies. The relationships between angular displacement, angular velocity and angular acceleration are essential in dealing with the work and energy and the impulse and momentum of rigid bodies.

This chapter is divided into three main sections, similar to those in Chapter 3: force and acceleration, work and energy, and impulse and momentum. The force and acceleration method is a direct application of Newton's second law, while the work and energy method and the impulse and momentum method are derived from it. The application of the force and acceleration method is presented for problems involving wheels and gears and problems involving linkages and mechanisms. This chapter also introduces some new principles related to the rotation of rigid bodies. For example, Section 4.3 introduces the concepts of rotational kinetic energy and the work of a couple. Section 4.4 introduces the principle of angular impulse and momentum, and the conservation of angular momentum.

4.2 FORCE AND ACCELERATION

4.2.1 Equations of motion of a rigid body

Consider a rigid body that consists of an infinite number of particles, as shown in Figure 4.1. The centre of gravity of the rigid body is located at point G, which has the co-ordinates (x_G, y_G). The ith particle has the general co-ordinates (x_i, y_i). The external forces F_n are acting on the rigid body as shown in Figure 4.1 and the internal forces are equal and opposite to each other at each particle so that they are not considered in writing Newton's second law.

Applying Newton's second law to each particle and summing for all particles:

$$\sum F = \sum m_i a_i \tag{4.1}$$

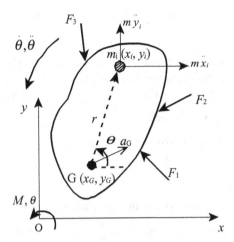

Figure 4.1: Forces acting on a rigid body

Writing Equation (4.1) in the x and y directions gives:

$$\sum F_x = \sum m_i \ddot{x}_i \tag{4.2}$$

$$\sum F_y = \sum m_i \ddot{y}_i \tag{4.3}$$

From the definition of the centre of gravity, the mass m of the rigid body ($m = \sum m_i$) and its acceleration components, the centre of gravity G, $\ddot{x}_G$ and $\ddot{y}_G$ are related to the particle's masses and acceleration by:

$$m\ddot{x}_G = \sum m_i \ddot{x}_i \tag{4.4}$$

$$m\ddot{y}_G = \sum m_i \ddot{y}_i \tag{4.5}$$

Substituting Equations (4.4) and (4.5) into Equations (4.2) and (4.3) gives:

$$\sum F_x = m \ddot{x}_G \tag{4.6}$$

$$\sum F_y = m \ddot{y}_G \tag{4.7}$$

And in a general form:

$$\sum F = m a_G \tag{4.8}$$

It should be noted that Equation (4.8) is identical to Newton's second law applied to a particle of mass m, moving with acceleration a_G and subjected to an external force $\sum F$ (see Equation (3.4)). Thus, Equation (4.8) indicates that, for a rigid body that consists of an infinite number of particles, there is a centre of gravity or centre of mass (point G), in which all masses of the system and all external forces acting on the system can be concentrated.

The summation of moment about the origin O for all particles can be obtained by applying the moment equation equivalent to Newton's second law, i.e. the sum of the moment due to external forces equals the sum of the moment due to inertia forces, which gives:

$$\sum M_O = \sum m_i \ddot{y}_i x_i - \sum m_i \ddot{x}_i y_i \tag{4.9}$$

The co-ordinates (x_i, y_i) can be written in terms of the centre of mass co-ordinates (x_G, y_G) and the relative co-ordinates $(x_{i/G}, y_{i/G})$:

$$x_i = x_{i/G} + x_G, \quad y_i = y_{i/G} + y_G \tag{4.10}$$

Differentiating Equation (4.10) twice with respect to time gives:

$$\ddot{x}_i = \ddot{x}_{i/G} + \ddot{x}_G, \quad \ddot{y}_i = \ddot{y}_{i/G} + \ddot{y}_G \tag{4.11}$$

Substituting Equations (4.10) and (4.11) into Equation (4.9) yields:

$$\sum M_O = \sum m_i x_{i/G} \ddot{y}_{i/G} + x_G \sum m_i \ddot{y}_{i/G} + \ddot{y}_G \sum m_i x_{i/G} + x_G \ddot{y}_G \sum m_i$$
$$- \sum m_i \ddot{x}_{i/G} y_{i/G} - y_G \sum m_i \ddot{x}_{i/G} - \ddot{x}_G \sum m_i y_{i/G} - y_G \ddot{x}_G \sum m_i \tag{4.12}$$

The terms $\sum m_i x_{i/G}$ and $\sum m_i y_{i/G}$ are the moment of mass about the centre of gravity; $\sum m_i x_{i/G} = m \, x_G$ and, since the origin is at G, $x_G = 0$. Consequently, $\sum m_i \ddot{x}_{i/G}$ and $\sum m_i \ddot{y}_{i/G}$ are also zero and Equation (4.12) becomes:

$$\sum M_O = \sum m_i x_{i/G} \ddot{y}_{i/G} + x_G \ddot{y}_G \sum m_i - \sum m_i \ddot{x}_{i/G} y_{i/G} - y_G \ddot{x}_G \sum m_i \tag{4.13}$$

Writing $(x_{i/G}, y_{i/G})$ in terms of the polar co-ordinates r and θ yields:

$$x_{i/G} = r_i \cos \theta \tag{4.14}$$
$$y_{i/G} = r_i \sin \theta \tag{4.15}$$

Differentiating Equations (4.14) and (4.15) twice with respect to time gives:

$$\ddot{x}_{i/G} = -r_i \ddot{\theta} \sin \theta - r_i \dot{\theta}^2 \cos \theta \tag{4.16}$$
$$\ddot{y}_{i/G} = r_i \ddot{\theta} \cos \theta - r_i \dot{\theta}^2 \sin \theta \tag{4.17}$$

Substituting Equations (4.14) to (4.17) into Equation (4.13) gives:

$$\sum M_O = m \ddot{y}_G x_G - m \ddot{x}_G y_G + I_G \ddot{\theta} \tag{4.18}$$

where I_G is equal to $\sum m_i r_i^2$ and is known as the mass moment of inertia of the body about the centre of gravity, G.

If the bending moment is taken about the centre of gravity, G, Equation (4.18) reduces to:

$$\sum M_G = I_G \ddot{\theta} \tag{4.19}$$

Equation (4.19) is the moment or rotational equation equivalent to Newton's second law applied at the centre of gravity of a rigid body, which describes the relationship between the moments ($\sum M_G$) due to the external forces and the inertia moment ($I_G \ddot{\theta}$) both calculated about the centre of gravity, G.

4.2.2 Mass moment of inertia

As can be seen from the derivation in Section 4.2.1, the mass moment of inertia of a body is a measure of its resistance to angular acceleration ($\sum M_G = I_G \ddot{\theta}$). Thus, the mass moment of inertia in a rotational motion performs the same function as the mass (m) in a translational motion ($\sum F = ma_G$). From Equation (4.18), the mass moment of inertia ($\sum m_i r_i^2$) can be defined as the integral of the second moment of all mass elements (dm) in the body. For the body shown in Figure 4.2, the mass moment of inertia can be written as:

$$I = \int_m r^2 dm \tag{4.20}$$

where r is the perpendicular distance from the axis $z - z$ to the mass element dm. The axis $z - z$ is often chosen to pass through the centre of gravity. The mass moment of inertia is a positive quantity and has the units of kg.m^2.

The mass element (dm) is written in terms of the density (ρ) and the element's volume (dV) as:

$$dm = \rho dV \tag{4.21}$$

Substituting Equation (4.21) into Equation (4.20) and keeping ρ outside the integration as it is a constant quantity, the mass moment of inertia becomes:

$$I = \rho \int_V r^2 dV \tag{4.22}$$

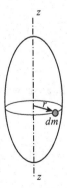

Figure 4.2: Mass moment of inertia

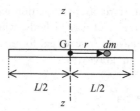

Figure 4.3: Calculating the mass moment of a uniform bar

Consider a bar of length L as shown in Figure 4.3. Its element's volume is given by:

$$dV = dA \times dr \tag{4.23}$$

where dA is the area of the element's cross-section. Substituting Equation (4.23) into Equation (4.22) gives:

$$I = \rho \int_A \int_r r^2 dA dr \tag{4.24}$$

For a uniform bar with constant cross-section area A, Equation (4.24) becomes:

$$I = \rho A \int_r r^2 dr \tag{4.25}$$

As the axis $z - z$ passes through the centre of gravity of the bar, the integration is performed from $-L/2$ to $+L/2$:

$$I_G = \rho A \int_{-L/2}^{L/2} r^2 dr \tag{4.26}$$

Integrating Equation (4.26) gives:

$$I_G = \frac{\rho A L^3}{12} \tag{4.27}$$

Since $\rho = \frac{m}{V} = \frac{m}{AL}$, the mass moment of a uniform bar becomes:

$$I_G = \frac{mL^2}{12} \tag{4.28}$$

For a cylinder of length L and radius R, as shown in Figure 4.4, the element's volume is given by:

$$dV = (2\pi r) \times L \times dr \tag{4.29}$$

Substituting Equation (4.29) into Equation (4.22) gives:

$$I = 2\pi L \rho \int_r r^3 dr \tag{4.30}$$

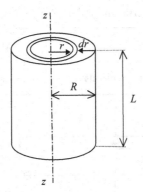

Figure 4.4: Calculating the mass moment of a cylinder

Integrating Equation (4.30) from 0 to R, the mass moment of inertia about the axis passing through the centre of gravity G is given by:

$$I_G = 2\pi L\rho \int_0^R r^3 dr = \frac{\pi L\rho R^4}{2} \tag{4.31}$$

Since $\rho = \frac{m}{V} = \frac{m}{\pi R^2 L}$, the mass moment of a cylinder becomes:

$$I_G = \frac{mR^2}{2} \tag{4.32}$$

The mass moment of inertia can also be defined using the radius of gyration, k_G. The radius of gyration has units of length and is related to the body mass and its mass moment of inertia by:

$$I_G = mk_G^2 \tag{4.33}$$

Table 4.1 summarizes the mass moment of inertia and the radius of gyration of some common bodies.

4.2.3 Application of Newton's second law to rigid bodies

The steps that were followed in applying Newton's second law to particles in Section 3.3 are also applicable to rigid bodies:

Step 1. Define and identify the rigid bodies or systems whose motion should be described. In Figure 4.5(a), two rigid bodies, of masses m_1 and m_2 and mass moment of inertia I_1 and I_2, are identified.

Step 2. Isolate the identified rigid bodies from all other bodies in the system. In Figure 4.5(b), the two rigid bodies are surrounded by an imaginary envelope. The bodies lying outside this envelope are replaced with forces and moments in Step 4.

Table 4.1: Mass moment of inertia and radius of gyration

Body	Dimensions	Mass moment of inertia (I_G) about z-z axis	Radius of gyration (k_G)
Uniform bar		$\dfrac{mL^2}{12}$	$\dfrac{L}{\sqrt{12}}$
Cylinder or disc		$\dfrac{mR^2}{2}$	$\dfrac{R}{\sqrt{2}}$
Sphere or ball		$\dfrac{2mR^2}{5}$	$\dfrac{\sqrt{2}R}{\sqrt{5}}$
Block		$\dfrac{m(L_1^2 + L_2^2)}{12}$	$\sqrt{\dfrac{(L_1^2 + L_2^2)}{12}}$

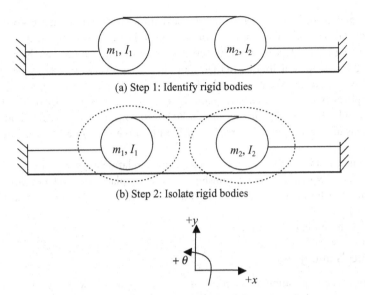

(a) Step 1: Identify rigid bodies

(b) Step 2: Isolate rigid bodies

(c) Step 3: Define co-ordinate system and sign convention

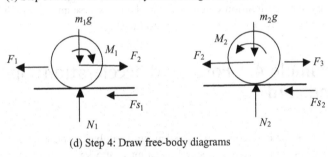

(d) Step 4: Draw free-body diagrams

A rigid body

$$\sum F_x = m\ddot{x}_G, \quad \sum F_y = m\ddot{y}_G, \quad \sum M = I\ddot{\theta}$$

(e) Step 5: Apply Newton's second law

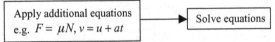

Apply additional equations
e.g. $F = \mu N$, $v = u + at$ → Solve equations

(f) Steps 6 and 7: Apply additional equations and solve the system of equations

Figure 4.5: Applying Newton's second law to rigid bodies

Step 3. Define a co-ordinate system and sign conventions as shown in Figure 4.5(c).

Step 4. Draw a free-body diagram for each rigid body, Figure 4.5(d), by identifying and drawing all the forces and moments acting on the rigid bodies at their centres of gravity and contributing to their motion. This is a very important first step in analyzing the kinetics of a rigid body.

Step 5. Apply Newton's second law (Equations (4.6) and (4.7) for translation and Equations (4.19) or (4.18) for rotation), taking into consideration the free-body diagrams and identifying all forces contributing to the motion as shown in Figure 4.5(e).

Step 6. Define any additional equations using frictional relationships, Equation (3.12), or kinematic equations, i.e. relationships between displacement, velocity and acceleration or between translation and rotation (see Chapter 2), as shown in Figure 4.5(f).

Step 7. Solve the system of equations.

4.2.4 Wheels and gears

In this section, the steps summarized in Figure 4.5 are applied to wheels and gears. The kinematic equations that are required to deal with wheel and gear problems are mainly the velocity and acceleration components in the transverse and radial directions. These equations were derived as Equations (2.8) to (2.10).

Example 4.1 Force and acceleration: Spool and cord

In Figure 4.6, a light cord is wrapped around a hub of a spool, which moves only in the vertical direction. The spool has a mass of 5 kg and mass moment inertia

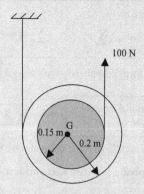

Figure 4.6: Representation of Example 4.1

of 0.0375 kg.m^2. When no external force is applied, the spool is stationery. If a vertical force of 100 N is applied to the cord on the right-hand side, the spool moves upwards. Determine:

a) the angular acceleration of the spool;
b) the tension in the cord on the left-hand side.

Solution

The mass moment of inertia is $I_G = 0.0375$ kg.m^2. From the free-body diagram in Figure 4.7, applying Newton's second law in the y direction gives:

$$\sum F_y = ma_y \Rightarrow 100 + T - 5g = 5a_y \tag{E4.1a}$$

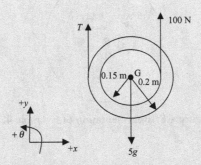

Figure 4.7: Free-body diagram for Example 4.1

From kinematics, $a_y = a_t = 0.2\ddot{\theta}$ (see Equation (2.10)), Equation (E4.1a) becomes:

$$T = -100 + 5g + 5 \times 0.2\ddot{\theta} \Rightarrow T = -50.95 + \ddot{\theta} \tag{E4.1b}$$

Applying the moment equation equivalent to Newton's second law about the centre of gravity G, Equation (4.19) gives:

$$\sum M_G = I_G\ddot{\theta} \Rightarrow 100 \times 0.15 - T \times 0.2 = 0.0375\ddot{\theta}$$
$$\Rightarrow T = 75 - 0.1875\ddot{\theta} \tag{E4.1c}$$

a) Equating Equations (E4.1b) and (E4.1c), the angular acceleration of the spool is obtained as:

$$T = -50.95 + \ddot{\theta} = 75 - 0.1875\ddot{\theta}, \text{ from which } \ddot{\theta} = 106.06 \text{ rad/s}^2$$

b) Using Equation (E4.1b), the tension in the cord is calculated as:

$$T = 75 - 0.1875 \times 106.06 = 55.11 \text{ N}$$

Example 4.2 Force and acceleration: Flywheel

A flywheel of mass 24 kg is unbalanced so that its centre of gravity G is shifted by 0.12 m from its centre O, as shown in Figure 4.8. The radius of gyration of the flywheel is equal to 0.2 m. If the flywheel rotates about O with an angular velocity of 10 rad/s clockwise due to an external torque of 115 N.m, determine (at the instant shown):

a) the flywheel angular acceleration;
b) the components of reaction at O.

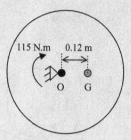

Figure 4.8: Representation of Example 4.2

Solution

From Equation (4.33) and Table 4.1, the mass moment of inertia is:

$$I_G = mk_G^2 = 24 \times 0.2^2 = 0.96 \, \text{kg.m}^2$$

a) Considering the free-body diagram in Figure 4.9 and taking the moment about G, Equation (4.19) gives:

$$\sum M_G = I_G \ddot{\theta} \Rightarrow -115 - 0.12 \times R_y = 0.96 \ddot{\theta} \qquad \text{(E4.2a)}$$

From Equations (2.9) and (2.10), the acceleration components are (at the instant shown):

$$a_r = \ddot{x}_G = -r\dot{\theta}^2 = -0.12 \times 10^2 = -12 \, \text{m/s}^2, \quad a_t = \ddot{y}_G = r\ddot{\theta} = 0.12\ddot{\theta}$$

Taking force summation in the y direction, Equation (4.7) yields:

$$\sum F_y = m\ddot{y}_G \Rightarrow R_y - 24g = 24 \times 0.12\ddot{\theta} \qquad \text{(E4.2b)}$$

Substituting Equation (E4.2b) into Equation (E4.2a) and solving for $\ddot{\theta}$ gives:

$$\ddot{\theta} = -109.72 \, \text{rad/s}^2$$

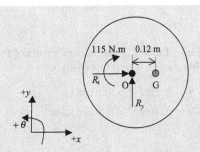

Figure 4.9: Free-body diagram for Example 4.2

b) From Equation (E4.2b), the reaction force in the y direction, R_y, is calculated as:

$$R_y = 24g + 24 \times 0.12 \times (-109.72) = -80.56\,\text{N}$$

Taking force summation in the x direction, Equation (4.6) gives the reaction force in the x-direction, R_x, as:

$$\sum F_x = m\ddot{x}_G \Rightarrow R_x = 24 \times (-12) = -288\,\text{N}$$

Example 4.3 Force and acceleration: Gears

Two gears are in mesh as shown in Figure 4.10. Gear 1 has a radius of 80 mm, a radius of gyration of 56.56 mm and a mass of 4 kg; gear 2 has a radius of 120 mm, a radius of gyration of 84.85 mm and a mass of 8 kg. If gear 1 has an angular acceleration of 3 rad/s^2, determine:

a) the angular acceleration of gear 2;
b) the contact force between the two gears;
c) the external moment M_1 applied to gear 1.

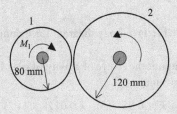

Figure 4.10: Representation of Example 4.3

Solution

From Equation (4.33) and Table 4.1, the mass moments of inertia for gears 1 and 2 (using $k_{G1} = 0.05656$ m and $k_{G2} = 0.08485$ m) are:

$$I_{G1} = mk_{G1}^2 = 4 \times 0.05656^2 = 0.0128 \text{ kg.m}^2$$
$$I_{G2} = mk_{G2}^2 = 8 \times 0.08485^2 = 0.0576 \text{ kg.m}^2$$

a) From kinematics, Equation (2.11), the relationship between the angular acceleration of gear 1 $\ddot{\theta}_1$ and the angular acceleration of gear 2, $\ddot{\theta}_2$ is given by:

$$r_1\ddot{\theta}_1 = r_2\ddot{\theta}_2$$

From which, the angular acceleration of gear 2 is calculated as:

$$\ddot{\theta}_2 = \frac{r_1\ddot{\theta}_1}{r_2} = \frac{0.08 \times 3}{0.12} = 2 \text{ rad/s}^2$$

b) Considering the free-body diagram in Figure 4.11 and applying the moment equation, Equation (4.19), to gear 2 gives:

$$\sum M_{G2} = I_{G2}\ddot{\theta}_2 \Rightarrow F \times 0.12 = 0.0576 \times \ddot{\theta}_2 \Rightarrow F = 0.48\ddot{\theta}_2$$

From which the contact force between the two gears F is calculated as:

$$F = 0.48\ddot{\theta}_2 = 0.48 \times 2 = 0.96 \text{ N}$$

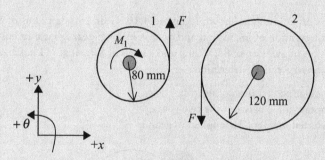

Figure 4.11: Free-body diagram for Example 4.3

c) Taking the moment about the centre of gravity of gear 1, the external moment applied at gear 1 is given by:

$$\sum M_{G1} = I_{G1}\ddot{\theta}_1 \Rightarrow F \times 0.08 - M_1 = 0.0128 \times (-3) \Rightarrow M_1 = 0.1152 \text{ N.m}$$

4.2.5 Linkage and mechanisms

The kinetics of linkages and mechanisms is concerned with the relationships between the forces acting on the different links in a linkage or mechanism and their velocities and accelerations. The forces acting on a mechanism or on machine elements can be due to various causes, e.g. gas pressure on pistons, torque on rotating links and reactions at bearings. Because there are cyclical and harmonic motion in the different parts in a linkage or mechanism, inertia forces produced by masses and accelerations should be taken into account when analyzing forces on any part of the mechanism. This section presents the forces and accelerations method, which is based on using Newton's second law to relate the forces acting on a linkage or mechanism to the acceleration.

Sign convention for moments due to inertia forces

When applying the bending moment equation equivalent to Newton's second law at a point other than the centre of gravity of the rigid body, Equation (4.18), the sign of the bending moment due to inertia forces should be taken into account.

For the rigid bar shown in Figure 4.12, the centre of gravity of the body is located at point G. The positive bending moment is in the anticlockwise sense, and thus the term $I_G \ddot{\theta}$ is always positive (positive $\ddot{\theta}$ is anticlockwise as defined in Figure 4.1). If the moment is taken about point O, $m\ddot{y}_G$ is making a positive (anti-clockwise) bending moment, $m\ddot{x}_G$ is making a negative (clockwise) bending moment and the moment equation, Equation (4.18), is obtained. However, if the moment is taken about point B, both $m\ddot{y}_G$ and $m\ddot{x}_G$ are making positive (anti-clockwise) bending moments and Equation (4.18) becomes:

$$\sum M_B = +m\ddot{y}_G x_G + m\ddot{x}_G y_G + I_G\ddot{\theta} \tag{4.34}$$

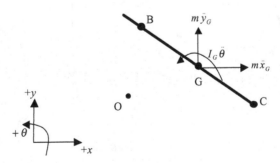

Figure 4.12: Sign convention for inertia moments

Similarly, taking moment about C, both $m\ddot{y}_G$ and $m\ddot{x}_G$ are making negative (clockwise) bending moment and Equation (4.18) becomes:

$$\sum M_C = -m\ddot{y}_G x_G - m\ddot{x}_G y_G + I_G\ddot{\theta} \tag{4.35}$$

Thus, a general form for the moment equation equivalent to Newton's second law at any point in a rigid body is:

$$\sum M = \pm m\ddot{y}_G x_G \pm m\ddot{x}_G y_G + I_G\ddot{\theta} \tag{4.36}$$

where the positive and negative signs depend upon the point about which the moment is taken. When the moment is taken about the centre of gravity, G, i.e. $x_G = y_G = 0$, Equation (4.36) is simply reduced to Equation (4.19) ($\sum M_G = I_G\ddot{\theta}$).

Acceleration components of a rotating rigid body

The acceleration components at the centre of gravity of a rotating rigid body in the Cartesian coordinate system can be obtained using the kinematic equations, Equations (2.9) and (2.10). In Figure 4.13(a), if the rigid bar is rotating about point O with angular velocity $\dot{\theta}$ and angular acceleration $\ddot{\theta}$, the radial acceleration is $a_r = -r\dot{\theta}^2$, Equation (2.9), and the transverse acceleration is $a_\theta = r\ddot{\theta}$, Equation (2.10). Resolving a_r and a_θ in the x and y directions, as shown in Figure 4.13(b) gives:

$$\ddot{x}_G = -r\dot{\theta}^2\sin\theta + r\ddot{\theta}\cos\theta \tag{4.37}$$

and

$$\ddot{y}_G = r\dot{\theta}^2\cos\theta + r\ddot{\theta}\sin\theta \tag{4.38}$$

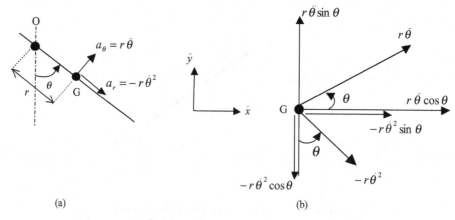

(a)   (b)

Figure 4.13: Acceleration components of a rotating rigid body

If the bar is in a vertical position ($\theta = 0$), $\ddot{x}_G$ and $\ddot{y}_G$ become $\ddot{x}_G = r\ddot{\theta}$ and $\ddot{y}_G = r\dot{\theta}^2$, respectively. If the bar is in a horizontal position ($\theta = 90°$), $\ddot{x}_G$ and $\ddot{y}_G$ become $\ddot{x}_G = -r\dot{\theta}^2$ and $\ddot{y}_G = r\ddot{\theta}$, respectively.

Example 4.4 Force and acceleration: Rigid rod I

A rigid rod, AB, of 0.9 m length, L, is rotating about its centre of gravity, point G, as shown in Figure 4.14. The bar has a mass, m, of 4 kg and an initial horizontal force, P, of −75 N is applied at a bar end, point B. Determine:

a) the angular acceleration of the rod;
b) the reaction components at point G.

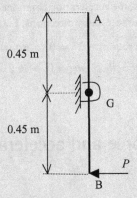

Figure 4.14: Representation of Example 4.4

Solution

The free-body diagram in Figure 4.15 shows R_{Gx} and R_{Gy} as the components of reaction at G in the x and y directions, respectively.

From Table 4.1 and Equation (4.28), the mass moment of inertia of a rigid rod is $I_G = \frac{ml^2}{12} = \frac{4 \times 0.9^2}{12} = 0.27\,\text{kg.m}^2$.

a) From the moment equation equivalent to Newton's second law, Equation (4.19), taking moment about the centre of gravity, G, the angular acceleration of the rod is calculated as:

$$\sum M_G = I_G\ddot{\theta} \Rightarrow -75 \times 0.45 = 0.27\ddot{\theta} \Rightarrow \ddot{\theta} = 125\,\text{rad/s}^2$$

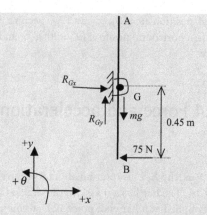

Figure 4.15: Free-body diagram for Example 4.4

b) Since the rod is supported at its centre of gravity, i.e. point G does not move in the x and y directions, the acceleration components of G are zero. Applying Newton's second law in the x and y directions, Equations (4.6) and (4.7) give:

$$\sum F_x = ma_x = 0 \Rightarrow R_{Gx} - 75 = 0 \Rightarrow R_{Gx} = 75\,\text{N}$$
$$\sum F_y = ma_y = 0 \Rightarrow R_{Gy} - mg = 0 \Rightarrow R_{Gy} = mg = 4 \times 9.81 = 39.24\,\text{N}$$

Example 4.5 Force and acceleration: Rigid rod II

A uniform rod ABC shown in Figure 4.16 has a mass of 50 kg and is connected to two frictionless collars of negligible mass, which slide on smooth horizontal rods. The rod ABC is making angle $\theta = 30°$ with the horizontal axis. A spring of stiffness k is connected to the rod at point B and a force P of 20 N is applied at point C. If the spring is in its un-deformed state at the start of the motion, determine the acceleration of the rod and the reaction at points B and C for the following cases:

a) the start of the motion;
b) when the spring is contracted by a displacement equivalent to $\frac{2P}{k}$.

Solution

The forces acting on the rod ABC are shown in the free-body diagram in Figure 4.17. The force exerted on the rod ABC by the spring is kx, where x is the

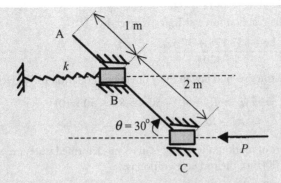

Figure 4.16: Representation of Example 4.5

deformation of the spring. The inertia forces $m\ddot{x}_G$ and $m\ddot{y}_G$ and the inertia moment $I_G\ddot{\theta}$ are acting at the centre of gravity G. Using Equation (4.6), the acceleration of the rod in the x direction can be found as:

$$\sum F_x = m\ddot{x}_G \Rightarrow -kx - 20 = 50 \times \ddot{x}_G \Rightarrow \ddot{x}_G = \frac{-kx - 20}{50} \qquad \text{(E4.5a)}$$

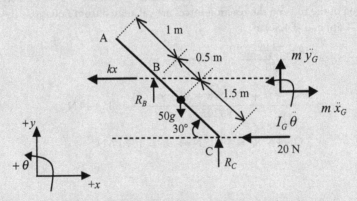

Figure 4.17: Free-body diagram for Example 4.5

Taking moment about C (Equation (4.35)) and using $\ddot{y}_G = \ddot{\theta} = 0$, as there is neither vertical motion nor rotation, leads to:

$$\sum M_C = -m\ddot{y}_G x_G - m\ddot{x}_G y_G + I_G\ddot{\theta}$$
$$\Rightarrow kx \times 2 \times \sin 30 - R_B \times 2 \times \cos 30 + 50 \times 9.81 \times 1.5 \times \cos 30$$
$$= -0 - 50 \times (\ddot{x}_G) \times 1.5 \times \sin 30 + 0$$

from which the reaction force at B, R_B is given by:

$$R_B = \frac{kx + 637.178 + 37.5\ddot{x}_G}{1.73205}$$ (E4.5b)

Applying Newton's second law in the y direction, Equation (4.7) gives:

$$\sum F_y = m\ddot{y}_G \Rightarrow R_B + R_C - 50 \times 9.81 = 50 \times (0)$$
$$\Rightarrow R_B + R_C = 490.5\,\text{N}$$ (E4.5c)

a) At the start of motion, the spring is in its un-deformed state, i.e. $x = 0$. From Equation (E4.5a), the rod acceleration is:

$$\ddot{x}_G = \frac{-0 - 20}{50} = -0.4\,\text{m/s}^2$$

From Equation (4.5b), the reaction R_B is;

$$R_B = \frac{0 + 637.178 + 37.5 \times (-0.4)}{1.73205} = 359.22\,\text{N}$$

And from Equation (4.5c), the reaction R_C is;

$$359.22 + R_C = 490.5\,\text{N} \Rightarrow R_C = 131.28\,\text{N}$$

b) If the spring is contracted by a displacement $\frac{2P}{k}$ (i.e. $x = -\frac{2P}{k}$), the acceleration of the rod and the reaction forces are calculated from Equations (E4.5a), (E4.5b) and (E4.5c) as:

$$\ddot{x}_G = \frac{-k \times \frac{-2\times20}{k} - 20}{50} = 0.4\,\text{m/s}^2$$

$$R_B = \frac{k \times \frac{-2\times20}{k} + 637.178 + 37.5 \times (0.4)}{1.73205} = 353.44\,\text{N}$$

$$353.44 + R_C = 490.5\,\text{N} \Rightarrow R_C = 137.06\,\text{N}$$

Example 4.6 Force and acceleration: Rigid rod III

A rod ABC has a mass of 10 kg and is supported at point C as shown in Figure 4.18. If a horizontal force P of magnitude 50 N is applied at B, determine the angular acceleration of the rod and the components of the reaction at C at the start of the motion.

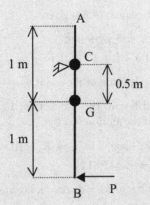

Figure 4.18: Representation of Example 4.6

Solution

The forces acting on the rod ACB are shown in the free-body diagram in Figure 4.19, where R_x and R_y are the reaction forces acting at point C in the x and y directions, respectively. The inertia forces $m\ddot{x}_G$ and $m\ddot{y}_G$ and the inertia moment $I_G\ddot{\theta}$ are acting at the centre of gravity G. From Equations (4.37) and (4.38) for $\theta = 0$, the acceleration components of the centre of gravity G are $\ddot{x}_G = r\ddot{\theta}$, $\ddot{y}_G = r\dot{\theta}^2$. From Table 4.1, the mass moment of inertia of the rod is $I_G = mL^2/12 = 3.333$ kg.m^2.

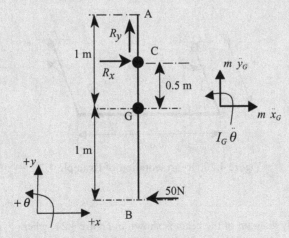

Figure 4.19: Free-body diagram for Example 4.6

Taking moment about C, Equation (4.34) gives:

$$\sum M_C = m\ddot{y}_G x_G + m\ddot{x}_G y_G + I_G\ddot{\theta}$$
$$\Rightarrow -50 \times 1.5 = 0 + 10 \times (0.5 \times \ddot{\theta}) \times 0.5 + 3.333 \times (\ddot{\theta})$$

from which the angular acceleration of the rod is $\ddot{\theta} = -12.857\,\text{rad/s}^2$.

Applying Newton's second law in the y direction, Equation (4.7) gives ($\ddot{y}_G = r\dot{\theta}^2 = 0$ since at the start of the motion $\dot{\theta} = 0$):

$$\sum F_y = m\ddot{y}_G \Rightarrow R_y - mg = m\ddot{y}_G \Rightarrow R_y - 10 \times 9.81$$
$$= 10 \times (0) \Rightarrow R_y = 981\,\text{N}$$

Applying Newton's second law in the x direction, Equation (4.6) gives ($\ddot{x}_G = r\ddot{\theta} = 0.5 \times -12.857\,\text{m/s}^2$):

$$\sum F_x = m\ddot{x}_G \Rightarrow R_x - 50 = 10 \times (-0.5 \times 12.857) \Rightarrow R_x = -14.28\,\text{N}$$

Example 4.7 Force and acceleration: Linkage mechanism

A linkage mechanism consists of a uniform beam of length 0.6 m and mass 80 kg, which is hung from two rods of 0.4 m length as shown in Figure 4.20. At the position shown, the rods are rotating with an angular velocity of 4 rad/s. Determine the angular acceleration of the rods and the force in each rod. Neglect the mass of the rods.

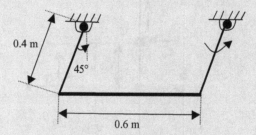

Figure 4.20: Representation of Example 4.7

Solution

The free-body diagram of the beam is shown in Figure 4.21, where F_1 and F_2 are the forces in the left and right rods, respectively. From Equations (4.37) and (4.38), the acceleration components of the centre of gravity of the beam ($\theta = 0$) are

$$\ddot{x}_G = r\ddot{\theta}, \quad \ddot{y}_G = r\dot{\theta}^2 = 0.4 \times 4^2 = 6.4\,\text{m/s}^2$$

where r is the rod length and $\ddot{\theta}, \dot{\theta}$ are the angular acceleration and angular velocity of the rods, respectively.

From Equation (4.6), Newton's second law in the x direction is:

$$\sum F_x = m\ddot{x}_G \Rightarrow mg\sin 45 = mr\ddot{\theta}$$

Solving for the angular acceleration $\ddot{\theta}$ gives:

$$\ddot{\theta} = \frac{g\sin 45}{r} = \frac{9.81\sin 45}{0.4} = 17.34\,\text{rad/s}^2$$

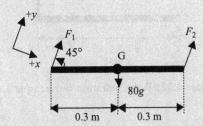

Figure 4.21: Free-body diagram for Example 4.7

Taking moment about G (i.e. applying Equation (4.19) with $\ddot{\theta} = 0$, since the beam does not rotate) gives:

$$\sum M_G = I_G\ddot{\theta} \Rightarrow -F_1 \times 0.3\cos 45$$
$$+ F_2 \times 0.3\cos 45 = 0$$

from which

$$F_1 = F_2 \tag{E4.7a}$$

Applying Newton's second law in the y direction, Equation (4.7) gives:

$$\sum F_y = m\ddot{y}_G \Rightarrow F_1 + F_2 - mg\cos 45 = mr\dot{\theta}^2$$
$$\Rightarrow F_1 + F_2 - 80 \times 9.81 \times \cos 45 = 80 \times 6.4$$

from which

$$F_1 + F_2 = 1066.94 \tag{E4.7b}$$

Solving Equations (E4.7a) and (E4.7b), the forces in the rods are calculated as:

$$F_1 = F_2 = 533.43\,\text{N}$$

Example 4.8 Force and acceleration: Suspended rod mechanism

A uniform rod of length 2.5 m and mass 5 kg is supported by a spring at A and a cord at B as shown in Figure 4.22. At the instant when cord is cut, determine the angular acceleration of the rod and the vertical reaction at A.

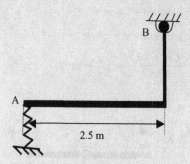

Figure 4.22: Representation of Example 4.8

Solution

The free-body diagram of the rod when the cord is cut is shown in Figure 4.23, where R_A is the reaction force at A.

From Equations (4.37) and (4.38), the acceleration components of the rod's centre of gravity are $\ddot{y}_G = r\ddot{\theta}, \ddot{x}_G = r\dot{\theta}^2 = 0$ (since at the instant when the rod is cut, i.e. the start of the motion, $\theta = 0$) and from Table 4.1 the rod's mass moment of inertia is $I_G = mL^2/12 = 2.60417\,\mathrm{kg.m^2}$.

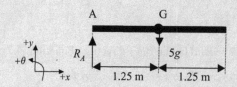

Figure 4.23: Representation of Example 4.8

Taking moment about A, Equation (4.34) gives:

$$\sum M_A = m\ddot{y}_G x_G + m\ddot{x}_G y_G + I_G \ddot{\theta} \Rightarrow -5 \times 9.81 \times 1.25$$
$$= 5 \times (1.25 \times \ddot{\theta}) \times 1.25 + 2.60417 \times (\ddot{\theta})$$

from which $\ddot{\theta} = -5.89 \, \text{rad/s}^2$.

Applying Newton's second law in the y direction, Equation (4.7) gives ($\ddot{y}_G = r\ddot{\theta} = 1.25 \times -5.89 \, \text{m/s}^2$):

$$\sum F_y = m\ddot{y}_G \Rightarrow R_A - mg = m\ddot{y}_G \Rightarrow R_A - 5 \times 9.81 = 5 \times 1.25 \times -5.89$$

From which the reaction force at A is obtained as:

$$R_A = 12.26 \, \text{N}$$

4.3 WORK AND ENERGY

The method of work and energy for particles was presented in Section 3.4. In addition to the work and energy equations derived for translational motion in Section 3.4, which are also applicable to rigid bodies, the rotational component of kinetic energy and work done by a couple has to be considered when studying the work and energy of rigid bodies. The derivation of the rotational kinetic energy is also based on Newton's second law.

4.3.1 Rotational kinetic energy

As mentioned in Chapter 2, rigid bodies can undergo both translation and rotation. The translational component of the kinetic energy of a rigid body is similar to that of a particle, which was derived in Equation (3.17), if the motion of the mass centre of the rigid body is considered. The rotational component of the kinetic energy can be obtained by integrating the rotational form of Equation (3.13). Recalling the moment equation equivalent to Newton's second law, Equation (4.19), and replacing $\ddot{\theta}$ by $\dot{\theta}\frac{d\dot{\theta}}{d\theta}$ (similar to $a = v\frac{dv}{ds}$ in a rectilinear motion) gives:

$$M_G = I_G\ddot{\theta} = I_G\dot{\theta}\frac{d\dot{\theta}}{d\theta} \tag{4.39}$$

Integrating Equation (4.39) over angular position θ and angular velocity $\dot{\theta}$ yields:

$$\int M_G d\theta = \int I_G\dot{\theta}d\dot{\theta} \tag{4.40}$$

For a constant mass moment of inertia I_G, Equation (4.40) becomes:

$$\int M_G d\theta = I_G \int \dot{\theta}d\dot{\theta} \tag{4.41}$$

If the body rotates from angular position θ_1 to angular position θ_2, with angular velocities $\dot{\theta}_1$ to $\dot{\theta}_2$, respectively, the integration of Equation (4.41) becomes:

$$\int_{\theta_1}^{\theta_2} M_G d\theta = I_G \int_{\theta_1}^{\theta_2} \dot{\theta} d\dot{\theta} \tag{4.42}$$

and performing the integration yields:

$$\int_{\theta_1}^{\theta_2} M_G d\theta = \frac{1}{2} I_G \dot{\theta}_2^2 - \frac{1}{2} I_G \dot{\theta}_1^2 \tag{4.43}$$

where the term on the left-hand side $\int_{\theta_1}^{\theta_2} M_G d\theta$ in Equation (4.43) is the work of the moment M_G and the term on the right-hand side $\frac{1}{2} I_G \dot{\theta}_2^2 - \frac{1}{2} I_G \dot{\theta}_1^2$ is the change in the rotational kinetic energy of a body of mass moment of inertia I_G. The change of the rotational kinetic energy is the difference between the kinetic energy at an initial position 1 and at a final position 2. The result of the integral $\int_{\theta_1}^{\theta_2} M_G d\theta$ depends upon the relationship between the external moment M_G and the rotation $d\theta$. The total kinetic energy of a rigid body that undergoes both translation and rotation is simply calculated by adding the translational and rotational kinetic energies:

$$KE = \frac{1}{2} m v_G^2 + \frac{1}{2} I_G \dot{\theta}^2 \tag{4.44}$$

where v_G^2 is the velocity of the centre of gravity of the rigid body.

4.3.2 Work of a couple

A rigid body subjected to a couple undergoes rotation. If a couple M_G is applied to a rigid body as shown in Figure 4.24, it can be equivalent to two equal and opposite forces, F, separated from one another by a distance r. Due to these forces, the body rotates by an angle $d\theta$. The displacement ds, can be calculated as $d\theta = \tan^{-1} \frac{ds}{r/2}$ and, for a small angle

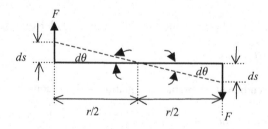

Figure 4.24: Work of a couple

$d\theta$, $\tan^{-1}\frac{ds}{r/2} \approx \frac{ds}{r/2}$ and $d\theta$ is given by:

$$d\theta = \frac{ds}{r/2} \Rightarrow ds = r/2\, d\theta \tag{4.45}$$

The work done by the two forces or the couple, dU_θ, is equal to:

$$dU_\theta = F \times r/2\, d\theta + F \times r/2\, d\theta = Fr\, d\theta = M_G\, d\theta \tag{4.46}$$

The work done by a rotation from angular position θ_1 to angular position θ_2 is obtained by integrating Equation (4.46) as:

$$U_\theta = \int_{\theta_1}^{\theta_2} M_G\, d\theta \tag{4.47}$$

For a constant couple M_G, the integration of Equation (4.47) yields:

$$U_\theta = M_G(\theta_2 - \theta_1) \tag{4.48}$$

The work is positive when the couple and the angular position have the same direction and is negative when they have opposite directions.

4.3.3 Summary of energies for rigid bodies

The conservation of energy law presented in Section 3.4.2, Equation (3.26), is also applicable to rigid bodies but the work done by a couple U_θ, derived in Section 4.3.2, should be added to the left-hand side in Equation (3.26). Therefore, the conservation of energy law of a rigid body is given by:

$$PE_1 + KE_1 + SE_1 + W_e + U_\theta = PE_2 + KE_2 + SE_2 \tag{4.49}$$

A summary of the work and energy terms of a rigid body is given in Figure 4.25, where G refers to the centre of gravity of a rigid body.

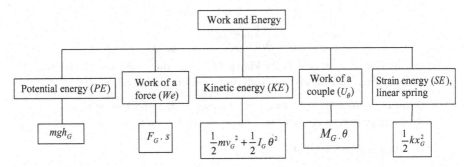

Figure 4.25: Summary of work and energy terms of a rigid body

Example 4.9 Work and energy: Lifting a mass

A motor applies a torque of 3.2 kN.m to a wheel in order to lift a sand bucket of mass 800 kg as shown in Figure 4.26. The wheel has a mass of 55 kg, an outer radius of 0.4 m and a radius of gyration about its centre of gravity of 0.27 m. Determine the speed of the bucket when it is lifted by 2 m starting from rest.

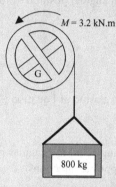

Figure 4.26: Representation of Example 4.9

Solution

Taking datum at position 1, before lifting the sand bucket as shown in Figure 4.27, the energies are calculated as follows:

$PE_1 = 0$ (datum passes through the sand bucket)

$KE_1 = 0$ (motion starts)

$SE_1 = 0$ (no elastic bodies)

In position 2:

$$PE_2 = mgh_G = 800 \times 9.81 \times 2 = 15696 \,\text{N.m}$$

$$KE_2 = \left(\frac{1}{2}mv^2\right)_{bucket} + \left(\frac{1}{2}mv_G^2 + \frac{1}{2}I_G\dot{\theta}^2\right)_{wheel}$$

where the first term represents the kinetic energy of the sand bucket, while the second and the third terms represent the kinetic energy of the wheel.

Since the centre of rotation is the centre of gravity of the wheel, point G, which is fixed, the velocity v_G is zero and the kinetic energy becomes:

$$KE_2 = \frac{1}{2}mv^2 + \frac{1}{2}I_G\dot{\theta}^2$$

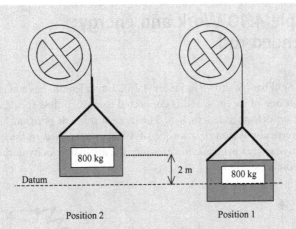

Figure 4.27: Analysis of Example 4.9

From Equation (2.8), the velocity of the sand bucket (the outer edge of the wheel) is $v = r\dot{\theta} = 0.4\dot{\theta}$ from which the angular velocity is $\dot{\theta} = \frac{v}{0.4}$. From Table 4.1 and Equation (4.33), the mass moment of inertia of the wheel is $I_G = mk_G^2 = 55 \times 0.27^2 = 4\,\text{kg.m}^2$. Therefore, the kinetic energy at position 2 is given by:

$$KE_2 = \frac{1}{2} \times 800 \times v^2 + \frac{1}{2} \times 4 \times \left(\frac{v}{0.4}\right)^2 = 412.5v^2$$

$SE_2 = 0$ (no elastic bodies)

The angular displacement of the wheel due to lifting the bucket a distance s is:

$$\theta = \frac{s}{2\pi r}\ \text{revolutions} = \frac{s}{r}\ \text{rad}$$

where r is the radius of the wheel. For $s = 2$ m, the angular displacement of the wheel is:

$$\theta = \frac{s}{r} = \frac{2}{0.4} = 5\,\text{rad}$$

The work of external couples is $U_\theta = M_G\theta = 3200 \times 5 = 16000\,\text{N.m}$.

The work due to external forces is $W_e = 0$ (no external forces).

Applying the conservation of energy principle, Equation (4.49) gives:

$$0 + 16000 = 15696 + 412.5v^2$$

From which the final velocity is calculated as:

$$v = 0.858\,\text{m/s}$$

Example 4.10 Work and energy: Suspended rod

A uniform, rigid bar (as shown in Figure 4.28), 1 m in length, has a mass of 20 kg, rotates about one of its ends, and is connected to a weightless spring of stiffness 100 N/m of un-deformed length 1 m. The motion starts at position 1, where the bar is in a vertical position. If a torque of 10 N.m is applied to the bar, use the conservation of energy principle to determine the angular velocity of the bar when it is in horizontal position 2.

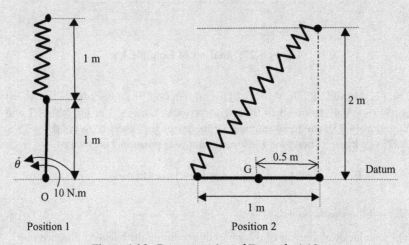

Figure 4.28: Representation of Example 4.10

Solution

Taking datum as shown in Figure 4.28 for position 1, where motion starts, the energies are calculated as:

$$PE_1 = mg\,L/2 = 20 \times 9.81 \times 0.5 = 98.1\,\text{N.m}$$
$$KE_1 = 0 \text{ (motion starts)}$$
$$SE_1 = 0 \text{ (spring is in its un-deformed state)}$$

In position 2:

$$PE_2 = 0 \text{ (datum passes through the centre of gravity of the bar)}$$
$$KE_2 = \frac{1}{2}mv_G^2 + \frac{1}{2}I_G\dot{\theta}^2$$

From Equation (2.8), the velocity of the centre of gravity of the rigid bar is $v_G = r\dot{\theta} = 0.5\dot{\theta}$, where r is $L/2 = 1/2 = 0.5$ m (the distance between the centre of rotation O and the centre of gravity G). From Table 4.1 and Equation (4.28), the mass moment of inertia of the bar is $I_G = mL^2/12 = 20 \times 1^2/12 = 1.667$ kg.m^2. Therefore, the kinetic energy at position 2 is given by:

$$KE_2 = \frac{1}{2} \times 20 \times (0.5\dot{\theta})^2 + \frac{1}{2} \times 1.667 \times \dot{\theta}^2 = 3.33\dot{\theta}^2$$

The strain energy due to the deformation of the spring is:

$$SE_2 = \frac{1}{2}ku^2$$

where u is the deformed length in position 2 and can be calculated as:

$$u = \sqrt{1^2 + 2^2} - 1 = 1.236 \, \text{m}$$

Therefore, the strain energy at position 2 is given by:

$$SE_2 = \frac{1}{2} \times 100 \times 1.236^2 = 76.39 \, \text{N.m}$$

The work due to external forces is $W_e = 0$ (no external forces).

The bar rotates 90° from position 1 to position 2 so that the angular displacement of the bar is:

$$\theta = \frac{\pi}{2} = 1.5708 \, \text{rad}$$

The work of external couples is $U_\theta = M_G\theta = 10 \times \frac{\pi}{2} = 15.708 \, \text{N.m}$.

Applying the conservation of energy principle, Equation (4.49) gives:

$$0 + 98.1 + 0 + 15.708 = 0 + 3.33\dot{\theta}^2 + 76.39 \Rightarrow \dot{\theta} = 3.35 \, \text{rad/s}$$

Example 4.11 Work and energy: Linkage mechanism I

A linkage mechanism shown in Figure 4.29 consists of a link AB of mass 5 kg and length 0.4 m, a link BC of mass 18 kg and length 2 m and a slider of 3 kg at C. If the slider C moves with a velocity of 1 m/s, determine the kinetic energy of the mechanism.

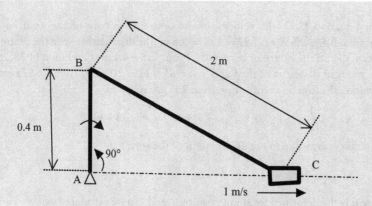

Figure 4.29: Representation of Example 4.11

Solution

The total kinetic energy of the mechanism has three contributions, from link AB, from link BC and from slider C:

$$KE = KE_{AB} + KE_{BC} + KE_C$$

Referring to Figure 4.30, these three terms can be calculated as:

$$KE_{AB} = \frac{1}{2}m_{AB}v_G^2 + \frac{1}{2}I_{AB}\dot{\theta}_{A/B}^2 = \frac{1}{2} \times 5 \times \left(\frac{0.4}{2} \times \dot{\theta}_{A/B}\right)^2 + \frac{1}{2} \times \frac{5 \times 0.4^2}{12}\dot{\theta}_{A/B}^2$$
$$= 0.1333\dot{\theta}_{A/B}^2$$

$$KE_{BC} = \frac{1}{2}m_{BC}v_D^2 + \frac{1}{2}I_{BC}\dot{\theta}_{C/B}^2 = \frac{1}{2} \times 18 \times v_D^2 + \frac{1}{2} \times \frac{18 \times 2^2}{12}\dot{\theta}_{C/B}^2$$
$$= 9v_D^2 + 3\dot{\theta}_{C/B}^2$$

$$KE_C = \frac{1}{2}m_Cv_C^2 = \frac{1}{2} \times 3 \times 1^2 = 1.5\,\text{N.m}$$

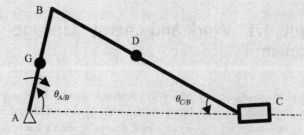

Figure 4.30: Analysis of energies in Example 4.11

To calculate the angular velocity $\dot{\theta}_{A/B}$, apply the kinematic equation, Equation (2.42). $\dot{\theta}_{A/B}$ is obtained as (using $\omega = \dot{\theta}_{A/B}$):

$$\dot{x}_C = -r\omega\left(\sin\theta + \frac{r}{2L}\sin 2\theta\right) \Rightarrow 1 = -0.4 \times \dot{\theta}_{A/B} \times \left(\sin 90 + \frac{0.4}{2 \times 2}\sin 180\right)$$

from which the angular velocity of the link AB is calculated as:

$$\dot{\theta}_{A/B} = -2.5\,\text{rad/s}$$

From Equations (2.51) and (2.52), the velocity components of point D are given by:

$$\dot{x}_D = \dot{x}_C + \frac{r^2\omega b}{2L^2}\sin 2\theta = 1 + \frac{(0.4)^2 \times (-2.5) \times 1}{2 \times (2)^2}\sin 180 = 1\,\text{m/s}$$

$$\dot{y}_D = \frac{r\omega b}{L}\cos\theta = \frac{0.4 \times (-2.5) \times 1}{2}\cos 90 = 0$$

$$v_D = \sqrt{\dot{x}_D^2 + \dot{y}_D^2} = 1\,\text{m/s}$$

From Figure 4.30, the relationship between $\theta_{A/B}$ and $\theta_{C/B}$ can be deduced as:

$$0.4\sin\theta_{A/B} = 2\sin\theta_{C/B}$$

Differentiating with respect to time gives:

$$0.4\dot{\theta}_{A/B}\cos\theta_{A/B} = 2\dot{\theta}_{C/B}\cos\theta_{C/B}$$

For $\theta_{A/B} = 90°$, the angular velocity of the link BC is $\dot{\theta}_{C/B} = 0$.

The kinetic energies of link AB and BC become:

$$KE_{AB} = 0.1333\dot{\theta}_{A/B}^2 = 0.1333 \times 2.5^2 = 0.833\,\text{N.m}$$

$$KE_{BC} = 9v_D^2 + 3\dot{\theta}_{C/B}^2 = 9 \times 1^2 + 0 = 9\,\text{N.m}$$

The total kinetic energy is given by:

$$KE = 0.833 + 9 + 1.5 = 11.33\,\text{N.m}$$

Example 4.12 Work and energy: Linkage mechanism II

In the linkage mechanism shown in Figure 4.31, two rods of length 2 m and mass 10 kg each, are used to support a beam of length 3 m and mass 12 kg. When the rods are in a vertical position, they rotate with an angular velocity of 3 rad/s. A couple of 100 N.m and a horizontal force of 150 N are applied to the linkage. Determine the angular acceleration of the rods at the position shown.

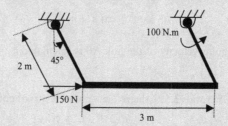

Figure 4.31: Representation of Example 4.12

Solution

Take datum at position 1, when the rods are in a vertical position, as shown in Figure 4.32.

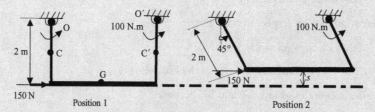

Figure 4.32: Analysis of energies in Example 4.12

For position 1, the energies are calculated as:

$PE_1 = 2 \times (mgh)_{rod} = 2 \times 10 \times 9.81 \times 1 = 196.2$ N.m (datum passes through the 12 kg beam)

$$KE_1 = 2 \times \left(\frac{1}{2}mv_C^2 + \frac{1}{2}I_C\dot{\theta}_C^2\right)_{rod} + \left(\frac{1}{2}mv_G^2 + \frac{1}{2}I_G\dot{\theta}_G^2\right)_{beam}$$

where the first term represents the kinetic energy of the rods, while the second term represents the kinetic energy of the beam. From Equation (2.8), the velocity of the centre of gravity of the rods is $v_C = r\dot{\theta}_C = 1 \times 3$ m/s, where r is $L/2 = 2/2 = 1$ m (the distance between the centre of rotation O, or O', and the centre of gravity of the rods C, or C'). Similarly, the velocity of the centre of gravity of the beam is $v_G = r\dot{\theta}_C = 2 \times 3$ m/s, where r is the rod length $L = 2$ m (the distance between the centre of rotation O, or O', and the centre of gravity of the beam, G). Since the beam does not rotate, the angular velocity of the centre of gravity of the beam is $\dot{\theta}_G = 0$. Using Table 4.1 and Equation (4.28) to calculate the mass moment of inertia of the rods $I_G = mL^2/12$, the kinetic energy becomes:

$$KE_1 = 2 \times \left(\frac{1}{2} \times 10 \times (1 \times 3)^2 + \frac{1}{2} \times \frac{10 \times 2^2}{12} \times 3^2 \right) + \left(\frac{1}{2} \times 12 \times (2 \times 3)^2 \right)$$

$$= 336 \, \text{N.m}$$

$SE_1 = 0$ (no elastic bodies)

In position 2, the beam moves upwards by a distance $s = 2 - 2 \times \cos 45$ as can be deduced from Figure 4.32.

$$PE_2 = 2 \times (mgh)_{rod} + (mgs)_{beam} = 2 \times 10 \times 9.81 \times (1 \times \cos 45 + 2 - 2$$

$$\times \cos 45) + 12 \times 9.81 \times (2 - 2 \times \cos 45) = 322.62 \, \text{N.m}$$

$$KE_2 = 2 \times \left(\frac{1}{2}mv^2 + \frac{1}{2}I_C\dot{\theta}^2 \right)_{rod} + \left(\frac{1}{2}mv_G^2 \right)_{beam}, \quad \text{where } \dot{\theta} \text{ denotes the final angular velocity of the rods.}$$

$$KE_2 = 2 \times \left(\frac{1}{2} \times 10 \times (1 \times \dot{\theta})^2 + \frac{1}{2} \times \frac{10 \times 2^2}{12} \times \dot{\theta}^2 \right) + \left(\frac{1}{2} \times 12 \times (2 \times \dot{\theta})^2 \right)$$

$$= 37.333\dot{\theta}^2;$$

$$SE_2 = 0 \quad \text{(no elastic bodies)}$$

The work due to external forces is $W_e = Fs = 150 \times 2 \times \sin 45 = 212.132 \, \text{N.m}$.

The rods rotate $45°$ from position 1 to position 2 so that the angular displacement of the rods is:

$$\theta = \frac{\pi}{4} = 0.7854 \, \text{rad}$$

The work of external couples is $U_\theta = M_G\theta = 100 \times \frac{\pi}{4} = 78.54 \, \text{N.m}$.

Applying the conservation of energy principle, Equation (4.49) gives:

$$196.2 + 336 + 212.132 + 78.54 = 322.62 + 37.333\dot{\theta}^2$$

from which the final angular velocity is calculated as:

$$\dot{\theta} = 3.66 \, \text{m/s}$$

4.4 IMPULSE AND MOMENTUM

The impulse and momentum of particles was presented in Section 3.5. This section considers the application of the principle of impulse of momentum to rigid bodies. The equation of linear impulse and momentum, Equation (3.31), and the conservation of linear momentum equation, Equation (3.33), are directly applicable to rigid bodies when the translation of the centre of gravity of the rigid bodies is considered.

Since rigid bodies undergo both translation and rotation, an angular momentum equation is required to deal with the rotational motion. In order to derive the angular momentum equation, a procedure similar to that followed in Section 3.5.1 is applied. Thus, the angular impulse and moment equation can be obtained by integrating the rotational form of Equation (3.27), i.e. the moment equation equivalent to Newton's second law, with respect to time:

$$M_G = I_G\ddot{\theta} = I_G\frac{d\dot{\theta}}{dt} \Rightarrow M_G dt = I_G\dot{\theta} \tag{4.50}$$

Integrating Equation (4.50) over time t and angular velocity $\dot{\theta}$ yields:

$$\int M_G dt = \int I_G d\dot{\theta} \tag{4.51}$$

If the angular velocity of the rigid body changes from $\dot{\theta}_1$ to $\dot{\theta}_2$ in the time interval from t_1 to t_2, Equation (4.51) can be integrated as:

$$\int_{t_1}^{t_2} M_G dt = \int_{\theta_1}^{\theta_2} I_G d\dot{\theta}$$

For a constant moment of inertia I_G, this gives:

$$\int_{t_1}^{t_2} M_G dt = I_G\dot{\theta}_2 - I_G\dot{\theta}_1 \tag{4.52}$$

which can be written as:

$$I_G\dot{\theta}_1 + \int_{t_1}^{t_2} M_G dt = I_G\dot{\theta}_2 \tag{4.53}$$

It can be seen that the angular impulse and momentum, Equation (4.52), are the moments of the linear impulse and momentum, Equation (3.30), respectively. The term on the right-hand side in Equation (4.52) is the time rate of change in angular momentum, while the term on the left-hand side is the sum of the moment about the centre of mass G due to all external forces acting on the body. Similarly to Equation (3.33), the conservation of angular momentum can be obtained when the angular impulses are equal to zero $\int_{t_1}^{t_2} M_G dt = 0$. Thus the conservation of angular momentum equation is:

$$\sum I_G\dot{\theta}_1 = \sum I_G\dot{\theta}_2 \tag{4.54}$$

or

$$\sum H_{G1} = \sum H_{G2} \tag{4.55}$$

where H_{G1} is the angular momentum at time t_1 and H_{G2} is the angular momentum at time t_2. Figure 4.33 illustrates the linear momentum, denoted as L_G, and the angular moment,

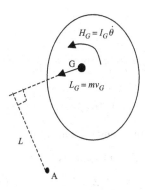

Figure 4.33: Linear and angular momentum of a rigid body

denoted as H_G, acting at the centre of gravity of a rigid body. They are given by:

$$L_G = mv_G \qquad (4.56)$$
$$H_G = I_G\dot{\theta} \qquad (4.57)$$

where v_G is the velocity of the centre of gravity of the rigid body. The angular momentum at any point A (H_A) in Figure 4.33, can be obtained by taking moment about A due to the linear and angular momentum at the centre of gravity L_G and M_G:

$$H_A = H_G + L \times L_G = I_G\dot{\theta} + L \times mv_G \qquad (4.58)$$

Example 4.13 Impulse and momentum: Rotating disk I

A disk of mass 8 kg and a radius 0.2 m rotates about point O on its edge with an angular velocity of 5 rad/s as shown in Figure 4.34. Determine the angular momentum of the disk about its centre of gravity, G, and about point O.

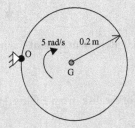

Figure 4.34: Representation of Example 4.13

Solution

From Table 4.1 and Equation (4.32), the mass moment of inertia of the disk about its centre of gravity, G, is:

$$I_G = \frac{mr^2}{2} = \frac{8 \times 0.2^2}{2} = 0.16 \, \text{kg.m}^2$$

Applying Equation (4.57), the angular momentum about G is calculated as:

$$H_G = I_G\dot\theta = 0.16 \times 5 = 0.8 \, \text{kg.m}^2/\text{s}$$

The velocity of G is $v_G = r \times \dot\theta = 0.2 \times 5 = 1$ m/s from Equation (2.8) and, from Equation (4.58), the angular momentum about point O is:

$$H_O = I_G\dot\theta + r \times mv_G = 0.8 + 0.2 \times 8 \times 1 = 2.4 \, \text{kg.m}^2/\text{s}$$

Example 4.14 Impulse and momentum: Rotating disk II

A disk of a mass 8 kg and of a radius 0.2 m rotates about its centre of gravity G as shown in Figure 4.35. A couple of 3 N.m and a force of 40 N acting at the outer edge of the disk are applied to the disk as shown. If the disk is initially at rest, determine the disk's angular velocity and the reaction at G after 2 seconds.

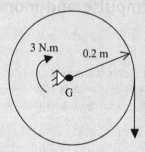

Figure 4.35: Representation of Example 4.14

Solution

The mass moment of inertia of the disk (from Table 4.1) about its centre of gravity, G, is:

$$I_G = \frac{mr^2}{2} = \frac{8 \times 0.2^2}{2} = 0.16 \, \text{kg.m}^2$$

Considering the free-body diagram in Figure 4.36 and applying the principle of angular impulse and momentum, Equation (4.53) gives:

$$I_G\dot{\theta}_1 + \int_{t_1}^{t_2} M_G dt = I_G\dot{\theta}_2$$

where the initial angular momentum is zero ($I_G\dot{\theta}_1 = 0$) since the disk is initially at rest, the final angular momentum is $I_G\dot{\theta}_2 = 0.16\dot{\theta}_2$ kg.m^2/s and the angular impulse during the time interval $0 \leq t \leq 2$ is $\int_0^2 M_G dt = \int_0^2 (M + F \cdot r)dt = 3 \times 2 + 40 \times 0.2 \times 2$ kg.m^2/s. The equation of angular impulse and momentum becomes:

$$0 + 3 \times 2 + 40 \times 0.2 \times 2 = 0.16 \times \dot{\theta}_2$$

from which the final angular velocity is calculated as:

$$\dot{\theta}_2 = 137.5 \, \text{rad/s}$$

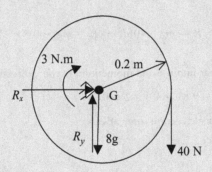

Figure 4.36: Free-body diagram for Example 4.14

Applying the principle of linear impulse and momentum, Equation (3.31), in the x direction gives:

$$mv_{x1} + \int_{t_1}^{t_2} F_x dt = mv_{x2}$$

Since there is no motion of the centre of gravity in the x direction, both initial and final velocities are zeros and therefore the initial and final linear momentums are zeros, i.e. $mv_{x1} = mv_{x2} = 0$. The linear impulse in the x direction during the time interval $0 \leq t \leq 2$ is $\int_{t_1}^{t_2} F_x dt = \int_0^2 R_x dt = R_x \times 2$ N.s. The equation of linear impulse and momentum in the x direction becomes:

$$0 + R_x \times 2 = 0$$

from which the horizontal reaction force at G is:

$$R_x = 0$$

Similarly, applying the principle of linear impulse and momentum, Equation (3.31), in the y direction gives:

$$mv_{y1} + \int_{t_1}^{t_2} F_y dt = mv_{y2}$$

Since there is no motion of the centre of gravity in the y direction, both initial and final velocities in the y direction are zero and therefore the initial and final linear momentums are zero, i.e. $mv_{y1} = mv_{y2} = 0$. The linear impulse in the y direction during the time interval $0 \leq t \leq 2$ is

$$\int_{t_1}^{t_2} F_y dt = \int_0^2 (R_y - mg - 40)dt = (R_y - mg - 40) \times 2 \, \text{N.s}$$

The equation of linear impulse and momentum in the y direction becomes:

$$0 + (R_y - 8 \times 9.81 - 40) \times 2 = 0$$

from which the vertical reaction force at G is:

$$R_y = 118.5 \, \text{N}$$

Example 4.15 Impulse and momentum: Rigid rod

A rod ABC of length 1 m and mass 4 kg is supported by a pin at its end A and is initially at rest as shown in Figure 4.37. If a bullet of 5 g is fired towards the rod, strikes it at an angle of 45° with the horizontal and penetrates into it, determine the angular velocity of the rod and the bullet just after the strike.

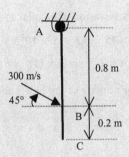

Figure 4.37: Representation of Example 4.15

Solution

Applying the equation of conservation of angular momentum about point A, Equation (4.55) gives:

$$\sum H_{A1} = \sum H_{A2}$$

The angular momentum of the bullet about point A just before the strike is $m_b \times v_1 \times \cos 45 \times l_b$ kg.m^2/s, where m_b is the bullet's mass, v_1 is the bullet's velocity just before the strike and l_b is the distance between the bullet and the centre of rotation, A, $l_b = 0.8$ m. The rod's angular momentum about point A just before the strike is zero (the rod is initially at rest).

The bullet's angular momentum at point A just after the strike is $m_b \times v_{B2} \times l_b$, where v_{B2} is the final velocity of the bullet. The angular momentum of the rod about point A just after the strike is $m_r \times v_{G2} \times l_G + I_G \times \dot\theta$, where m_r is the rod's mass, v_{G2} is the final velocity of the rod's centre of gravity, l_G is the distance between the rod's centre of gravity and the centre of rotation A and is 0.5 m, and I_G is the mass moment of inertia of the rod $I_G = \frac{1}{12} \times m_r \times (1)^2$. The equation of conservation of angular momentum about point A, becomes:

$$m_b \times v_1 \times \cos 45 \times 0.8 = m_b \times v_{B2} \times 0.8 + m_r \times v_{G2} \times 0.5 + \frac{1}{12}$$
$$\times m_r \times (1)^2 \times \dot\theta$$

From kinematics, Equation (2.8), the final velocities of the bullet and the rod's centre of gravity are given by:

$$v_{B2} = 0.8 \times \dot\theta \quad \text{and} \quad v_{G2} = 0.5 \times \dot\theta$$

Rewriting the equation of conservation of angular momentum gives:

$$5 \times 10^{-3} \times 300 \times \cos 45 \times 0.8 = 5 \times 10^{-3} \times 0.8 \times 0.8 \times \dot\theta$$
$$+ 4 \times 0.5 \times 0.5 \times \dot\theta + \frac{1}{12} \times 4 \times (1)^2 \times \dot\theta$$

from which, the final angular velocity of the bullet and the rod is obtained as:

$$\dot\theta = 0.635 \, \text{rad/s}$$

Example 4.16 Impulse and momentum: Gears

A gear A has an initial angular velocity of 3 rad/s before it contacts with gear B, which is initially at rest as shown in Figure 4.38. Gear A has a mass moment of inertia of 0.05 kg.m^2 and a radius of 0.15 m, while gear B has a mass moment of inertia of 0.15 kg.m^2 and a radius of 0.2 m. Determine the final angular velocities of gears A and B when they are in contact.

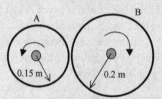

Figure 4.38: Representation of Example 4.16

Solution

The free-body diagrams of gears A and B are shown in Figure 4.39, where R_{xB} and R_{yB} are the reaction forces in the x and y directions, respectively, at the centre of gravity of gear B. Similarly, R_{xA} and R_{yA} are the reaction forces in the x and y directions, respectively, at the centre of gravity of gear A.

Applying the principle of angular impulse and momentum, Equation (4.53), to gear A gives:

$$I_A \dot\theta_{A1} + \int_{t_1}^{t_2} M_A dt = I_A \dot\theta_{A2}$$

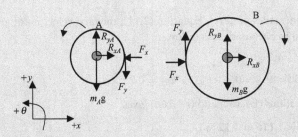

Figure 4.39: Free-body diagrams for Example 4.16

where the initial angular momentum of gear A is $I_A\dot{\theta}_{A1} = 0.05 \times 3$ kg.m^2/s, the final angular momentum of gear A is $I_A \times \dot{\theta}_{A2} = 0.05 \times \dot{\theta}_{A2}$ kg.m^2/s and the angular impulse applied to gear A during the time interval 0 to t is $\int_0^t M_A dt = \int_0^t -F_y \times r_A dt = -F_y \times 0.15 \times t$ kg.m^2/s (negative because the moment is clockwise). The equation of angular impulse and momentum of gear A becomes:

$$I_A\dot{\theta}_{A1} - F_y \times r_A \times t = I_A \times \dot{\theta}_{A2}$$
$$0.05 \times 3 - F_y \times 0.15 \times t = 0.05 \times \dot{\theta}_{A2}$$

from which

$$\dot{\theta}_{A2} = 3 - 3 \times F_y \times t \qquad \text{(E4.16a)}$$

Similarly, applying the principle of angular impulse and momentum, Equation (4.53), to gear B gives:

$$I_B\dot{\theta}_{B1} + \int_{t_1}^{t_2} M_B dt = I_A\dot{\theta}_{B2}$$

where the initial angular momentum of gear B is $I_B\dot{\theta}_{B1} = 0$ kg.m^2/s (because gear B is initially at rest), the final angular momentum of gear B is $I_B \times \dot{\theta}_{B2} = -0.15 \times \dot{\theta}_{B2}$ kg.m^2/s (negative because gear B rotates clockwise) and the angular impulse applied to gear B during the time interval 0 to t is $\int_0^t M_B dt = \int_0^t -F_y \times r_B dt = -F_y \times 0.2 \times t$ kg.m^2/s (negative because the moment is clockwise). The equation of angular impulse and momentum of gear B becomes:

$$I_B\dot{\theta}_{B1} - F_y \times r_B \times t = -I_B \times \dot{\theta}_{B2}$$
$$0.15 \times 0 - F_y \times 0.2 \times t = -0.15 \times \dot{\theta}_{B2} \Rightarrow F_y \times t = 0.75 \times \dot{\theta}_{B2} \qquad \text{(E4.16b)}$$

Eliminating $F_y \times t$ from Equations (E4.16a) and (E4.16b) leads to:

$$\dot{\theta}_{A2} = 3 - 2.25 \times \dot{\theta}_{B2} \qquad \text{(E4.16c)}$$

From the kinematics of gears, Equation (2.11), the final velocities of gears A and B are related to one another as follows:

$$r_A \times \dot{\theta}_{A2} = r_B \times \dot{\theta}_{B2}$$
$$0.15 \times \dot{\theta}_{A2} = 0.2 \times \dot{\theta}_{B2} \Rightarrow \dot{\theta}_{B2} = 0.75 \times \dot{\theta}_{A2} \qquad \text{(E4.16d)}$$

Solving Equations (E4.16c) and (E4.16d) gives:

$$\dot{\theta}_{A2} = 1.116 = 1.12 \,\text{rad/s}$$
$$\dot{\theta}_{B2} = 0.847 = 0.84 \,\text{rad/s}$$

4.5 Tutorial Sheet

4.5.1 Force and acceleration

Q4.1   A light cord is wrapped around a hub of a spool, which moves only in the vertical direction as shown in Figure 4.40. The spool has a mass of 10 kg and mass moment inertia of 0.15 kg.m². If the spool is moving upwards with angular acceleration $\ddot{\theta} = 10.06 \,\text{rad/s}^2$, determine:

a) the vertical force P in the cord on the right-hand side;

[80 N]

b) the tension in the cord on the left-hand side.

[48.3 N]

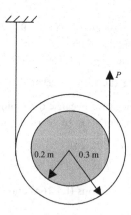

Figure 4.40: Representation of Question 4.1

Q4.2   A car of mass 1.8×10^3 kg has its centre of gravity at a distance of 1.23 m from its rear wheels and a distance of 0.74 m from its front wheels as shown in Figure 4.41. The vertical distance between the road and the car's centre of gravity, CG, is 0.3 m. The coefficient of friction between the wheels and the road is 0.25. If the rear (driving) wheels slip, the front wheels rotate and the mass of the wheels is negligible, determine:

a) the vertical reaction forces at the front and rear wheels;

[10.76 N, 6.89 N]

b) the car's acceleration.

[0.96 m/s^2]

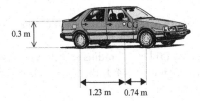

0.3 m

1.23 m   0.74 m

Figure 4.41: Representation of Question 4.2

Q4.3   A wheel, shown in Figure 4.42, has a mass of 8 kg and a radius of gyration of 0.2 m. An external couple, which is defined as $M = 4t$, where M is in N.m and t is the time in seconds, is applied anti-clockwise to the wheel starting from rest. When $t = 2$ s, determine:

a) the angular velocity of the wheel;

[25 rad/s]

b) the components of reaction at the centre of gravity of the wheel, G.

[0, 78.48 N]

$M = 4t$

Figure 4.42: Representation of Question 4.3

Q4.4   A fan blade, shown in Figure 4.43, is subjected to a moment given by $M = 3.2\,(1 - e^{-0.2t})$, where M is in N.m and t is time in seconds. The fan blade has a mass of 2.1 kg and a radius of gyration of 0.276 m. When $t = 3$ s, determine the angular velocity of the fan blade.

[14.88 rad/s]

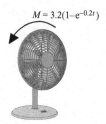

$M = 3.2(1 - e^{-0.2t})$

Figure 4.43: Representation of Question 4.4

Q4.5   A disk of mass 60 kg and radius of 1.2 m is supported by a pin at point O as shown in Figure 4.44. If the disk is released from rest at the position shown, determine:

a) the initial angular acceleration;

[−5.45 rad/s²]

b) the components of reaction at point O.

[0, 196.2 N]

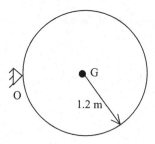

G

O

1.2 m

Figure 4.44: Representation of Question 4.5

Q4.6   Two gears, 1 and 2, are in mesh as shown in Figure 4.45. Gear 1 has a radius of 80 mm, a radius of gyration of 50 mm and a mass of 4.5 kg, while gear 2 has a radius of 120 mm, a radius of gyration of 80 mm and a mass of 9 kg. If an external moment of 0.5 N.m is applied to gear 1, determine the angular acceleration of gear 2.

[9.05 rad/s^2]

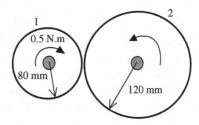

Figure 4.45: Representation of Question 4.6

Q4.7   A uniform rod ABC of mass 8 kg is connected to two collars of negligible mass which slide on smooth horizontal rods as shown in Figure 4.46. If a force of magnitude 40 N to the left is applied at C, determine:

a) the acceleration of the rod;

[−5 m/s^2]

b) the reaction at B and C.

[41.54 N, 36.94 N]

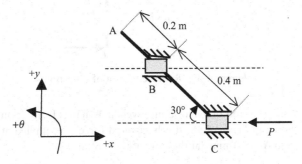

Figure 4.46: Representation of Question 4.7

Q4.8   If the reaction at B in Figure 4.46 is 45 N upwards, determine:

a) the acceleration of the rod;

[−4 m/s^2]

b) the force P;

[−32 N]

c) the reaction force at C.

[33.48 N]

Q4.9   A uniform rod of length 0.9 m and mass 4 kg is supported by a pin at point C as shown in Figure 4.47. A horizontal force P of magnitude 75 N is applied at end B. The distance between the support, C, and the rod's centre of gravity, G, is r. For $r = 0.225$ m, determine (at the start of the motion):

a) the angular acceleration of the rod;

[−107.14 rad/s²]

b) the components of the reaction at C.

[−21.43 N, 39.24 N]

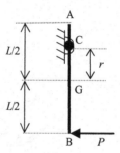

Figure 4.47: Representation of Question 4.9

Q4.10   A uniform rod of length L and weight W hangs freely from a hinge A as shown in Figure 4.48. If a horizontal force P is applied at a distance h from A, determine (at the start of the motion):

a) the angular acceleration of the rod;

$[-\frac{3Phg}{WL^2}]$

b) the components of the reaction at A.

$[\frac{2PL-3Ph}{2L}, W]$

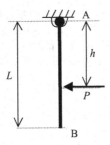

Figure 4.48: Representation of Question 4.10

Q4.11 In Figure 4.48, if the horizontal force P is applied at a distance $h = L$ (i.e. at B), determine (at the start of the motion):

a) the angular acceleration of the rod;

$$[-\tfrac{3Pg}{WL}]$$

b) the components of the reaction at A.

$$[-\tfrac{P}{2}, W]$$

Q4.12 A uniform beam of length 0.7 m and mass 90 kg hangs from two rods of 0.45 m in length as shown in Figure 4.49. The forces in the rods are identical and equal to 600 N. Determine the angular velocity and the angular acceleration of the rods (ignoring the mass of the rods).

[3.77 rad/s, 15.42 rad/s²]

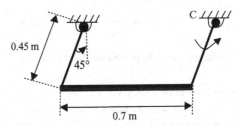

Figure 4.49: Representation of Question 4.12

Q4.13 A diving board of mass 24 kg and length 3 m is supported at point B by a spring of stiffness 6800 N/m as shown in Figure 4.50. When someone jumps off the diving board at point C, the spring is compressed by 0.19 m and motion just starts. Determine the angular acceleration of the board.

[22.01 rad/s^2]

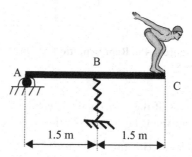

Figure 4.50: Representation of Question 4.13

Q4.14 A uniform rod of mass 15 kg and length 2 m rotates about point A as shown in Figure 4.51. When the bar is in a horizontal position, the angular velocity of the rod is 3 rad/s and a couple of 45 N.m is applied. Determine:

a) the angular acceleration of the rod;

[−9.61 rad/s^2]

b) the horizontal and vertical reaction components at A.

[−135 N, 3 N]

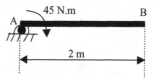

Figure 4.51: Representation of Question 4.14

Q4.15 An aeroplane of mass 240×10^3 kg travels with an acceleration a during take-off. The distances between the plane's centre of gravity and the front and rear wheels are 21.8 m and 5.2 m, respectively, in the horizontal direction, while the distance between the plane's centre of gravity and the engines is 2.1 m in the vertical direction, as shown in Figure 4.52. If the thrust of the engine is equal to 680 kN, determine:

a) the acceleration a of the plane;

[2.83 m/s^2]

b) the vertical components of the reaction at the wheels assuming that the horizontal components are negligible.

[400.6 kN, 1953.8 kN]

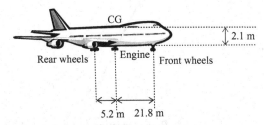

Figure 4.52: Representation of Question 4.15

Q4.16 A uniform rod of length $2L$ and weight W hangs freely from a hinge A as shown in Figure 4.53. If a horizontal force P is applied at a distance $3L/4$ from A, show that:

a) the angular acceleration of the rod is $-9Pg/16WL$;
b) the components of the reaction at A are $7P/16$ and W.

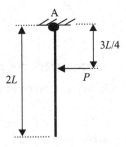

Figure 4.53: Representation of Question 4.16

Q4.17 A uniform rod AB, shown in Figure 4.54, has a mass of 5 kg and a length of 3 m and is simply supported at point A. If a horizontal force $P = 10$ N is applied at end B, determine:

a) the angular acceleration of the rod;

$$[-3.8 \text{ rad/s}^2]$$

b) the components in the x and y directions of the reaction at A.

$$[-10.2 \text{ N}, 28.8 \text{ N}]$$

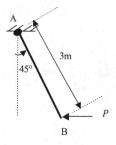

Figure 4.54: Representation of Question 4.17

4.5.2 Work and energy

Q4.18 A disk has a mass of 32 kg and a radius of 1.2 m. It rolls with an angular velocity of 4 rad/s and its centre of gravity has a velocity of 8 m/s at the instant shown in Figure 4.55. Determine the kinetic energy of the disk.

$$[1.21 \text{ kN.m}]$$

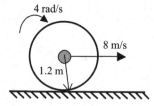

Figure 4.55: Representation of Question 4.18

Q4.19 A wheel has a mass of 18 kg, a radius of 0.24 m and a radius of gyration of 0.156 m. It is attached to a spring of stiffness 150 N/m as shown in Figure 4.56. When the spring is not stretched and the wheel is at rest, a torque of 20 N.m is applied so that the wheel rolls without slipping. Determine the angular velocity of the wheel when its centre of gravity moves a distance of 0.12 m.

[2.78 rad/s]

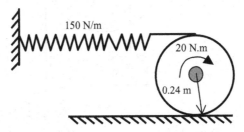

Figure 4.56: Representation of Question 4.19

Q4.20 A wheel has a mass of 24 kg and a radius of 0.2 m and is attached at its centre of gravity to a spring of stiffness 120 N/m as shown in Figure 4.57. When the spring is not stretched and the wheel is at rest, a torque of 32 N.m is applied so that the wheel rolls without slipping. Determine the distance that the wheel would travel before stopping.

[2.67 m]

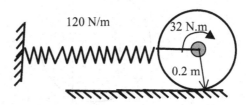

Figure 4.57: Representation of Question 4.20

Q4.21 Two masses, m_1 and m_2, of 2 kg each are attached to a compound disk pulley as shown in Figure 4.58. The radii of the inner and outer rims are 0.03 m and 0.09 m, respectively. The disk pulley has a mass of 4 kg and a radius of gyration of 0.05 m. Determine the speed of mass m_1 after it has descended 0.3 m, starting from rest.

[1.51 m/s]

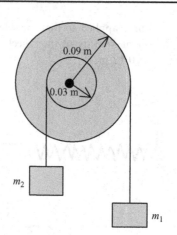

Figure 4.58: Representation of Question 4.21

Q4.22 A 15 kg mass is attached to a spool of mass 40 kg, a radius of gyration of 0.22 m and inner radius of 0.2 m as shown in Figure 4.59. If the mass is released from rest, determine the distance that it would have descended when the angular velocity of the spool reaches 4 rad/s.

[0.14 m]

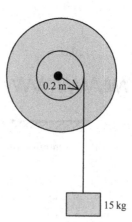

Figure 4.59: Representation of Question 4.22

Q4.23 A uniform, rigid bar, of 1 m in length, has a mass of 15 kg. It rotates about one of its ends as shown in Figure 4.60 and is connected to a weightless spring of stiffness k and an un-deformed length of 0.7 m. The motion starts at position 1 where the bar is in a vertical position. If an anti-clockwise torque of 12 N.m is applied to the bar, the bar's angular velocity when it is in a horizontal position (position 2) is 3 rad/s. Determine:

a) the potential energy at position 1 (take datum as shown);

[73.57 N.m]

b) the kinetic energy at position 2;

[22.5 N.m]

c) the strain energy at position 2 (use the conservation of energy method);

[69.92 N.m]

d) the stiffness of the spring.

[86.39 N/m]

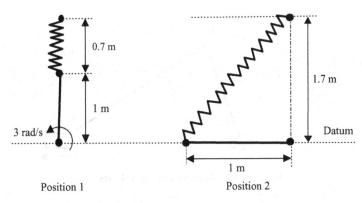

Position 1   Position 2

Figure 4.60: Representation of Question 4.23

Q4.24 A rod AB of mass 8 kg and length 0.6 m is restricted to move vertically at A and horizontally at B as shown in Figure 4.61. When $\theta = 0°$, the rod is at rest. A horizontal force of 45 N is applied at B. Determine the angular velocity of the rod at the position shown.

[8.56 rad/s]

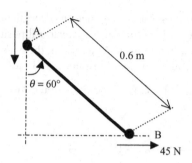

Figure 4.61: Representation of Question 4.24

Q4.25 Determine the kinetic energy of the mechanism shown in Figure 4.62 if the slider moves with a velocity of 1 m/s. Link AB has a mass of 6 kg, link BC has a mass of 15 kg and slider C has a mass of 3 kg.

[9.57 N.m]

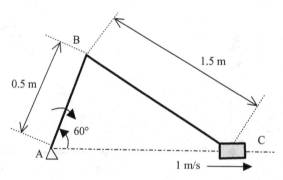

Figure 4.62: Representation of Question 4.25

Q4.26 A uniform bar, AB, of mass 3 kg and length 2 m is free to rotate about A. At the position shown in Figure 4.63, the bar has an angular velocity of 4 rad/s. A couple of 20 N.m and a force of 10 N, which is always perpendicular to the bar, are applied at B. Determine the angular velocity when the bar has rotated 360°.

[11.9 rad/s]

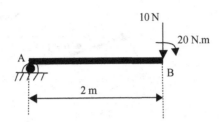

Figure 4.63: Representation of Question 4.26

Q4.27 For the linkage mechanism shown in Figure 4.64, determine the horizontal force F that should be applied to the beam in order to achieve an angular velocity of the rods of 3.5 rad/s. Each of the two rods has a length of 2 m and mass of 9 kg, while the beam has a length of 3 m and mass of 12 kg. When the rods are vertical, they rotate with an angular velocity of 3 rad/s.

[168.1 N]

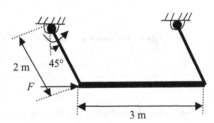

Figure 4.64: Representation of Question 4.27

Q4.28 A rod AB of mass 12 kg and length 0.5 m is attached to a spring of stiffness 1200 N/m as shown in Figure 4.65. Point A is constrained to move horizontally, while point B is constrained to move vertically. If the rod is released from rest, determine the angular acceleration of the rod when it is horizontal.

[2.2 rad/s]

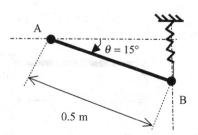

Figure 4.65: Representation of Question 4.28

Q4.29 A rod of mass 12 kg and length 2 m rotates about its end A and is attached to a spring of stiffness k and length 1.5 m at the other end B, as shown in Figure 4.66. The motion starts when the bar is horizontal and the spring is not stretched. Determine the spring's stiffness k if the bar comes to rest after rotating 90° clockwise.

[36.75 N/m]

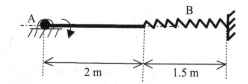

Figure 4.66: Representation of Question 4.29

Q4.30 A uniform, rigid bar, AB has a mass of 10 kg, rotates about one of its ends A and is connected to a weightless spring of stiffness k and undeformed length of 2 m at the other end, B, as shown in Figure 4.67. If the motion starts at position 1 when the angle $\theta = 30°$ and a couple of 250 N.m is applied anti-clockwise, use the conservation of energy principle to determine the spring's stiffness k so that the angular velocity of the bar when it is horizontal (position 2) is 30 rev/min.

[95.56 N/m]

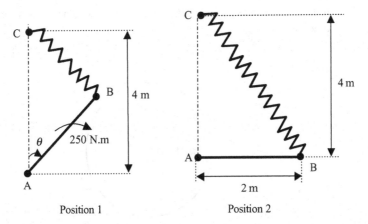

Position 1   Position 2

Figure 4.67: Representation of Question 4.30

Q4.31 A ball of 2 kg is attached to a string of 1 m length as shown in Figure 4.30. If the ball is released from rest in position 1, use the conservation of energy principle to determine the ball's velocity when it is in position 2 (ignore elastic deformation in the string).

[3.13 m/s]

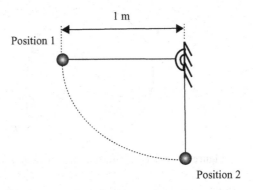

Figure 4.68: Representation of Question 4.31

4.5.3 Impulse and momentum

Q4.32 A disk with mass of 8 kg and a radius of 0.2 m rotates about point O at a distance of 0.1 m from its centre of gravity, G, as shown in Figure 4.69.

If the angular velocity is 5 rad/s, determine the angular momentum of the disk about point O.

[1.2 kg.m²/s]

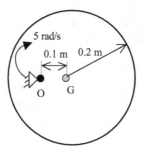

Figure 4.69: Representation of Question 4.32

Q4.33 A disk with mass of 15 kg and a radius of 0.3 m rotates about its centre of gravity, G, as shown in Figure 4.70. The disk is initially at rest. Determine the force *F* acting at the outer edge of the disk, which should be applied so that the angular velocity of the disk after 2 seconds is 100 rad/s.

[112.5 N]

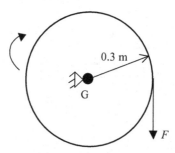

Figure 4.70: Representation of Question 4.33

Q4.34 A rod ABC of length 1 m and mass 4 kg is supported by a pin at its end A as shown in Figure 4.71. A bullet of mass 5 g is fired towards the rod, strikes it at an angle of θ with the horizontal and penetrates it. If the angular velocity of the rod and the bullet just after the strike is 0.778 rad/s, determine the angle of strike θ.

[30°]

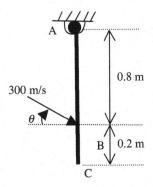

Figure 4.71: Representation of Question 4.34

Q4.35 A gear A has an initial angular velocity of $\dot{\theta}_{A1}$ before it contacts with gear B, which is initially at rest, as shown in Figure 4.72. Gear A has a mass moment of inertia of 0.05 kg.m^2 and a radius of 0.1 m; gear B has a mass moment of inertia of 0.15 kg.m^2 and a radius of 0.2 m. If the final angular velocity of gear B is 1 rad/s, determine the initial angular velocity of gear A.

[3.5 rad/s]

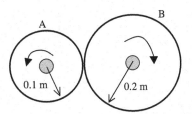

Figure 4.72: Representation of Question 4.35

CHAPTER 5

Balancing of Machines

5.1 INTRODUCTION

As has been seen in Chapters 3 and 4, a body that possesses a mass and undergoes motion with acceleration produces inertia forces and inertia moments. In machines, the inertia forces and moments produced by the motions of the various links are transferred to the bearings and then to the foundations. Therefore, it is important to consider inertia forces when designing these components. The acceleration of the slider in a slider–crank mechanism, Equation (2.44), is proportional to the square of the angular velocity. Consequently, the inertia force produced by the piston ($m\ddot{x}_A$) is proportional to the square of the angular velocity so that, for high-speed machines, large forces are produced causing premature failure to machine elements. If all links and masses in a machine are arranged in such a way that the inertia forces self-balance, the bearings and foundations are freed from these forces. Removing the effect of the unbalanced mass, or adding an equal mass in such a way that the unbalanced effect is cancelled, eliminates the undesirable unbalanced force. In this chapter, the balancing of machines is considered. The chapter is divided into two main sections that deal with the balancing of rotating masses and of reciprocating engines.

5.2 BALANCING OF ROTATING MASSES

In Figure 5.1, three masses are attached to a shaft (AB) rotating with an angular velocity ω about the shaft, i.e. the fixed axis AB.

If mass m_i is located at a distance r_i from the axis AB, the inertia force on the shaft produced by m_i is:

$$F_i = m_i a_i = m_i r_i \omega^2 \tag{5.1}$$

where a_i is the radial acceleration of particle i, which is equal to $a_i = r_i \omega^2$ and acting towards the shaft (see Equation (2.9)). The sum of forces produced by all masses is:

$$\sum F = \sum m r \omega^2$$

The reactions introduced at bearings A and B, R_A and R_B, respectively, should be equal to the sum of the forces:

$$R_A + R_B = \sum F \tag{5.2}$$

The right hand side of Equation (5.1) can be obtained using a vector summation, taking into account both magnitudes and directions. A simple way to do so, is to resolve the forces in the x and y directions as shown in Figure 5.2.

$$\sum F_x = \sum m r \omega^2 \cos\theta \tag{5.3}$$

$$\sum F_y = \sum m r \omega^2 \sin\theta \tag{5.4}$$

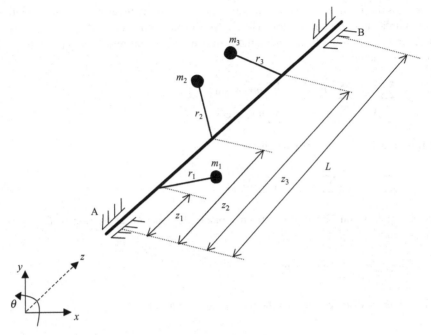

Figure 5.1: Rotating masses

The total out-of-balance force is then calculated as:

$$\sum F = \sqrt{\left(\sum F_x\right)^2 + \left(\sum F_y\right)^2} \tag{5.5}$$

As the masses are attached to the shaft at different positions, bending moments take place. The magnitude of these bending moments depends on the chosen position of a reference plane, i.e. the plane about which the bending moment is taken. The resultant bending moment is:

$$\sum M = \sum mrz\,\omega^2 \tag{5.6}$$

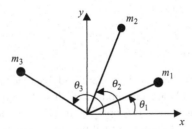

Figure 5.2: Angular diagram of rotating masses

where z is the distance along the shaft's axis measured from the reference plane. Equation (5.6) implies a vector summation. The analytical solution is obtained by resolving M in the x and y directions (F_x is making moment about the y axis and F_y is making moment about the x axis):

$$\sum M_x = \sum F_y \times z = \sum mrz\omega^2 \sin\theta \tag{5.7}$$

$$\sum M_y = \sum F_x \times z = \sum mrz\omega^2 \cos\theta \tag{5.8}$$

The total out-of-balance moment is:

$$\sum M = \sqrt{\left(\sum M_x\right)^2 + \left(\sum M_y\right)^2} \tag{5.9}$$

If bearing A is considered as a reference plane, the reaction at bearing B (R_B) can be calculated as:

$$R_B \times L = \sum M_A \Rightarrow R_B = \frac{\sum M_A}{L} \tag{5.10}$$

where L is the distance between the two bearings, A and B.

5.2.1 Static balance

The condition for static balance is that the summation of all inertia forces is equal to zero, $\sum F = 0$ (Equation (5.5)). If all masses are in the same transverse plane, the static balance condition should be sufficient as the inertia moment at bearing A would also be equal to zero, $\sum M = L \times \sum F = 0$, and neither bearing would have a reaction force. If the masses are in different planes, an additional condition, dynamic balance, is required.

5.2.2 Dynamic balance

The condition for dynamic balance is that the summation of all inertia moments is equal to zero, $\sum M = 0$ (Equation (5.9)).

5.2.3 Complete balance

In order to achieve complete balance, both static and dynamic balance conditions should be satisfied, i.e. the force and moment summations should be equal to zero, regardless of the location of the reference plane. It is possible, therefore, to achieve complete balance by selecting a combination of mass m, radial position r, position along the shaft z and relative angular position θ so that both force and moment summations vanish.

Example 5.1 Balancing of rotating masses I

A shaft is supported by two bearings, A and B, and carries three pulleys, 1, 2 and 3 as shown in Figure 5.3. The angular speed of the shaft is $\omega = 600$ rev/min anti-clockwise. The angular positions of planes 2 and 3 relative to plane 1 are 260° and 120°, respectively, measured anticlockwise.

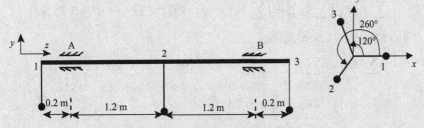

Figure 5.3: Representation of Example 5.1

The pulleys have the characteristics given in Table 5.1 (where m is the mass, r is the radial position, θ is the angular position and the reference plane is taken at bearing A).

Table 5.1: Data for Example 5.1

Plane	mr (kg.m)	z (m)	mrz (kg.m^2)	θ (°)
1	0.03	−0.2	−0.006	0
2	0.04	1.2	0.048	260
3	0.045	2.6	0.117	120

Determine the out-of-balance force and moment and the reaction force on each bearing.

Solution

The angular speed is $\omega = 600 \times \frac{2\pi}{60} = 62.832$ rad/s.

From Equation (5.3), the summation of forces in the x direction is:

$$\sum F_x = \sum mr\omega^2 \cos\theta = \omega^2 \times \sum mr\cos\theta$$
$$= (62.832)^2 \times [0.03 \times \cos 0 + 0.04 \times \cos 260 + 0.045 \times \cos 120]$$
$$= 2.187\,\text{N}$$

From Equation (5.4), the summation of forces in the y direction is:

$$\sum F_y = \sum mr\omega^2 \sin\theta = \omega^2 \times \sum mr\sin\theta$$
$$= (62.832)^2 \times [0.03 \times \sin 0 + 0.04 \times \sin 260 + 0.045 \times \sin 120]$$
$$= -1.663\,\text{N}$$

From Equation (5.5), the total out-of-balance force is:

$$\sum F = \sqrt{\left(\sum F_x\right)^2 + \left(\sum F_y\right)^2} = \sqrt{(2.187)^2 + (-1.663)^2} = 2.75\,\text{N}$$

From Equation (5.7), M_x about A is:

$$\sum M_{Ax} = \sum F_y \times z = \sum mrz\omega^2 \sin\theta = \omega^2 \times \sum mrz \sin\theta$$
$$= (62.832)^2 \times [-0.006 \times \sin 0 + 0.048 \sin 260 + 0.117$$
$$\times \sin 120] = 213.4\,\text{N.m}$$

and M_y from Equation (5.8) is:

$$\sum M_{Ay} = \sum F_x \times z = \sum mrz\omega^2 \cos\theta = \omega^2 \times \sum mrz \cos\theta$$
$$= (62.832)^2 \times [-0.006 \times \cos 0 + 0.048 \times \cos 260 + 0.117 \times \cos 120]$$
$$= -287.543\,\text{N.m}$$

The total out-of-balance moment (Equation (5.9)) is:

$$\sum M_A = \sqrt{\left(\sum M_{Ax}\right)^2 + \left(\sum M_{Ay}\right)^2} = \sqrt{(213.4)^2 + (-287.543)^2}$$
$$= 358.08\,\text{N.m}$$

The component of the reaction at bearing B in the y direction, $R_B \sin\theta_B$, is obtained from Equation (5.10) using $\sum M_{Ax}$:

$$R_B \sin\theta_B = \frac{\sum M_{Ax}}{L} = \frac{213.4}{2.4} = 88.917\,\text{N}$$

Similarly, the component of the reaction at B in the x direction, $R_B \cos\theta_B$, is obtained from Equation (5.10) using $\sum M_{Ay}$:

$$R_B \cos\theta_B = \frac{\sum M_{Ay}}{L} = \frac{-287.543}{2.4} = -119.81\,\text{N}$$

The magnitude of the reaction at B is the vector summation of the x and y components:

$$R_B = \sqrt{(88.917)^2 + (-119.81)^2} = 149.2\,\text{N}$$

To calculate the reaction at A, Equation (5.2) gives $R_A + R_B = \sum F$, which implies a vector summation is used. Applying Equation (5.2) in the x direction gives:

$$\sum F_x = R_A \cos\theta_A + R_B \cos\theta_B = 2.187 \Rightarrow R_A \cos\theta_A - 119.81 = 2.187$$

from which the component of the reaction at A in the x direction, $R_A \cos\theta_A$, is calculated as:

$$R_A \cos\theta_A = 122\,\text{N}$$

Similarly, applying Equation (5.2) in the y direction gives:

$$\sum F_y = R_A \sin\theta_A + R_B \sin\theta_B = -1.663 \Rightarrow R_A \sin\theta_A + 88.917 = -1.663$$

from which the component of the reaction at A in the y direction, $R_A \sin\theta_A$, is calculated as:

$$R_A \sin\theta_A = -90.58\,\text{N}$$

The magnitude of the reaction at A is the vector summation of the x and y components:

$$R_A = \sqrt{(122)^2 + (-90.58)^2} = 151.95\,\text{N}$$

Example 5.2 Balancing of rotating masses II

A shaft carries two masses at planes 1 and 2 as shown in Figure 5.4. Determine the magnitude and the angular position of two masses that should be added at planes 3 and 4, each at a radial position of 0.25 m in order to achieve complete balance.

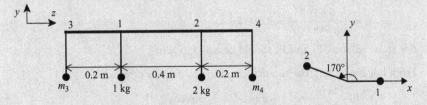

Figure 5.4: Representation of Example 5.2

The planes have the characteristics given in Table 5.2 (where m is the mass, r is the radial position, θ is the angular position and the reference plane is taken at plane 3).

Table 5.2: Data for Example 5.2

Plane	m (kg)	r (m)	θ (°)	z (m)	mr (kg.m)	mrz (kg.m²)
1	1	0.2	0	0.2	0.2	0.04
2	2	0.4	170	0.6	0.8	0.48
3	m_3	0.25	θ_3	0	$0.25\,m_3$	0
4	m_4	0.25	θ_4	0.8	$0.25\,m_4$	$0.2\,m_4$

Solution

For dynamic balance, M_x and M_y should be equal to zero. Using Equation (5.7) and equating M_x to zero gives:

$$\omega^2 \times [0.04 \times \sin 0 + 0.48 \times \sin 170 + 0.2m_4 \sin \theta_4] = 0$$

from which

$$m_4 \sin \theta_4 = -0.41675 \tag{E5.2a}$$

Using Equation (5.8) and equating M_y to zero gives:

$$\omega^2 \times [0.04 \times \cos 0 + 0.48 \cos 170 + 0.2m_4 \cos \theta_4] = 0$$

from which

$$m_4 \cos \theta_4 = 2.1635 \tag{E5.2b}$$

Dividing Equation (E5.2a) by Equation (E5.2b), gives:

$$\tan \theta_4 = \frac{-0.41675}{2.1635} \Rightarrow \theta_4 = -10.903° = 349.097°$$

From Equation (E5.2a), the mass m_4 is calculated as:

$$m_4 = \frac{-0.41675}{\sin 349.097} = 2.203 \, \text{kg}$$

For static balance, F_x and F_y should be equal to zero.

From Equation (5.3), the summation of forces in the x direction is:

$$\sum F_x = \sum mr\omega^2 \cos \theta = \omega^2 \times \sum mr \cos \theta$$

Equating F_x to zero, gives:

$$\omega^2 \times [0.2 \times \cos 0 + 0.8 \times \cos 170 + 0.25m_3\cos \theta_3 + 0.25m_4 \cos \theta_4] = 0$$

$$\omega^2 \times [0.2 \times \cos 0 + 0.8 \times \cos 170 + 0.25m_3 \cos \theta_3$$
$$+ 0.25 \times 2.203 \times \cos 349.097] = 0$$

from which

$$m_3 \cos \theta_3 = 0.18815 \qquad \text{(E5.2c)}$$

From Equation (5.4), the summation of forces in the y direction is:

$$\sum F_y = \sum mr\omega^2 \sin \theta = \omega^2 \times \sum mr \sin \theta$$

Equating F_y to zero, gives:

$$\omega^2 \times [0.2 \times \sin 0 + 0.8 \times \sin 170 + 0.25 m_3 \sin \theta_3 + 0.25 m_4 \sin \theta_4] = 0$$

$$\omega^2 \times [0.2 \times \sin 0 + 0.8 \times \sin 170 + 0.25 m_3 \sin \theta_3 + 0.25 \times 2.203$$

$$\times \sin 349.097] = 0$$

from which

$$m_3 \sin \theta_3 = -0.13898 \qquad \text{(E5.2d)}$$

Dividing Equation (E5.2d) by Equation (E5.2c), gives:

$$\tan \theta_3 = \frac{-0.13898}{0.18815} \Rightarrow \theta_3 = -36.453° = 323.547°$$

From Equation (E5.2d), the mass m_3 is calculated as:

$$m_3 = \frac{-0.13898}{\sin 323.547} = 0.234 \, \text{kg}$$

Example 5.3 Balancing of rotating masses III

A shaft of 5.5 m length is supported by two bearings, A and B, and carries four pulleys in planes 1, 2, 3 and 4 as shown in Figure 5.5.

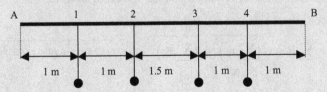

Figure 5.5: Representation of Example 5.3

The pulleys have the characteristics given in Table 5.3 (where m is the mass, r is the radial position, θ is the angular position and z is measured from A).

Table 5.3: Data for Example 5.3

Plane	m (kg)	r (mm)	mr (kg.m)	z (m)	mrz (kg.m^2)	θ (°)
1	2	12	0.024	1	0.024	0
2	3	15	0.045	2	0.09	30
3	4	10	0.04	3.5	0.14	60
4	3	15	0.045	4.5	0.2025	120

If the shaft speed is 200 rev/min, determine:

a) the total out-of-balance force;
b) the total out-of-balance moment about bearing A;
c) the radial positions of two additional masses of 2 kg and 8 kg that should be added in planes 2 and 3 at angular positions 150° and 240°, respectively, in order to achieve static balance.

Solution

a) The angular speed $\omega = 200$ rev/min $= 200 \times 2\pi/60 = 20.944$ rad/s.
The summation of forces in the x direction is:

$$\sum F_x = \sum mr\omega^2 \cos\theta = \omega^2 \times \sum mr \cos\theta$$
$$= (20.944)^2 \times [0.024 \times \cos 0 + 0.045 \times \cos 30$$
$$+ 0.04 \times \cos 60 + 0.045 \times \cos 120] = 26.5257\,\text{N}$$

The summation of forces in the y direction is:

$$\sum F_y = \sum mr\omega^2 \sin\theta = \omega^2 \times \sum mr \sin\theta$$
$$= (20.944)^2 \times [0.024 \times \sin 0 + 0.045 \times \sin 30$$
$$+ 0.04 \times \sin 60 + 0.045 \times \sin 120] = 42.1597\,\text{N}$$

The total out-of-balance force is:

$$\sum F = \sqrt{\left(\sum F_x\right)^2 + \left(\sum F_y\right)^2} = \sqrt{(26.5257)^2 + (42.1597)^2}$$
$$= 49.81\,\text{N}$$

b) M_{Ax} is:

$$\sum M_{Ax} = \sum F_y \times z = \sum mrz\omega^2 \sin\theta = \omega^2 \times \sum mrz \sin\theta$$
$$= (20.944)^2 \times [0.024 \times \sin 0 + 0.09 \times \sin 30$$
$$+ 0.14 \times \sin 60 + 0.2025 \times \sin 120] = 149.84\,\text{N.m}$$

M_{Ay} is:

$$\sum M_{Ay} = \sum F_x \times z = \sum mrz\omega^2 \cos\theta = \omega^2 \times \sum mrz \cos\theta$$
$$= (20.944)^2 \times [0.024 \times \cos 0 + 0.09 \times \cos 30$$
$$+ 0.14 \times \cos 60 + 0.2025 \times \cos 120] = 31.0 \, \text{N.m}$$

The total out-of-balance moment is:

$$\sum M_A = \sqrt{\left(\sum M_{Ax}\right)^2 + \left(\sum M_{Ay}\right)^2} = \sqrt{(149.84)^2 + (31.0)^2}$$
$$= 153.0 \, \text{N.m}$$

c) The summation of forces in the x direction should be equal to zero in order to achieve static balance:

$$\sum F_x = \sum mr\omega^2 \cos\theta = \omega^2 \times \sum mr \cos\theta = 0$$
$$= 26.5257 + (20.944)^2 \times [2 \times r_2 \times \cos 150 + 8 \times r_3 \times \cos 240] = 0$$

from which

$$-1.732 r_2 - 4 r_3 = 16.5368 \tag{E5.3a}$$

The summation of forces in the y direction should be equal to zero in order to achieve static balance:

$$\sum F_y = \sum mr\omega^2 \sin\theta = \omega^2 \times \sum mr \sin\theta = 0$$
$$= 42.1597 + (20.944)^2 \times [2 \times r_2 \times \sin 150 + 8 \times r_3 \times \sin 240] = 0$$

from which

$$r_2 - 6.928 r_3 = 10.4045 \tag{E5.3b}$$

Solving Equations (E5.3a) and (E5.3b) in two unknowns, r_2 and r_3, gives:

$$r_2 = 0.0025 \, \text{m},$$
$$r_3 = 0.014 \, \text{m}$$

Example 5.4 Balancing of rotating masses IV

A shaft AB carries two unbalanced pulleys in the transverse planes 1 and 2 as shown in Figure 5.6. The shaft length is 6 m and the pulleys are 2 m apart.

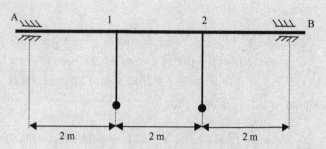

Figure 5.6: Representation of Example 5.4

The pulleys have the characteristics given in Table 5.4 (where m is the mass, r is the radial position, θ is the angular position and z is measured from A).

Table 5.4: Data for Example 5.4

Plane	m (kg)	r (mm)	z (m)	θ (°)
1	5	30	2	45
2	10	20	4	135

If the shaft rotates with a constant angular speed of 200 rev/min, determine:

a) the out-of-balance force;
b) the out-of balance moment at bearing A;
c) the reaction forces at bearings A and B.

Solution

The angular speed $\omega = 200$ rev/min $= 200 \times 2\pi/60 = 20.944$ rad/s
a) The out-of-balance forces are:

$$\sum F_x = \omega^2 \sum mr \cos\theta$$
$$= (20.944)^2 \left[5 \times \frac{30}{1000} \times \cos 45 + 10 \times \frac{20}{1000} \times \cos 135 \right]$$
$$= -15.5087\,\mathrm{N}$$

$$\sum F_y = \omega^2 \sum mr \sin\theta$$
$$= (20.944)^2 \left[5 \times \frac{30}{1000} \times \sin 45 + 10 \times \frac{20}{1000} \times \sin 135 \right]$$
$$= 108.5606\,\mathrm{N}$$

The total out-of-balance force is:

$$\sum F = \sqrt{\left(\sum F_x\right)^2 + \left(\sum F_y\right)^2} = 109.7\,\text{N}$$

b) The out-of balance moments are:

$$\sum M_{Ax} = \sum F_y \times z = \omega^2 \sum mrz\sin\theta$$

$$= (20.944)^2 \left[5 \times \frac{30}{1000} \times 2 \times \sin 45 + 10 \times \frac{20}{1000} \times 4 \times \sin 135\right]$$

$$= 341.19\,\text{N.m}$$

$$\sum M_{Ay} = \sum F_x \times z = \omega^2 \sum mrz\cos\theta$$

$$= (20.944)^2 \left[5 \times \frac{30}{1000} \times 2 \times \cos 45 + 10 \times \frac{20}{1000} \times 4 \times \cos 135\right]$$

$$= -155.086\,\text{N.m}$$

The total out-of-balance moment is:

$$\sum M_A = \sqrt{\left(\sum M_{Ax}\right)^2 + \left(\sum M_{Ay}\right)^2} = 374.8\,\text{N.m}$$

c) The component of the reaction at B in the y direction, $R_B \sin\theta_B$, is obtained from Equation (5.10) using $\sum M_{Ax}$:

$$R_B \sin\theta_B = \frac{\sum M_{Ax}}{6} = \frac{341.19}{6} = 56.865\,\text{N}$$

Similarly, the component of the reaction at B in the x direction, $R_B \cos\theta_B$, is obtained Equation (5.10) using $\sum M_{Ay}$:

$$R_B \cos\theta_B = \frac{\sum M_{Ay}}{6} = \frac{-155.086}{6} = -25.84767\,\text{N}$$

The magnitude of the reaction at B is the vector summation of the x and y components:

$$R_B = \sqrt{(56.865)^2 + (-25.84767)^2} = 62.46\,\text{N}$$

To calculate the reaction at A, apply Equation (5.2) in the x direction to give:

$$\sum F_x = R_A \cos\theta_A + R_B \cos\theta_B = -15.5087 \Rightarrow R_A \cos\theta_A - 25.84767$$

$$= -15.5087$$

The component of the reaction at A in the x direction, $R_A \cos\theta_A$, is calculated as:

$$\Rightarrow R_A \cos\theta_A = 10.34\,\text{N}$$

Similarly, applying Equation (5.2) in the y direction, gives:

$$\sum F_y = R_A \sin\theta_A + R_B \sin\theta_B = 108.5606 \Rightarrow R_A \sin\theta_A + 56.865$$
$$= 108.5606$$

The component of the reaction at A in the y direction, $R_A \sin\theta_A$, is calculated as:

$$\Rightarrow R_A \sin\theta_A = 51.6956\,\text{N}$$

The magnitude of the reaction at A is the vector summation of the x and y components:

$$R_A = \sqrt{(10.34)^2 + (51.6956)^2} = 52.72\,\text{N}$$

5.3 BALANCING OF RECIPROCATING ENGINES

5.3.1 Balancing of a single-cylinder engine

Equation (2.44) in Chapter 2 shows that the acceleration of the piston of an engine is approximately $r\omega^2(\cos\theta + 1/n\,\cos 2\theta)$, where n is the ratio of the length of the connecting-rod to the radius of the crank ($n = \frac{L}{r}$), at a crank angular velocity ω and a crank angle θ. This produces an inertia force, F_A, which is equal to the mass times the acceleration:

$$F_A = mr\omega^2 \cos\theta + \frac{mr\omega^2}{n}\cos 2\theta \qquad (5.11)$$

where m is the mass of the reciprocating parts (the piston and part of the connecting-rod mass). The first term in Equation (5.11), $mr\omega^2 \cos\theta$, is called the primary force and the second term, $\frac{mr\omega^2}{n}\cos 2\theta$, is called the secondary force.

Both the primary and secondary forces act along the line of stroke. These inertia forces are unbalanced and are transmitted to the bearings and foundations. The primary force could be balanced by attaching two masses equal to half the reciprocating mass at a radius r from the crank axis as shown in Figure 5.7(a). One of these two masses rotates in the same direction as the crank, while the second rotates in the opposite direction. Both primary force and moment balances are achieved for any crank angle θ. In a similar way, the secondary force could be balanced by attaching two masses equal to $(\frac{m}{2n})$ as shown in Figure 5.7(b). However, in this case the balance masses should be driven at twice the crank speed. In practice, it is difficult to attach the secondary balance masses to the crank and make them rotate at twice the crank speed and thus complete balance is difficult to achieve for a single-cylinder engine.

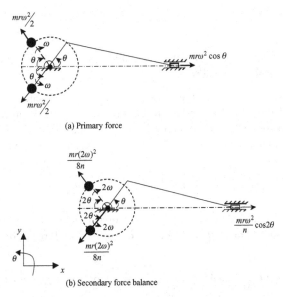

(a) Primary force

(b) Secondary force balance

Figure 5.7: Balancing of a single-cylinder engine

Example 5.5 Balancing of a single-cylinder engine

A single-cylinder engine has a reciprocating mass of 5 kg and is running at a speed of $\omega = 650$ rev/min. The crank radius is 0.15 m and the connecting rod is of length $L = 0.5$ m. Two masses of 2.5 kg are attached to the crank as shown in Figure 5.8. Determine the total out-of-balance force when the crank is right at the top ($\theta = 0$), with and without the attached masses.

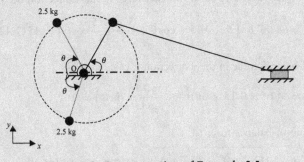

Figure 5.8: Representation of Example 5.5

Solution

The ratio of the connecting rod to the crank is $n = 0.5/0.15 = 3.33$ and $\omega = 650 \times 2\pi/60 = 68.0678$ rad/s.

The out-of-balance force in the x direction with the attached masses is the sum of the primary and secondary forces produced by the reciprocating mass $\omega^2 \left(mr \cos\theta + \frac{mr}{n} \cos 2\theta \right)$ (see Equation (5.11)) and the out-of-balance force produced by the attached mass $2 \times m_1 r \cos\theta_1 \times \omega^2$. Thus, the out-of-balance force in the x direction is:

$$F_x = \omega^2 \left(mr \cos\theta + \frac{mr}{n} \cos 2\theta + 2 \times m_1 r \cos\theta_1 \right)$$

where m_1 is the attached mass at angle θ_1 measured anticlockwise. When the crank is right at the top ($\theta = 0$), $\theta_1 = 180°$. The out-of-balance force in the x direction becomes:

$$F_x = 68.0678^2 \left(5 \times 0.15 \times \cos 0 + \frac{5 \times 0.15}{3.333} \cos 0 + 2 \times 2.5 \times 0.15 \times \cos 180 \right)$$
$$= 1042.6 \, \text{N}$$

The out-of-balance force in the y direction is calculated as:

$$F_y = \omega^2 \left(5 \times 0.15 \times \sin 0 + \frac{5 \times 0.15}{3.333} \sin 0 + 2 \times 2.5 \times 0.15 \times \sin 180 \right) = 0$$

The total out-of-balance force is:

$$\sum F = \sqrt{\left(\sum F_x \right)^2 + \left(\sum F_y \right)^2} = 1042.6 \, \text{N} = 1.04 \, \text{kN}$$

The out-of-balance force without the attached masses is the sum of only the primary and secondary forces produced by the reciprocating mass, $\omega^2 \left(mr \cos\theta + \frac{mr}{n} \cos 2\theta \right)$. The out-of-balance force in the x direction becomes:

$$F_x = 68.0678^2 \left(5 \times 0.15 \times \cos 0 + \frac{5 \times 0.15}{3.333} \cos 0 \right) = 4517.4 \, \text{N}$$

The out-of-balance force in the y direction is calculated as:

$$F_y = \omega^2 \left(5 \times 0.15 \times \sin 0 + \frac{5 \times 0.15}{3.333} \sin 0 \right) = 0$$

The total out-of-balance force is:

$$\sum F = \sqrt{\left(\sum F_x \right)^2 + \left(\sum F_y \right)^2} = 4517.4 \, \text{N} = 4.5 \, \text{kN}$$

5.3.2 Balancing of a multi-cylinder engine

For multi-cylinder engines, a number of slider–crank mechanisms (cylinders) are connected to the same crankshaft. If the lines of stroke of all the cylinders are parallel, as in Figure 5.9, the engine is called an in-line engine.

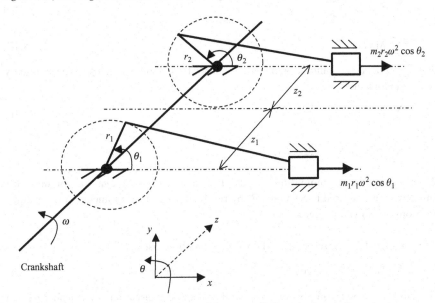

Figure 5.9: In-line engine with two cylinders (primary forces)

It is possible to balance the primary and secondary reciprocating inertia forces and moments by selecting appropriate radii for the cranks, relative angular positions and distances between cranks along the shafts in the axial (z) direction.

To balance the primary force in the line of stroke direction, the following condition should be satisfied:

$$\sum F_x = \omega^2 \sum mr \cos \theta = 0 \tag{5.12}$$

To balance the primary moments:

$$\sum M_y = \omega^2 \sum mrz \cos \theta = 0 \tag{5.13}$$

where z is measured from a convenient reference plane.

Similarly, the balancing of secondary force and moment requires:

$$\sum F_x = \omega^2 \sum \frac{mr}{n} \cos 2\theta = 0 \tag{5.14}$$

$$\sum M_y = \omega^2 \sum \frac{mrz}{n} \cos 2\theta = 0 \tag{5.15}$$

Equations (5.12) to (5.15) represent the instantaneous balancing in the direction of the line of stroke. For an overall balancing, the vector summation of the forces and moments should vanish. For overall balancing of the primary force and moment, the vector summation of the forces and moments should be equal to zero:

$$\sum F = \omega^2 \sum_{\theta} mr = 0 \qquad\qquad (5.16)$$

$$\sum M = \omega^2 \sum_{\theta} mrz = 0 \qquad\qquad (5.17)$$

where $\sum_{\theta}$ is a vector summation over θ. Similarly, for overall balancing of the secondary force and moment:

$$\sum F = \omega^2 \sum_{2\theta} \frac{mr}{n} = 0 \qquad\qquad (5.18)$$

$$\sum M = \omega^2 \sum_{2\theta} \frac{mrz}{n} = 0 \qquad\qquad (5.19)$$

where $\sum_{2\theta}$ is a vector summation over 2θ. The analytical solution can be obtained, in a similar way to that used in Section 5.2, by resolving forces and moments in the x and y directions as shown in Figure 5.10.

Equation (5.16) can be resolved in the x and y directions as:

$$\sum F_x = \omega^2 \sum mr \cos\theta, \qquad \sum F_y = \omega^2 \sum mr \sin\theta \qquad\qquad (5.20)$$

The primary moment can also be resolved in the x and y directions by considering that F_x is making moment about the y axis and F_y is making moment about the x axis:

$$\sum M_x = \sum F_y \times z = \omega^2 \sum mrz \sin\theta \qquad \sum M_y = \sum F_x \times z = \omega^2 \sum mrz \cos\theta \qquad\qquad (5.21)$$

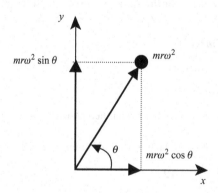

Figure 5.10: Resolving the primary force in the x and y directions

Similarly, the secondary force and moment in the x and y directions are obtained as:

$$\sum F_x = \omega^2 \sum \frac{mr}{n} \cos 2\theta, \qquad \sum F_y = \omega^2 \sum \frac{mr}{n} \sin 2\theta \qquad (5.22)$$

$$\sum M_x = \sum F_y \times z = \omega^2 \sum \frac{mrz}{n} \sin 2\theta, \qquad \sum M_y = \sum F_x \times z = \omega^2 \sum \frac{mrz}{n} \cos 2\theta \qquad (5.23)$$

Example 5.6 Balancing of a multi-cylinder engine I

A petrol engine has four cylinders in line as shown in Figure 5.11. Each cylinder has a crank radius of $r = 0.16$ m, a connecting-rod length of $L = 0.48$ m and reciprocating mass of $m = 30$ kg. When crank 1 is right at the top ($\theta_1 = 0$), the angular positions of cranks 2, 3 and 4 are 90°, 180° and 270°, respectively. If the engine runs at speed $\omega = 650$ rev/min, determine the primary and secondary out-of-balance forces and moments.

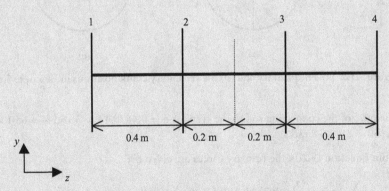

Figure 5.11: Representation of Example 5.6

The values required to calculate the out-of-balance forces and moments are given in Table 5.5 (where m is the mass, r is the radial position, θ is the angular position and z is measured from the middle plane of the shaft).

Table 5.5: Data for Example 5.6

Crank	m (kg)	r (m)	mr (kg m)	z (m)	mrz (kg m^2)	θ (°)	2θ (°)
1	30	0.16	4.8	−0.6	−2.88	0	0
2	30	0.16	4.8	−0.2	−0.96	90	180
3	30	0.16	4.8	0.2	0.96	180	0
4	30	0.16	4.8	0.6	2.88	270	180

Solution

Figure 5.12(a) shows the (real) primary crank positions when crank 1 is right at the top (i.e. $\theta_1 = 0$). The (imaginary) secondary crank positions are obtained by doubling all angles relative to the top, as shown in Figure 5.12(b).

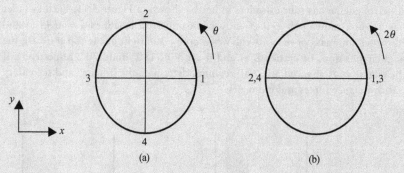

Figure 5.12: (a) The primary and (b) the secondary crank positions in Example 5.6

The ratio of the connecting rod to the crank is $n = 0.48/0.16 = 3$ and $\omega = 650 \times 2\pi/60 = 68.0678$ rad/s.

From Equation (5.20), the primary forces are given by:

$$\sum F_x = \omega^2 \sum mr \cos\theta = mr \times \omega^2 \sum \cos\theta$$
$$= 4.8 \times \omega^2 [\cos 0 + \cos 90 + \cos 180 + \cos 270] = 0$$
$$\sum F_y = \omega^2 \sum mr \sin\theta = mr \times \omega^2 \sum \sin\theta$$
$$= 4.8 \times \omega^2 [\sin 0 + \sin 90 + \sin 180 + \sin 270] = 0$$

Thus, the total primary force balances.

From Equation (5.21), the primary moments are given by:

$$\sum M_x = \sum F_y \times z = \omega^2 \sum mrz \sin\theta$$
$$= 4.8 \times (68.0678)^2 [-0.6 \times \sin 0 - 0.2 \times \sin 90$$
$$+ 0.2 \times \sin 180 + 0.6 \times \sin 270]$$
$$= -17791.59 \, \text{N.m}$$

$$\sum M_y = \sum F_x \times z = \omega^2 \sum mrz \cos\theta$$
$$= 4.8 \times (68.0678)^2 [-0.6 \times \cos 0 - 0.2 \times \cos 90$$
$$+ 0.2 \times \cos 180 + 0.6 \times \cos 270]$$
$$= -17791.59 \, \text{N.m}$$

The total out-of-balance primary moment is:

$$\sum M = \sqrt{\left(\sum M_x\right)^2 + \left(\sum M_y\right)^2} = \sqrt{(-17791.59)^2 + (-17791.59)^2}$$
$$= 25161.11 \, \text{N.m}$$

From Equation (5.22), the secondary forces are given by:

$$\sum F_x = \omega^2 \sum \frac{mr}{n} \cos 2\theta = \frac{mr \times \omega^2}{n} \sum \cos 2\theta$$
$$= \frac{4.8 \times \omega^2}{3} [\cos 0 + \cos 180 + \cos 0 + \cos 180] = 0$$

$$\sum F_y = \omega^2 \sum \frac{mr}{n} \sin 2\theta = \frac{mr \times \omega^2}{n} \sum \sin 2\theta$$
$$= \frac{4.8 \times \omega^2}{3} [\sin 0 + \sin 180 + \sin 0 + \sin 180] = 0$$

Thus, the total secondary force balances.

From Equation (5.23), the secondary moments are given by:

$$\sum M_x = \sum F_y \times z = \omega^2 \sum \frac{mrz}{n} \sin 2\theta = \frac{mr \times \omega^2}{n} \sum z \sin 2\theta$$
$$= \frac{4.8 \times \omega^2}{3} [-0.6 \times \sin 0 - 0.2 \times \sin 180$$
$$+ 0.2 \times \sin 0 + 0.6 \times \sin 180] = 0$$

$$\sum M_y = \sum F_x \times z = \omega^2 \sum \frac{mrz}{n} \cos 2\theta = \frac{3 \times \omega^2}{3.333} \sum z \cos 2\theta$$
$$= \frac{4.8 \times (68.0678)^2}{3} [-0.6 \times \cos 0 - 0.2 \times \cos 180$$
$$+ 0.2 \times \cos 0 + 0.6 \times \cos 180]$$
$$= -5930.53 \, \text{N.m}$$

The total out-of-balance secondary moment is:

$$\sum M = \sqrt{\left(\sum M_x\right)^2 + \left(\sum M_y\right)^2} = \sqrt{(0)^2 + (-5930.53)^2} = 5930.53\,\text{N.m}$$

Example 5.7 Balancing of a multi-cylinder engine II

For the four cylinder in-line engine shown in Figure 5.13, the cylinders have identical characteristics: crank radius, r, ratio of connecting-rod length to crank radius, n, and reciprocating mass, m. The cranks are at equal spaces a. When crank 1 is right at the top ($\theta_1 = 0$), the angular positions of cranks 2, 3 and 4 are 180°, 180° and 360°, respectively. If the engine runs at speed ω rad/s, determine the primary and secondary out-of-balance forces and moments in terms of m, r, a, ω and n.

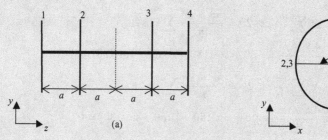

Figure 5.13: Representation of Example 5.7

Solution

From Equation (5.20), the primary forces are given by:

$$\sum F_x = \omega^2 \sum mr\cos\theta = mr \times \omega^2 \sum \cos\theta$$
$$= \omega^2 [\cos 0 + \cos 180 + \cos 180 + \cos 360] = 0$$
$$\sum F_y = \omega^2 \sum mr\sin\theta = mr \times \omega^2 \sum \sin\theta$$
$$= \omega^2 [\sin 0 + \sin 180 + \sin 180 + \sin 360] = 0$$

Thus, the total primary force balances.

From Equation (5.21), the primary moments are given by:

$$\sum M_x = \sum F_y \times z = \omega^2 \sum mrz\sin\theta$$
$$= mr\omega^2 [-2a \times \sin 0 - a \times \sin 180 + a \times \sin 180 + 2a \times \sin 360] = 0$$

$$\sum M_y = \sum F_x \times z = \omega^2 \sum mrz\cos\theta$$
$$= mr\omega^2 \left[-2a \times \cos 0 - a \times \cos 180 + a \times \cos 180 + 2a \times \cos 360\right] = 0$$

Thus, the total primary moment balances.

From Equation (5.22), the secondary forces are given by:

$$\sum F_x = \omega^2 \sum \frac{mr}{n}\cos 2\theta = \frac{mr \times \omega^2}{n}\sum \cos 2\theta$$
$$= \frac{mr \times \omega^2}{n}\left[\cos 0 + \cos 0 + \cos 0 + \cos 0\right] = \frac{4mr\omega^2}{n}$$

$$\sum F_y = \omega^2 \sum \frac{mr}{n}\sin 2\theta = \frac{mr \times \omega^2}{n}\sum \sin 2\theta$$
$$= \frac{mr \times \omega^2}{n}\left[\sin 0 + \sin 0 + \sin 0 + \sin 0\right] = 0$$

The total out-of-balance secondary force $= \frac{4mr\omega^2}{n}$.

From Equation (5.23), the secondary moments are given by:

$$\sum M_x = \sum F_y \times z = \omega^2 \sum \frac{mrz}{n}\sin 2\theta = \frac{mr \times \omega^2}{n}\sum z\sin 2\theta$$
$$= \frac{mr\omega^2}{n}\left[-2a \times \sin 0 - a \times \sin 0 + a \times \sin 0 + 2a \times \sin 0\right] = 0$$

$$\sum M_y = \sum F_x \times z = \omega^2 \sum \frac{mrz}{n}\cos 2\theta = \frac{mr\omega^2}{n}\sum z\cos 2\theta$$
$$= \frac{mr\omega^2}{n}\left[-2a \times \cos 0 - a \times \cos 0 + a \times \cos 0 + 2a \times \cos 0\right] = 0$$

Thus, the total secondary moment balances.

Example 5.8 Balancing of a multi–cylinder engine III

For the three-cylinder in-line engine shown in Figure 5.14, the reciprocating assemblies for each cylinder have the same mass (m), crank radius (r) and connecting-rod length (L). If the crank shaft rotates with a constant angular speed ω, show that

a) the primary and secondary forces are $mr\omega^2$ and $2.07\frac{mr}{n}\omega^2$, respectively;
b) the primary and secondary moments at crank 2 are $1.81mra\omega^2$ and $1.53\frac{mra}{n}\omega^2$, respectively.

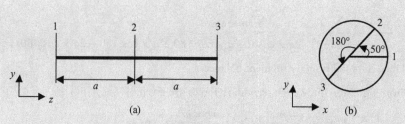

Figure 5.14: Representation of Example 5.8

The values required to calculate the forces and moments are given in Table 5.6 (where m is the mass, r is the radial position, θ is the angular position and z is measured from crank 2).

Table 5.6: Data for Example 5.8

Crank	m (kg)	r (m)	mr (kg m)	z (m)	mrz (kg m^2)	θ (°)	2θ (°)
1	m	r	mr	$-a$	$-mra$	0	0
2	m	r	mr	0	0	50	100
3	m	r	mr	a	mra	230	460

Solution

a) From Equation (5.20), the primary forces are given by:

$$\sum F_x = \omega^2 \sum mr \cos\theta = \omega^2 mr \left[\cos 0 + \cos 50 + \cos 230\right] = mr\omega^2$$
$$\sum F_y = \omega^2 \sum mr \sin\theta = \omega^2 mr \left[\sin 0 + \sin 50 + \sin 230\right] = 0$$

The total primary force is:

$$\sum F = \sqrt{\left(\sum F_x\right)^2 + \left(\sum F_y\right)^2} = mr\omega^2$$

From Equation (5.22), the secondary forces are given by:

$$\sum F_x = \omega^2 \sum \frac{mr}{n} \cos 2\theta = \frac{mr}{n}\omega^2 \left[\cos 0 + \cos 100 + \cos 460\right] = 0.652 \frac{mr}{n}\omega^2$$
$$\sum F_y = \omega^2 \sum \frac{mr}{n} \sin 2\theta = \frac{mr}{n}\omega^2 \left[\sin 0 + \sin 100 + \sin 460\right] = 1.969 \frac{mr}{n}\omega^2$$

The total secondary force is:

$$\sum F = \sqrt{\left(\sum F_x\right)^2 + \left(\sum F_y\right)^2} = 2.074 \frac{mr}{n}\omega^2$$

b) From Equation (5.21), the primary moments are given by:

$$\sum M_x = \sum F_y \times z = \omega^2 \sum mrz \sin\theta$$
$$= \omega^2 mr [-a \sin 0 + 0 \sin 50 + a \sin 230] = -0.766 mra\omega^2$$
$$\sum M_y = \sum F_x \times z = \omega^2 \sum mrz \cos\theta$$
$$= \omega^2 mr [-a \cos 0 + 0 \cos 50 + a \cos 230] = -1.642 mra\omega^2$$

The total out-of-balance primary moment is:

$$\sum M = \sqrt{\left(\sum M_x\right)^2 + \left(\sum M_y\right)^2} = 1.81 mra\omega^2$$

From Equation (5.23), the secondary moments are given by:

$$\sum M_x = \sum F_y \times z = \omega^2 \sum \frac{mrz}{n} \sin 2\theta$$
$$= \frac{mr}{n}\omega^2 [-a \sin 0 + 0 \sin 100 + a \sin 460] = 0.985\frac{mra}{n}\omega^2$$
$$\sum M_y = \sum F_x \times z = \omega^2 \sum \frac{mrz}{n} \cos 2\theta$$
$$= \frac{mr}{n}\omega^2 [-a \cos 0 + 0 \cos 100 + a \cos 460] = -1.173\frac{mra}{n}\omega^2$$

The total out-of-balance secondary moment is:

$$\sum M = \sqrt{\left(\sum M_x\right)^2 + \left(\sum M_y\right)^2} = 1.53\frac{mra}{n}\omega^2$$

5.4 Tutorial Sheet

5.4.1 Balancing of rotating masses

Q5.1   A shaft rotates at 250 rev/min and is supported by two bearings, A and B, 0.5 m apart, as shown in Figure 5.15. At a distance of 0.2 m from bearing A, three masses of 8, 10 and 12 kg are attached to the shaft at radii 25, 35 and 20 mm and angular positions of 0°, 60° and 240°, respectively. Determine:

a) the total out-of-balance force;

[186.6 N]

b) the total out-of balance moment at bearing A;

[37.32 N.m]

c) the reaction forces at bearings A and B.

[74.63 N, 111.9 N]

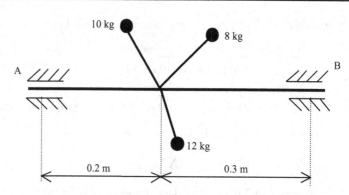

Figure 5.15: Representation of Question 5.1

Q5.2   In Figure 5.16, four masses are attached to a shaft in four planes 1.5 m apart. The masses (m), their radial positions (r) and their angular positions (θ) are summarized in Table 5.7. The shaft rotates with a constant angular velocity of 450 rev/min and is supported by two bearings, A and B, at a distance 1 m outside planes 1 and 4, respectively. Determine:

a) the total out-of-balance force;

[179.8 N]

b) the total out-of balance moment at bearing A;

[667.4 N.m]

c) the reaction forces at bearings A and B.

[135.5 N, 102.7 N]

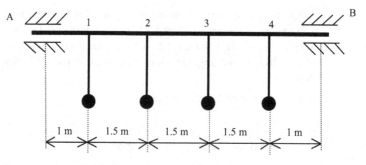

Figure 5.16: Representation of Question 5.2

Table 5.7: Data for Question 5.2

Plane	m (kg)	r (mm)	θ (°)
1	4	15	0
2	3	25	90
3	5	10	150
4	2	15	220

Q5.3   A shaft carries two masses at planes 1 and 2 as shown in Figure 5.17. The masses and their angular and radial positions are given in Table 5.8. If it is decided to add two masses, each of 1 kg in planes 3 and 4 in order to achieve the complete balancing of the shaft, determine the radial and the angular positions of these two additional masses.

[0.0585 m, 323.5°, 0.5508 m, 349.1°]

Table 5.8: Data for Question 5.3

Plane	m (kg)	r (m)	θ (°)
1	1	0.2	0
2	2	0.4	170

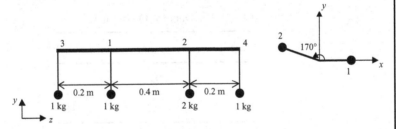

Figure 5.17: Representation of Question 5.3

Q5.4   A uniform disc has four holes at a radius of 0.12 m as shown in Figure 5.18. The holes have been drilled at the angles given in Table 5.9. The equivalent removed mass for each hole is also given. Determine the equivalent removed mass and the angular position of a fifth hole that should be drilled at a radius of 0.15 m in order to statically balance the disk.

[200 g, 283.9°]

Table 5.9: Data for Question 5.4

Hole	m (g)	θ (°)
1	120	0
2	120	60
3	160	120
4	160	180

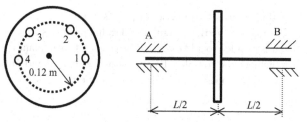

Figure 5.18: Representation of Question 5.4

Q5.5   A rim of a flywheel has a radius of 5 m. Attached at its outer edge are three masses, as shown in Figure 5.19, which have the characteristics given in Table 5.10. Determine the mass and the angular position of an additional mass that should be attached to the rim at its outer edge in order to statically balance the flywheel.

[1 kg, 306.87°]

Table 5.10: Data for Question 5.5

Mass	m (kg)	θ (°)
1	0.6	0
2	0.8	90
3	1.2	180

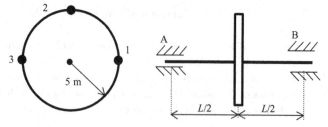

Figure 5.19: Representation of Question 5.5

Q5.6   Three masses are attached along a shaft at the locations shown in Figure 5.20. The values of the masses and their radial and angular positions are summarized in Table 5.11. The shaft is supported at its ends by two bearings, A and B. In order to dynamically balance the shaft, an additional mass of 2 kg at a radial position of 0.5 m is attached to the shaft. Determine the location along the shaft measured from bearing A and the angular position of this additional mass.

[1.54 m, 289.13°]

Table 5.11: Data for Question 5.6

Mass	m (kg)	r (m)	θ (°)
1	1.2	0.6	0
2	2.1	0.8	210
3	1.6	0.4	320

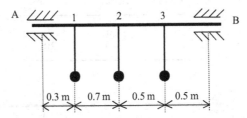

Figure 5.20: Representation of Question 5.6

Q5.7   A shaft AB, 5.5 m long, carries four pulleys in planes 1, 2, 3 and 4 as shown in Figure 5.21. The pulleys have the characteristics given in Table 5.12. If the shaft speed is 200 rev/min, determine the radial positions of two additional masses of 4 kg each that should be added in planes 2 and 3 at angular positions 150° and 240°, respectively, in order to achieve static balance.

[1.3 mm, 28 mm]

Table 5.12: Data for Question 5.7

Plane	m (kg)	r (mm)	θ (°)
1	2	12	0
2	3	15	30
3	4	10	60
4	3	15	120

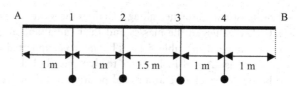

Figure 5.21: Representation of Question 5.7

Q5.8   Determine the reaction forces at bearings A and B of the shaft AB shown in Figure 5.22. The shaft rotates with a constant angular velocity of 350 rev/min and carries two unbalanced pulleys in the transverse planes 1 and 2 for which the data are given in Table 5.13.

[161.5 N, 191.3 N]

Table 5.13: Data for Question 5.8

Plane	m (kg)	r (mm)	θ (°)
1	5	30	45
2	10	20	135

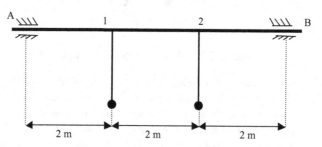

Figure 5.22: Representation of Question 5.8

5.4.2 Balancing of reciprocating engines

Q5.9   Figure 5.23 shows the crank arrangement of a four-cylinder in-line engine in which all the reciprocating assemblies have identical values: mass $m = 0.4$ kg, crank radius $r = 75$ mm, and ratio of connecting-rod length to crank radius $n = 4$. The crank spacing is constant and is equal to $a = 60$ mm. If the crankshaft rotates with a constant angular velocity of $\omega = 4500$ rev/min, determine:

a) the primary and secondary forces;

[0, 6.66 kN]

b) the primary and secondary moments at the mid-plane.

[799.4 N.m, 0]

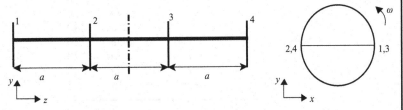

Figure 5.23: Representation of Question 5.9

Q5.10 Figure 5.24 shows the crank arrangement of a three-cylinder in-line engine in which all the reciprocating assemblies have identical values: mass $m = 0.4$ kg, crank radius $r = 75$ mm, and ratio of connecting-rod length to crank radius $n = 4$. The crank spacing is constant and is equal to $a = 60$ mm. If the crankshaft rotates with a constant angular velocity of $\omega = 4500$ rev/min, determine:

a) the primary and secondary forces;

[0, 0]

b) the primary and secondary moments at the mid-plane.

[692.3 N.m, 173 N.m]

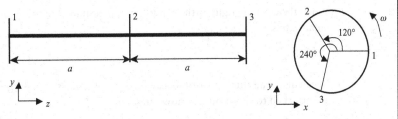

Figure 5.24: Representation of Question 5.10

Q5.11 Show that, for the six-cylinder engine shown in Figure 5.25, for which the reciprocating assemblies are summarized in Table 5.14, a complete balance is achieved. z is measured from the middle of the shaft.

Table 5.14: Data for Question 5.11

Cylinder	mr (kg)	z (mm)	θ (°)
1	mr	$-2.5a$	0
2	mr	$-1.5a$	240
3	mr	$-0.5a$	120
4	mr	$0.5a$	120
5	mr	$1.5a$	240
6	mr	$2.5a$	0

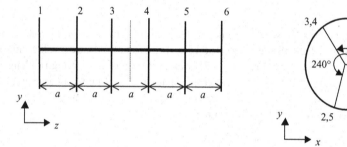

Figure 5.25: Representation of Question 5.11

Q5.12 A single-cylinder engine has a reciprocating mass of 6 kg and is running at a speed $\omega = 65$ rad/s. The crank radius is 0.15 m and the connecting rod length is $L = 0.45$ m.

a) When the crank is right at the top ($\theta = 0$), determine the total out-of-balance force.

[5.07 kN]

b) Recalculate the total out-of-balance force, if two masses of 3 kg each are attached to the crank as shown in Figure 5.26.

[1.27 kN]

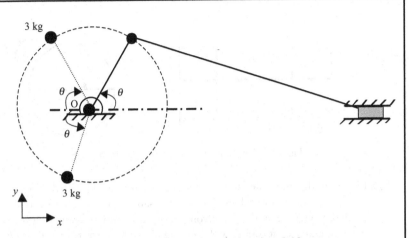

Figure 5.26: Representation of Question 5.12

Q5.13 In the two-cylinder engine shown in Figure 5.27, both cylinders have identical reciprocating assemblies, mass m, crank radius r and ratio of connecting-rod length to crank radius n. The crank shaft rotates with an angular speed ω. Show that:

  a) the primary force balances and the out-of-balance primary moment equals $2mra\omega^2$;

  b) the out-of-balance secondary force equals $\frac{2mr\omega^2}{n}$ and the secondary moment balances.

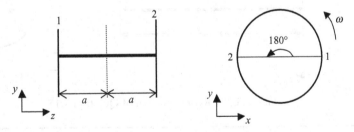

Figure 5.27: Representation of Question 5.13

Q5.14 Show that the primary force, primary moment and secondary moment balance in the four-cylinder in-line engine shown in Figure 5.28. If the cylinders have identical data $mr = 0.1$ kg.m and $n = 3$ and the engine runs with 360 rad/s, determine the out-of-balance secondary force.

[17.28 kN]

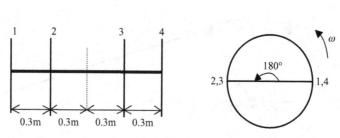

Figure 5.28: Representation of Question 5.14

Q5.15 For the three-cylinder in-line engine shown in Figure Q5.29, the reciprocating assemblies for each cylinder are the same: mass $m = 0.5$ kg, crank radius $r = 0.07$ m and ratio of connecting-rod length to crank radius $n = 4$. If the crank shaft rotates with a constant angular velocity of 3000 rev/min, determine:

a) the primary and secondary forces;

[3.45 kN, 1.79 kN]

b) the primary and secondary moments at crank 2.

[407 N.m, 86 N.m]

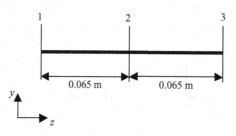

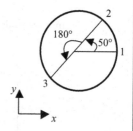

Figure 5.29: Representation of Question 5.15

PART II

Vibration

CHAPTER 6

Free Vibration of Systems with a Single Degree of Freedom

6.1 INTRODUCTION

Vibration can be regarded as a branch of dynamics that deals with periodic or oscillatory motion. Common examples of vibration problems are the response of civil engineering structures to dynamics loading, ambient conditions and earthquakes, vibration of unbalanced rotating machines and vibration of power lines due to wind excitation, and aircraft wings.

A periodic motion is a motion that repeats itself at equal time intervals. Oscillatory motion can be periodic or non-periodic. An example of a periodic motion is a simple pendulum, where the motion repeats regularly at equal time intervals. An example of a non-periodic motion would be the response of a structure to an earthquake, where the motion is irregular and can be repeated at unequal time intervals. A harmonic motion is a periodic motion that varies in a sinusoidal way. In a simple harmonic motion, the displacement, x, of the system varies as:

$$x = A \sin \omega_n t \tag{6.1}$$

where A is a constant, ω_n is the angular velocity of the system and t is the time. The velocity of the system is obtained by differentiating the displacement, Equation (6.1), with respect to time:

$$\dot{x} = \frac{dx}{dt} = A\omega_n \cos \omega_n t \tag{6.2}$$

Differentiating again with respect to time, the acceleration of the system is obtained as:

$$\ddot{x} = \frac{d^2 x}{dt^2} = -A\omega_n^2 \sin \omega_n t \tag{6.3}$$

The acceleration is therefore directly proportional to the displacement, i.e. $\ddot{x} = -\omega^2 x$, in a simple harmonic motion. Figure 6.1 shows typical variations over time of the displacement, velocity and acceleration of a simple harmonic motion.

We can define the following terms:

- A **cycle of vibration** is a measure of the motion of a vibrating system, from equilibrium ($x = 0$) to a maximum, then to equilibrium again, then to a minimum and back to equilibrium. In Figure 6.1(a), the cycle takes place in 8 seconds. A cycle is also equivalent to one revolution. So if the system is vibrating or rotating with an angular velocity of ω rad/s, the time for one cycle (period of oscillation τ) is equivalent to the time of one revolution, i.e. $\tau = \frac{2\pi}{\omega}$. For the response in Figure 6.1, the angular velocity or angular frequency is $\omega = \frac{2\pi}{8} = 0.7854$ rad/s.
- The **amplitude of vibration** is the maximum displacement of a vibrating system from its equilibrium position. In Figure 6.1(a), the amplitude of vibration is 0.1 m.
- The **period of oscillation** is the time taken to complete one cycle (τ). In Figure 6.1(a), the period of oscillation is 8 seconds.

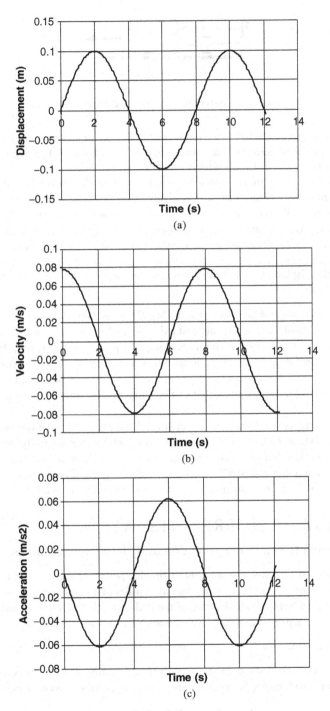

Figure 6.1: A simple harmonic motion

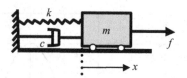

Figure 6.2: A system with a single degree of freedom

- The **frequency of oscillation** is the number of cycles per unit of time and is calculated as the inverse of the period of oscillation, i.e. $f' = \frac{1}{\tau}$. In Figure 6.1(a), the frequency of oscillation is $f' = \frac{1}{8} = 0.125$ cycles per second, or hertz (Hz).
- The **natural frequency** of a system is the frequency of oscillation at which the system is left to vibrate freely without the action of any external forces.
- The **phase angle** of two harmonic motions is the shift of the maximum values between the motions. As an example, for two displacement responses of $x_1 = A_1 \sin \omega t$ and $x_2 = A_2 \sin(\omega t + \phi)$, the phase angle between them is ϕ. The phase angle between the displacement in Figure 6.1(a) and the velocity in Figure 6.1(b) is $\frac{\pi}{2}$ (the time difference between the displacement amplitude and velocity amplitude is 2 seconds, thus $\phi = \frac{2}{8} \times 2\pi = \frac{\pi}{2}$).

A system with a single degree of freedom (SDOF) is the simplest vibrating system. In general, it consists of a mass, a spring and a damper as shown in Figure 6.2. The mass is allowed to move only in one direction, i.e. along the direction of the spring. Hence, only one degree of freedom is analyzed, the longitudinal displacement (x) for the system shown in Figure 6.2, where m is the mass, k is the spring's stiffness, c is the damping coefficient and f is an externally applied force that can be a function of time.

This chapter presents and discusses both undamped and damped free vibration of a system with a single degree of freedom. The equations of motion are basically derived using Newton's second law, but a brief comment on how to derive them using the energy method is also given. For a damped SDOF system, three damping conditions are considered: an overdamped system, a critically damped system and an underdamped system.

6.2 UNDAMPED FREE VIBRATION

6.2.1 Deriving the equation of motion using Newton's second law

For an undamped free SDOF system, both the damping coefficient c and the external force f are equal to zero and the system simply consists of a mass and a spring as shown in Figure 6.3(a). The free-body diagram of the undamped free SDOF is shown in Figure 6.3(b) and the equation of motion, derived from Newton's second law, is:

$$\sum F_x = m\ddot{x} \tag{6.4}$$

where $\sum F_x$ is the sum of the external forces in the x direction and $\ddot{x}$ is the acceleration in the x direction. From Figure 6.3(b), the only force acting on the mass is kx in the negative

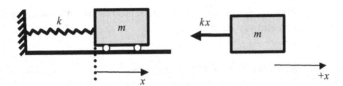

Figure 6.3: a) An undamped free SDOF system and b) its free-body diagram

x direction, thus:

$$-kx = m\ddot{x} \tag{6.5}$$

Rearranging Equation (6.5) gives:

$$m\ddot{x} + kx = 0 \tag{6.6}$$

or

$$\ddot{x} + \omega_n^2 x = 0 \tag{6.7}$$

where $\omega_n = \sqrt{\frac{k}{m}}$ is the angular frequency of the system in rad/s.

For the undamped free SDOF system shown in Figure 6.4(a), where a body of mass m is suspended by a spring of stiffness k, the free-body diagram is shown in Figure 6.4(b). In such a case, the weight of the body mg and the force due to static deflection $k\delta_o$ should be included in the equation of motion. Applying Newton's second law to the free-body diagram in Figure 6.4(b) gives:

$$mg - k(x + \delta_o) = m\ddot{x} \tag{6.8}$$

where δ_o is the static deflection.

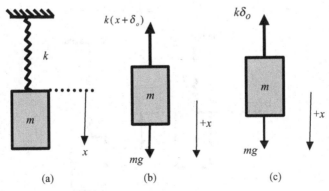

Figure 6.4: a) An undamped free SDOF system, b) its free-body diagram and c) its free-body diagram in static equilibrium

From the static equilibrium (see Figure 6.4(c)), the relationship between the weight of the body and the static defection is given by:

$$mg - k\delta_o = 0 \tag{6.9}$$

Substituting Equation (6.9) into Equation (6.8) gives:

$$-kx = m\ddot{x} \Rightarrow m\ddot{x} + kx = 0 \quad \text{or} \quad \ddot{x} + \omega_n^2 x = 0$$

This is exactly the same as Equation (6.7) for the undamped free SDOF system in Figure 6.3. Thus, the weight of the body does not have any effect on the equation of motion and its subsequent solution.

6.2.2 Deriving the equation of motion using the energy method

The equation of motion derived in Section 6.2.1 may also be obtained using the conservation of energy method (see Section 3.4 in Chapter 3 and Section 4.3 in Chapter 4). Since the SDOF system is undamped, the principle of conservation of energy is applicable and may lead to a simple derivation of the equation of motion. Consider the SDOF system in Figure 6.5, where a mass m is suspended by a spring of stiffness k. If the motion is measured from the equilibrium position, Figure 6.5(a), the total energy for the system in the general position, Figure 6.5(b), is given by:

$$KE + PE + SE = \frac{1}{2}m\dot{x}^2 - mgx + \frac{1}{2}k(x + \delta_o)^2 - \frac{1}{2}k\delta_o^2 \tag{6.10}$$

where KE is the kinetic energy $KE = \frac{1}{2}m\dot{x}^2$, PE is the potential energy $PE = -mgx$ (datum taken at the equilibrium position) and SE is the strain energy $SE = k(x + \delta_o)^2/2 - k\delta_o^2/2$ where δ_o is the static deflection. From Equation (6.9), the static deflection is $\delta_o = mg/k$, thus Equation (6.10) becomes:

$$KE + PE + SE = \frac{1}{2}m\dot{x}^2 + \frac{1}{2}kx^2 \tag{6.11}$$

The principle of conservation of energy states that the total energy at different positions is identical, thus:

$$KE_1 + PE_1 + SE_1 = KE_2 + PE_2 + SE_2 = KE_3 + PE_3 + SE_3 = \text{Constant}$$

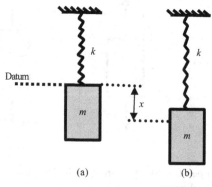

Figure 6.5: An undamped SDOF system in a) the equilibrium position and b) the general position

Therefore, the differentiation of Equation (6.11) with respect to time should be equal to zero:

$$\frac{d}{dt}\left(\frac{1}{2}m\dot{x}^2 + \frac{1}{2}kx^2\right) = 0 \tag{6.12}$$

Differentiating Equation (6.12) gives $m\dot{x}\ddot{x} + k\dot{x}x = 0$, from which

$$m\ddot{x} + kx = 0 \tag{6.13}$$

which is identical to the equation of motion, Equation (6.6).

In the case of simple harmonic motion, i.e. $x = A\sin\omega t$, the energy changes from maximum kinetic energy ($\frac{1}{2}m\dot{x}_{max}^2$, where $\dot{x}_{max}$ is the maximum velocity) and zero potential and strain energies at $x = 0$ (equilibrium position) to maximum potential and strain energies ($\frac{1}{2}kx_{max}^2$, where x_{max} is the maximum displacement) and zero kinetic energy. Applying the conservation of energy principle for these two positions gives:

$$KE_{max} = (PE + SE)_{max}$$

or

$$\frac{1}{2}m\dot{x}_{max}^2 = \frac{1}{2}kx_{max}^2 \tag{6.14}$$

From Equations (6.2) and (6.1), $\dot{x}_{max} = \omega_n A = \omega_n x_{max}$, thus Equation (6.14) becomes:

$$\frac{1}{2}m(\omega_n x_{max})^2 = \frac{1}{2}kx_{max}^2 \Rightarrow \omega_n = \sqrt{\frac{k}{m}} \tag{6.15}$$

which provides a direct determination of the natural frequency of the system.

6.2.3 Time response

The solution of the second-order differential equation, Equation (6.7), can be obtained using a Laplace transform as:

$$x = A\cos\omega_n t + B\sin\omega_n t \tag{6.16}$$

where A and B are constants and can be determined from the initial conditions.

$$\text{At } t = 0, \quad x = x_o \quad \text{and} \quad \dot{x} = x_o \tag{6.17}$$

Differentiating Equation (6.16) with respect to time and substituting the initial conditions from Equation (6.17) yields:

$$\dot{x} = -A\omega_n \sin\omega_n t + B\omega_n \cos\omega_n t$$
$$x_o = A \times \cos 0 + B\sin 0 \Rightarrow x_o = A$$
$$\dot{x}_o = -A\omega_n \sin 0 + B\omega_n \cos 0 \Rightarrow \dot{x}_o = B\omega_n \Rightarrow B = \frac{\dot{x}_o}{\omega_n} \tag{6.18}$$

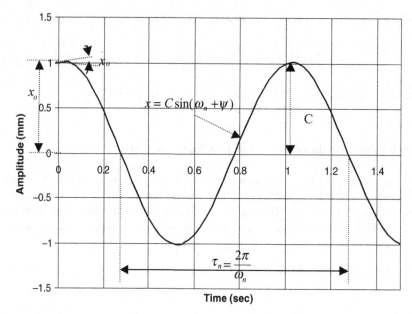

Figure 6.6: A typical time response of an undamped SDOF

Substituting A and B from Equation (6.18) into Equation (6.16) gives:

$$x = x_o \cos \omega_n t + \frac{\dot{x}_o}{\omega_n} \sin \omega_n t \tag{6.19}$$

Equation (6.19) can also be written as:

$$x = C \sin(\omega_n t + \psi) \tag{6.20}$$

where

$$C = \sqrt{x_o^2 + \left(\frac{\dot{x}_o}{\omega_n}\right)^2} \quad \text{and} \quad \psi = \tan^{-1} \frac{x_o}{\dot{x}_o/\omega_n} = \tan^{-1} \frac{x_o \omega_n}{\dot{x}_o}$$

C is the amplitude of vibration and ψ is the phase angle between the initial displacement x_o and initial velocity $\dot{x}_o$.

Figure 6.6 shows the typical time response for an undamped free vibration of an SDOF system. The following parameters can be identified:

$\tau_n = \frac{2\pi}{\omega_n}$ is the period of motion in seconds (s)

$f_n = \frac{1}{\tau_n} = \frac{\omega_n}{2\pi}$ is the natural frequency in Hz (1 Hz = 1 cycle per second)

C is the amplitude of vibration

x_o and $\dot{x}_o$ are the initial displacement and initial velocity, respectively.

6.2.4 Equivalent stiffness for structural systems

Many structural systems can be approximated to a system with a single degree of freedom in order to determine their fundamental natural frequency. In order to do so, the equivalent stiffness and the equivalent mass of the system are required.

The equivalent stiffness can be calculated by dividing the force F by the deflection u at the point where the force F is applied. The force must be applied at the point where the amplitude of the first mode of vibration of the structure takes place. The defection u is calculated from the basic principles of structural analysis and depends on the end conditions, the direction and position of the applied force. In Table 6.1, the formula for the deflection, u, as well as the stiffness, k, is presented for some common structural systems. The usefulness of this table is illustrated in Example 6.1 and Chapter 7.

In order to consider the system's weight, i.e. the uniform distributed load ($\rho A L$), the equivalent mass (m) for the SDOF system can be approximated by equating the kinetic energy of an SDOF system to that of the structural system. The equivalent masses are also given in Table 6.1.

Table 6.1: The SDOF equivalent stiffness and equivalent mass for structural systems

	Deflection u	Stiffness $k = F/u$	Mass m
Axial bar	$u = \dfrac{FL}{EA}$	$k = \dfrac{EA}{L}$	$m = 0.101\rho AL$
Simply supported beam	$u = \dfrac{FL^3}{48EI}$	$k = \dfrac{48EI}{L^3}$	$m = 0.4857\rho AL$
Fixed-free beam	$u = \dfrac{FL^3}{3EI}$	$k = \dfrac{3EI}{L^3}$	$m = 0.2357\rho AL$
Fixed-fixed beam	$u = \dfrac{FL^3}{192EI}$	$k = \dfrac{192EI}{L^3}$	$m = 0.37\rho AL$
Fixed-simply supported beam	$u = \dfrac{7FL^3}{768EI}$	$k = \dfrac{768EI}{7L^3}$	$m = 0.46\rho AL$

E is Young's modulus, A is the cross-section area, L is the bar or beam length, I is the area moment of inertia, F is the applied load, u is the displacement (deflection) at the loaded point and ρ is the density.

Example 6.1 Undamped free vibration: wind turbine

The tower of the wind turbine, Figure 6.7(a), has a height of 54 m and a circular hollow cross-section, Figure 6.7(b), that has an inner diameter of 0.3 m and an outer diameter of 0.5 m. The rotor and hub mass is 15×10^3 kg and the tower is made of steel that has a Young's modulus of 200 GPa. Considering the cases when the tower's weight is ignored and when it is taken into account, determine:

a) the natural frequency of transverse vibration of the system;
b) the time response due to initial transverse displacement $x_o = 0.1$ m;
c) the maximum values of velocity and acceleration.

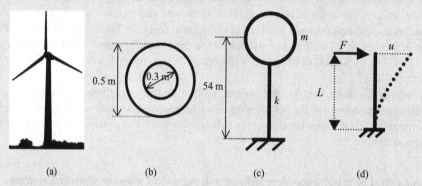

| (a) | (b) | (c) | (d) |

Figure 6.7: Example 6.1 a) a wind turbine, b) its lower cross-section, c) the idealized SDOF and d) its analysis

Solution

a) From structural analysis (Figure 6.7(d) and Table 6.1), the stiffness k of a cantilever beam is:

$$k = \frac{F}{u} = \frac{3EI}{L^3}$$

The area moment of inertia I about the x axis or the y axis (see Appendix B) is:

$$I = \frac{\pi}{64} \left(D_o^4 - D_i^4 \right) = \frac{\pi}{64} (0.5^4 - 0.3^4) = 2.67035 \times 10^{-3} \, \text{m}^4$$

Thus, k is calculated as:

$$k = \frac{3EI}{L^3} = \frac{3 \times 200 \times 10^9 \times 2.67035 \times 10^{-3}}{54^3} = 10175.1 \, \text{N/m}$$

The mass of the tower is:

$$M = \rho AL = 7800 \times \frac{\pi}{4}(0.5^2 - 0.3^2) \times 54 = 52929.55\,\text{kg}$$

If the tower's weight is ignored, $m = 15 \times 10^3$ kg and, from Equation (6.7), the angular frequency is:

$$\omega_n = \sqrt{\frac{k}{m}} = \sqrt{\frac{10175.1}{15 \times 10^3}} = 0.8236\,\text{rad/s}$$

The natural frequency in Hz is:

$$f_n = \frac{\omega_n}{2\pi} = 0.131\,\text{Hz}$$

If the tower's weight is taken into consideration, the total mass of the system is: $m + 0.2357\rho AL = 15 \times 10^3 + 0.2357 \times 52929.55 = 27475.495$ kg and, again from Equation (6.7), the angular frequency is:

$$\omega_n = \sqrt{\frac{k}{m}} = \sqrt{\frac{10175.1}{27475.495}} = 0.60855\,\text{rad/s}$$

The natural frequency in Hz is:

$$f_n = \frac{\omega_n}{2\pi} = 0.097\,\text{Hz}$$

b) From Equation (6.20), the time response is given by:

$$x = C \sin(\omega_n t + \psi)$$

where

$$C = \sqrt{x_o^2 + \left(\frac{\dot{x}_o}{\omega_n}\right)^2} = \sqrt{0.1^2 + (0)^2} = 0.1\,\text{m} \quad \text{and} \quad \psi = \tan^{-1}\frac{x_o}{\dot{x}_o/\omega_n}$$

$$= \tan^{-1}\frac{0.1 \times \omega_n}{0} = \frac{\pi}{2}$$

If the tower's weight is ignored, the time response, as shown in Figure 6.8(a), is:

$$x = 0.1\sin\left(0.8236t + \frac{\pi}{2}\right)\,\text{m}$$

If the tower's weight is taken into consideration, the time response, as shown in Figure 6.8(b), is:

$$x = 0.1\sin\left(0.60855t + \frac{\pi}{2}\right)\,\text{m}$$

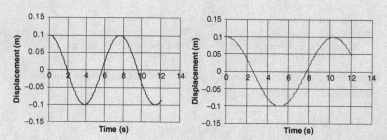

Figure 6.8: Time response for turbine a) ignoring and b) including weight of tower

c) The velocity and the acceleration are obtained by differentiating the displacement time response with respect to time.

If the tower's weight is ignored, the velocity (as shown in Figure 6.9(a)) is:

$$\dot{x} = C \times \omega_n \cos(\omega_n t + \psi) = 0.1 \times 0.8236 \cos\left(0.8236t + \frac{\pi}{2}\right)$$

The maximum velocity is:

$$\dot{x}_{max} = C \times \omega_n = 0.1 \times 0.8236 = 0.082\,\text{m/s}$$

The acceleration (as shown in Figure 6.9(b)) is obtained by further differentiating the velocity with respect to time:

$$\ddot{x} = -C \times \omega_n^2 \sin(\omega_n t + \psi) = -0.1 \times 0.8236^2 \sin\left(0.8236t + \frac{\pi}{2}\right)$$

The maximum acceleration is:

$$\ddot{x}_{max} = C \times \omega_n^2 = 0.1 \times 0.8236^2 = 0.068\,\text{m/s}^2$$

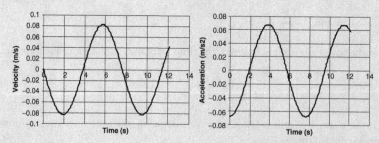

Figure 6.9: a) The velocity and b) the acceleration for turbine, ignoring weight of tower

If the tower's weight is taken into consideration, the velocity (as shown in Figure 6.10(a)) is:

$$\dot{x} = C \times \omega_n \cos(\omega_n t + \psi) = 0.1 \times 0.60855 \cos\left(0.60855t + \frac{\pi}{2}\right)$$

The maximum velocity is:

$$\dot{x}_{max} = C \times \omega_n = 0.1 \times 0.60855 = 0.061 \, \text{m/s}$$

The acceleration (as shown in Figure 6.10(b)) is obtained by further differentiating the velocity with respect to time:

$$\ddot{x} = -C \times \omega_n^2 \sin(\omega_n t + \psi) = -0.1 \times 0.60855^2 \sin\left(0.60855t + \frac{\pi}{2}\right)$$

The maximum acceleration is:

$$\ddot{x}_{max} = C \times \omega_n^2 = 0.1 \times 0.60855^2 = 0.037 \, \text{m/s}^2$$

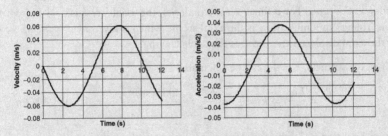

Figure 6.10: a) The velocity and b) the acceleration for turbine, including weight of tower

Example 6.2 Undamped free vibration: lift

A lift cage, shown in Figure 6.11, has a mass of 1100 kg and is suspended from a cable made of steel that has a Young's modulus of 200 GPa. At the position shown, the cable is stationary, has a length of 50 m and starts its motion with an initial velocity of 3 m/s downwards. If the natural frequency of the system should not exceed 2.5 Hz, determine the minimum cross-section area of the cable and the amplitude of vibration of the cage.

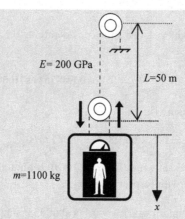

Figure 6.11: Representation of Example 6.2

Solution

For a cable in tension, the axial stress is $\sigma = \frac{P}{A}$, where P is the load and A is the cross-section area. For a linear elastic material, the axial stress is $\sigma = E\varepsilon$, where E is Young's modulus and ε is the axial strain, which is given by $\varepsilon = \frac{\delta_o}{L}$, where δ_o is the axial displacement and L is the cable length.

Thus, the stiffness, $k = \frac{P}{\delta_o}$, can be calculated as:

$$\sigma = \frac{P}{A} = \frac{E\delta_o}{L} \Rightarrow k = \frac{P}{\delta_o} = \frac{EA}{L} = \frac{200 \times 10^9 \times A}{50} = 4 \times 10^9 A \, \text{N/m}$$

From Equation (6.7), the angular frequency is given by:

$$\omega_n = \sqrt{\frac{k}{m}} \Rightarrow 2.5 \times 2\pi = \sqrt{\frac{4 \times 10^9 A}{1100}}$$

The cross-section area, A, is calculated as:

$$A = 67.85 \times 10^{-6} \, \text{m}^2 = 67.85 \, \text{mm}^2$$

From Equation (6.20), the time response is given by:

$$x = C\sin(\omega_n t + \psi)$$

where C is the amplitude of vibration and is calculated as:

$$C = \sqrt{x_o^2 + \left(\frac{\dot{x}_o}{\omega_n}\right)^2} = \sqrt{0^2 + \left(\frac{3}{2.5 \times 2\pi}\right)^2} = 0.19 \, \text{m}$$

Example 6.3 Undamped free vibration: car

A car has a natural frequency of 3.5 Hz without passengers and 3.25 Hz with passengers. If the mass of the car is 1.2×10^3 kg, determine the stiffness of the car and the mass of the passengers.

Solution

From Equation (6.7), the angular frequency of the car without passengers is:

$$\omega_n = \sqrt{\frac{k}{m}} \Rightarrow 3.5 \times 2\pi = \sqrt{\frac{k}{1200}}$$

from which the stiffness is calculated as:

$$k = 580333 \quad N = 580.3 \, kN$$

The angular frequency of the car with passengers of mass m_p is:

$$3.25 \times 2\pi = \sqrt{\frac{k}{m + m_p}} \Rightarrow 3.25 \times 2\pi = \sqrt{\frac{580333}{1200 + m_p}}$$

From which the mass of the passengers is calculated as:

$$m_p = 191.72 = 192 \, kg$$

Example 6.4 Undamped free vibration: beams

A cantilever beam of length L_1 is supported by a simply supported beam of length L_2 at its midpoint C, as shown in Figure 6.12. Both beams have identical Young's modulus E and area moment of inertia I. If the system carries a mass m at point C, determine the angular frequency of the system in terms of E, I, m, L_1 and L_2.

Solution

The stiffness of the system has a contribution from the cantilever beam and the simply supported beam:

$$k = k_1 + k_2$$

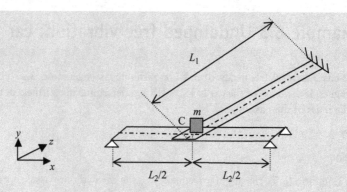

Figure 6.12: Representation of Example 6.4

where k is the overall beam stiffness, k_1 is the stiffness of the cantilever beam and k_2 is the stiffness of the simply supported beam. From Table 6.1, k_1 is:

$$k_1 = \frac{3EI}{L_1^3}$$

and k_2 is:

$$k_2 = \frac{48EI}{L_2^3}$$

Thus, the overall beam stiffness is $k = \frac{3EI}{L_1^3} + \frac{48EI}{L_2^3}$ and the angular frequency of the system is obtained as:

$$\omega_n = \sqrt{\frac{k}{m}} = \sqrt{\frac{\frac{3EI}{L_1^3} + \frac{48EI}{L_2^3}}{m}}$$

which gives

$$\omega_n = \sqrt{\frac{3EI}{m}\left(\frac{16L_1^3 + L_2^3}{L_1^3 \times L_2^3}\right)}$$

6.2.5 Systems with rotational degree of freedom

A pendulum

A simple pendulum can be analyzed as having a single degree of freedom, where the degree of freedom is the angle θ as shown in Figure 6.13. The moment equation equivalent to Newton's second law of motion (Equation (4.19)) is applicable as:

$$\sum M_o = I_o \ddot{\theta} \tag{6.21}$$

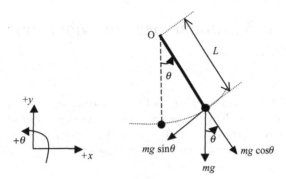

Figure 6.13: A pendulum

where $\sum M_o$ is the sum of moment about the pivot O, I_o is the mass moment of inertia about O and $\ddot{\theta}$ is the angular acceleration.

Taking moment about O and applying Equation (6.21) gives:

$$-mg\,L\sin\theta = I_0\ddot{\theta} \tag{6.22}$$

Rearranging Equation (6.22), and considering small oscillation, i.e., $\sin\theta \approx \theta$ yields:

$$I_0\ddot{\theta} + mg\,L\theta = 0 \tag{6.23}$$

or

$$\ddot{\theta} + \omega_n^2\theta = 0 \tag{6.24}$$

where

$$\omega_n = \sqrt{\frac{mg\,L}{I_o}} \text{ is the angular frequency in rad/s.}$$

If the mass moment of inertia can be written as $I_o = mk_g^2$, where k_g is the radius of gyration (see Equation (4.33)), the angular frequency becomes:

$$\omega_n = \sqrt{\frac{mg\,L}{mk_g^2}} = \sqrt{\frac{g\,L}{k_g^2}}$$

or

$$\frac{2\pi}{\tau_n} = \sqrt{\frac{g\,L}{k_g^2}} \tag{6.25}$$

where τ_n is the period of oscillation. For a simple pendulum, where the mass is concentrated at the end, $k_g = L$, the angular frequency becomes:

$$\omega_n = \sqrt{\frac{g}{L}} \tag{6.26}$$

Example 6.5 Undamped free vibration: bell crank

A bell crank ABC, shown in Figure 6.14, is supported at point A by a spring of stiffness k and at point B by a hinged pin support. A mass m is attached at point C and the crank is in static equilibrium when the link BC is in a horizontal position. Write the equation of motion for the bell crank for small oscillation angle θ. Ignore the crank weight.

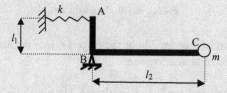

Figure 6.14: Representation of Example 6.5

Solution

The free-body diagram of the bell crank in its static equilibrium can be obtained as shown in Figure 6.15, where δ_o is the static deflection. Taking moment about point B:

$$\sum M_B = -mg \times l_2 - k\delta_o \times l_1 = 0$$

from which

$$\delta_o = -\frac{mg}{k} \times \frac{l_2}{l_1} \tag{E6.5a}$$

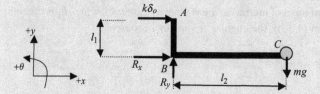

Figure 6.15: Free-body diagram of Example 6.5

The equation of motion can be obtained by considering the vibrating system and its free-body diagram in Figure 6.16, taking moment about B and applying Equation (6.21):

$$\sum M_B = -mg \times l_2 - k(l_1\theta + \delta_o) \times l_1 = I_B\ddot{\theta} \tag{E6.5b}$$

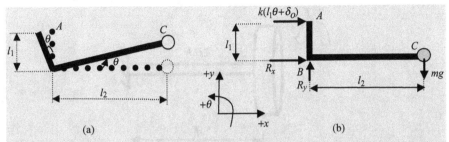

Figure 6.16: Example 6.5 a) SDOF system of bell crank and b) its free-body diagram

Substituting Equation (E6.5a) into Equation (E6.5b) gives:

$$-mg \times l_2 - k\left(l_1\theta - \frac{mg}{k} \times \frac{l_2}{l_1}\right) \times l_1 = I_B\ddot{\theta}$$

which can be rearranged to give:

$$\ddot{\theta} + \frac{kl_1^2}{I_B}\theta = 0 \qquad\qquad (E6.5c)$$

The mass moment of inertia of a particle, of mass m at point C, about point B is $I_B = ml_2^2$, thus the equation of motion, Equation (E6.5c) becomes:

$$\ddot{\theta} + \frac{kl_1^2}{ml_2^2}\theta = 0$$

or

$$\ddot{\theta} + \omega_n^2\theta = 0$$

where the angular frequency is:

$$\omega_n^2 = \frac{k}{m} \times \left(\frac{l_1}{l_2}\right)^2 \Rightarrow \omega_n = \frac{l_1}{l_2}\sqrt{\frac{k}{m}}$$

A rotor on a fixed shaft

The torsional vibration of a single rotor on a fixed shaft can be analyzed as a single degree of freedom, where the degree of freedom is again the angle θ as shown in Figure 6.17. Applying the moment equation equivalent to Newton's second law (Equation (4.19)) for the system in Figure 6.17 gives:

$$\sum T = J\ddot{\theta} \qquad\qquad (6.27)$$

where $\sum T$ is the sum of the torques, J is the mass moment of inertia of the rotor (or disc) and $\ddot{\theta}$ is the torsional angular acceleration.

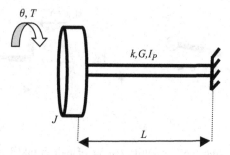

Figure 6.17: Torsional vibration of a shaft

From static torsional analysis, the stiffness k is given by:

$$k = \frac{T}{\theta} = \frac{GI_p}{L} \tag{6.28}$$

where G is the shear modulus of the shaft material and Ip is the polar moment of inertia of the shaft. From Figure 6.17, the sum of the torques is:

$$-k\theta = J\ddot{\theta} \tag{6.29}$$

Rearranging Equation (6.29) yields:

$$J\ddot{\theta} + k\theta = 0$$

or

$$\ddot{\theta} + \omega_n^2\theta = 0 \tag{6.30}$$

where $\omega_n = \sqrt{\frac{k}{J}}$ is the angular frequency in rad/s.

From Equation (6.28), using $k = \frac{GI_p}{L}$, the angular frequency ω_n can also be written as:

$$\omega_n = \sqrt{\frac{GI_p}{LJ}} \tag{6.31}$$

Example 6.6 Undamped free vibration: torsional vibration

The rotor of a wind turbine is mounted on a uniform hollow shaft of inner and outer diameters 0.2 m and 0.4 m, respectively. The shaft is made of steel and has a shear modulus of 77 GPa. The rotor has a mass of 12×10^3 kg and a radius of gyration of 18 m. If a brake is installed at a distance of 4 m behind the rotor, as shown in Figure 6.18, determine the period of free torsional vibration of the rotor when the brake is stopping the rotor rotation (ignore the shaft weight).

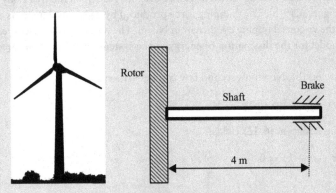

Figure 6.18: Representation of Example 6.6

Solution

The polar moment of inertia of the shaft (see Appendix B) is:

$$I_p = \frac{\pi}{32}\left(D_o^4 - D_i^4\right) = \frac{\pi}{32}(0.4^4 - 0.2^4) = 2.356 \times 10^{-3}\,\text{m}^4$$

From Table 4.1 and Equation (4.33), the rotor mass moment of inertia J is calculated as:

$$J = mk_g^2 = 12000 \times 18^2 = 3\,888\,000\,\text{kg.m}^2$$

The shear modulus of the shaft is:

$$G = 77\,\text{GPa} = 77 \times 10^9\,\text{N/m}^2$$

From Equation (6.31), the angular frequency can be calculated as:

$$\omega_n = \sqrt{\frac{GI_p}{LJ}} = \sqrt{\frac{77 \times 10^9 \times 2.356 \times 10^{-3}}{4 \times 3\,888\,000}} = 3.415\,\text{rad/s}$$

and the period of free vibration is:

$$\tau_n = \frac{2\pi}{\omega_n} = \frac{2\pi}{3.415} = 1.84\,\text{s}$$

6.3 VISCOUS DAMPED FREE VIBRATION

6.3.1 Deriving the equation of motion using Newton's second law

For a damped free vibration SDOF, the force f shown in Figure 6.2 is equal to zero and the system is reduced to that shown in Figure 6.19(a), for which the free-body diagram is drawn in Figure 6.19(b). The damping force produced by a viscous damper is equal to $c\dot{x}$, where c is the viscous damping coefficient in N.s/m. The damping coefficient, c, represents a simple model for the dissipation of energy in the system due to friction forces.

From Figure 6.9(b), Newton's second law can be written as:

$$-kx - c\dot{x} = m\ddot{x} \tag{6.32}$$

Rearranging Equation (6.32) yields:

$$m\ddot{x} + c\dot{x} + kx = 0 \tag{6.33}$$

or

$$\ddot{x} + 2\zeta\omega_n\dot{x} + \omega_n^2 x = 0 \tag{6.34}$$

where $\omega_n = \sqrt{\frac{k}{m}}$ is the angular frequency in rad/s and ζ is the damping ratio and is equal to:

$$\zeta = \frac{c}{2m\omega_n} \tag{6.35}$$

Defining the critical damping coefficient, c_c, as:

$$c_c = 2m\omega_n \tag{6.36}$$

Thus, ζ is:

$$\zeta = \frac{c}{c_c} \tag{6.37}$$

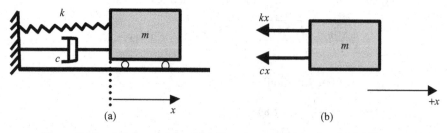

Figure 6.19: a) A damped free SDOF system and b) its free-body diagram

6.3.2 Time response

The solution of Equation (6.34) is obtained using the Laplace transform and has the following form:

$$x = Ae^{\lambda t} \tag{6.38}$$

Differentiating Equation (6.38) twice ($\dot{x} = A\lambda e^{\lambda t}$, $\ddot{x} = A\lambda^2 e^{\lambda t}$) and substituting into Equation (6.34) yields:

$$\lambda^2 + 2\zeta\omega_n\lambda + \omega_n^2 = 0 \tag{6.39}$$

The roots of the quadratic Equation (6.39) are:

$$\lambda_1 = \omega_n(-\zeta + \sqrt{\zeta^2 - 1}), \quad \lambda_2 = \omega_n(-\zeta - \sqrt{\zeta^2 - 1}) \tag{6.40}$$

By adding the two solutions together (superposition), Equation (6.38) becomes:

$$x = A_1 e^{\lambda_1 t} + A_2 e^{\lambda_2 t}$$

or

$$x = A_1 e^{(-\zeta + \sqrt{\zeta^2 - 1})\omega_n t} + A_2 e^{(-\zeta - \sqrt{\zeta^2 - 1})\omega_n t} \tag{6.41}$$

There are three categories of damped motion that can be defined:

- Overdamped systems ($\zeta > 1$)
- Critically damped systems ($\zeta = 1$)
- Underdamped systems ($\zeta < 1$)

Overdamped systems

An overdamped system is achieved when $\zeta > 1$, which means that λ_1 and λ_2 are real negative numbers. In such a case, the time response is in the form $x = Ae^{-at}$, where a is a positive real number. Thus, the motion decays so that x approaches zero when $t \Rightarrow \infty$, i.e. there is no oscillation.

The constants A_1 and A_2 in Equation (6.41) can be determined from initial conditions, at $t = 0$, $x = x_o$ and $\dot{x} = \dot{x}_o$. Substituting the first initial condition, $t = 0$, $x = x_o$, into Equation (6.41) gives:

$$x_o = A_1 + A_2 \tag{6.42}$$

Differentiating Equation (6.41) with respect to time and substituting the second initial condition, $t = 0$, $\dot{x} = \dot{x}_o$, yields:

$$\dot{x} = (-\zeta + \sqrt{\zeta^2 - 1})\omega_n A_1 e^{(-\zeta + \sqrt{\zeta^2 - 1})\omega_n t} + (-\zeta - \sqrt{\zeta^2 - 1})\omega_n A_2 e^{(-\zeta - \sqrt{\zeta^2 - 1})\omega_n t}$$

$$\dot{x}_o = (-\zeta + \sqrt{\zeta^2 - 1})\omega_n A_1 + (-\zeta - \sqrt{\zeta^2 - 1})\omega_n A_2 \tag{6.43}$$

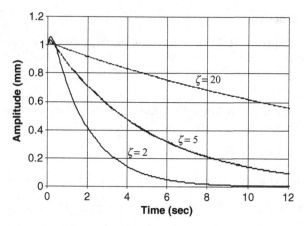

Figure 6.20: Displacement time response of an overdamped system

Solving Equations (6.42) and (6.43), in two unknowns, A_1 and A_2, gives:

$$A_1 = \frac{\dot{x}_o + x_o \omega_n (\zeta + \sqrt{\zeta^2 - 1})}{2\omega_n \sqrt{\zeta^2 - 1}}$$

$$A_2 = \frac{-\dot{x}_o - x_o \omega_n (\zeta - \sqrt{\zeta^2 - 1})}{2\omega_n \sqrt{\zeta^2 - 1}} \tag{6.44}$$

Substituting Equation (6.44) into Equation (6.41), the displacement time response for an overdamped SDOF system is obtained as:

$$x = \frac{\dot{x}_o + x_o \omega_n (\zeta + \sqrt{\zeta^2 - 1})}{2\omega_n \sqrt{\zeta^2 - 1}} e^{(-\zeta + \sqrt{\zeta^2 - 1})\omega_n t}$$

$$- \frac{\dot{x}_o + x_o \omega_n (\zeta - \sqrt{\zeta^2 - 1})}{2\omega_n \sqrt{\zeta^2 - 1}} e^{(-\zeta - \sqrt{\zeta^2 - 1})\omega_n t} \tag{6.45}$$

A typical displacement time response for an overdamped SDOF system is shown in Figure 6.20 for $\zeta = 2$, $\zeta = 5$ and $\zeta = 20$. It can be seen that the higher the damping ratio, ζ, the longer the time to decay. For heavily overdamped systems, ζ is much larger than 1, $\sqrt{\zeta^2 - 1} \approx \zeta$ and Equation (6.45) becomes:

$$x = \frac{\dot{x}_o + 2\zeta x_o \omega_n}{2\omega_n \zeta} - \frac{\dot{x}_o}{2\omega_n \zeta} e^{-2\zeta \omega_n t}$$

or

$$x = x_o + \frac{\dot{x}_o}{2\omega_n \zeta} \left(1 - \frac{\dot{x}_o}{2\omega_n \zeta} e^{-2\zeta \omega_n t} \right) \tag{6.46}$$

Critically damped systems

For a critically damped system, the damping ratio is equal to unity ($\zeta = 1$), which means that $\lambda_1 = \lambda_2 = -\omega_n$ (see Equation (6.40) for $\zeta = 1$) and the time response, Equation (6.41), is in the form (the solution for two equal roots):

$$x = (A_1 + A_2 t)e^{-\omega_n t} \tag{6.47}$$

Again the motion decays so that x approaches zero when $t \Rightarrow \infty$, i.e. there is no oscillation.

For critically damped systems, the constants A_1 and A_2 in Equation (6.47) can be determined from initial conditions, i.e. $t = 0$, $x = x_o$ and $\dot{x} = \dot{x}_o$. Substituting the first initial condition, $t = 0$, $x = x_o$ into Equation (6.47) gives:

$$A_1 = x_o \tag{6.48}$$

Differentiating Equation (6.47) with respect to time and substituting the second initial condition $t = 0$, $\dot{x} = \dot{x}_o$ yields:

$$\dot{x} = A_2 e^{-\omega_n t} - \omega_n(A_1 + A_2 t)e^{-\omega_n t}$$
$$\dot{x}_o = A_2 - \omega_n A_1 \tag{6.49}$$

Thus, A_2 is:

$$A_2 = \dot{x}_o + \omega_n A_1 \tag{6.50}$$

Substituting Equations (6.48) and (6.50) into Equation (6.47), the displacement time response for a critically damped system is obtained as:

$$x = (x_o + (\dot{x}_o + \omega_n x_o)t)e^{-\omega_n t} \tag{6.51}$$

A typical displacement time response for a critically damped SDOF system is shown in Figure 6.21.

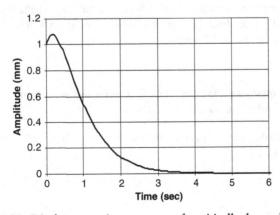

Figure 6.21: Displacement time response of a critically damped system

Underdamped systems

For an underdamped system, $\zeta < 1$, so that $\zeta^2 - 1 < 0$ and $\sqrt{\zeta^2 - 1} = i\sqrt{1 - \zeta^2}$, where $i = \sqrt{-1}$. Using the relationship $e^{a+b} = e^a e^b$, Equation (6.41) can be re-written as:

$$x = \left[A_1 e^{i\sqrt{1-\zeta^2}\omega_n t} + A_2 e^{-i\sqrt{1-\zeta^2}\omega_n t} \right] e^{-\zeta \omega_n t} \tag{6.52}$$

defining the damped natural frequency, ω_d as:

$$\omega_d = \omega_n \sqrt{1 - \zeta^2} \tag{6.53}$$

Substituting Equation (6.53) into Equation (6.52) gives:

$$x = \left[A_1 e^{i\omega_d t} + A_2 e^{-i\omega_d t} \right] e^{-\zeta \omega_n t} \tag{6.54}$$

Using Euler's formula ($e^{\pm ix} = \cos x \pm i \sin x$) and re-writing Equation (6.54) gives:

$$x = \left[A_1 (\cos \omega_d t + i \sin \omega_d t) + A_2 (\cos \omega_d t - i \sin \omega_d t) \right] e^{-\zeta \omega_n t}$$

or

$$x = \left[A_3 \cos \omega_d t + A_4 \sin \omega_d t \right] e^{-\zeta \omega_n t} \tag{6.55}$$

where A_3 and A_4 are new constants, i.e. $A_3 = A_1 + A_2$ and $A_4 = i(A_1 - A_2)$.

Another form of Equation (6.55) is:

$$x = D e^{-\zeta \omega_n t} \sin(\omega_d t + \theta) \tag{6.56}$$

where

$$D = \sqrt{A_3^2 + A_4^2} \quad \text{and} \quad \theta = \tan^{-1} \frac{A_3}{A_4}$$

In order to determine the constants A_3 and A_4 in Equation (6.55), the initial conditions at $t = 0$, $x = x_o$ and $\dot{x} = \dot{x}_o$ should be applied. Substituting the first initial condition, $t = 0$ and $x = x_o$ into Equation (6.55) gives:

$$A_3 = x_o \tag{6.57}$$

Differentiating Equation (6.55) with respect to time and substituting the second initial condition, $t = 0$ and $\dot{x} = \dot{x}_o$ yields:

$$\dot{x} = (-\omega_d A_3 \sin \omega_d t + \omega_d A_4 \cos \omega_d t) e^{-\zeta \omega_n t} - \zeta \omega_n (A_3 \cos \omega_d t + A_4 \sin \omega_d t) e^{-\zeta \omega_n t}$$

$$\dot{x}_o = \omega_d A_4 - \zeta \omega_n A_3 \tag{6.58}$$

Thus, A_4 is

$$A_4 = \frac{\dot{x}_o + \zeta \omega_n A_3}{\omega_d} \tag{6.59}$$

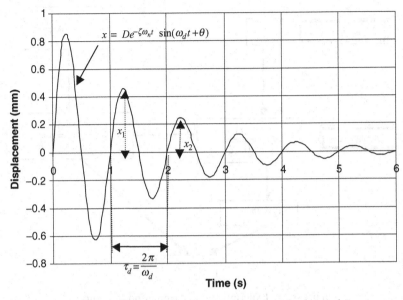

Figure 6.22: Displacement time response of an underdamped system

Substitute Equations (6.57) and (6.59) into Equation (6.55) to obtain the displacement time response for an underdamped system as:

$$x = \left[x_o \cos \omega_d t + \frac{\dot{x}_o + \zeta \omega_n x_o}{\omega_d} \sin \omega_d t \right] e^{-\zeta \omega_n t} \qquad (6.60)$$

An example of a displacement time response of an underdamped free vibration of an SDOF system is shown in Figure 6.22, where the period of motion in seconds is defined as:

$$\tau_d = \frac{2\pi}{\omega_d} \qquad (6.61)$$

The damping ratio ζ can be measured experimentally and hence the damping coefficient c can be evaluated. Considering Figure 6.22, the ratio of two successive amplitudes, x_1 and x_2 can be written (using Equation (6.56)) as:

$$\frac{x_1}{x_2} = \frac{D e^{-\zeta \omega_n t_1} \sin(\omega_d t_1 + \theta)}{D e^{-\zeta \omega_n (t_1 + \tau_d)} \sin(\omega_d (t_1 + \tau_d) + \theta)}$$

which yields:

$$\frac{x_1}{x_2} = e^{\zeta \omega_n \tau_d} \qquad (6.62)$$

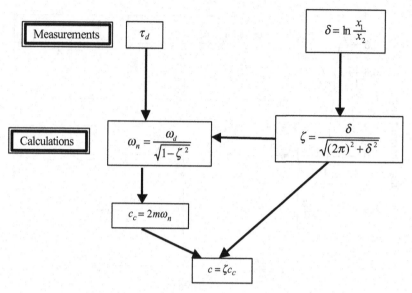

Figure 6.23: Experimental determination of ζ and c

defining a logarithmic decrement δ such that:

$$\delta = \ln \frac{x_1}{x_2} = \zeta \omega_n \tau_d \tag{6.63}$$

Substituting Equation (6.53) into Equation (6.61) and then into Equation (6.63), and solving for ζ gives:

$$\zeta = \frac{\delta}{\sqrt{(2\pi)^2 + \delta^2}} \tag{6.64}$$

A procedure for determining ζ and c from measurements and calculations is summarized in Figure 6.23.

Example 6.7 Damped free vibration I

A vehicle suspension, consisting of a spring, a shock absorber and a linkage, has to carry a mass of 300 kg. It is idealized to the SDOF shown in Figure 6.24. If the damped period of vibration is 1.2 s and the ratio between the two amplitudes of

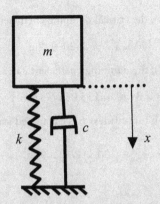

Figure 6.24: Representation of Example 6.7

vibration in two successive cycles is measured as $\frac{x_2}{x_1} = \frac{1}{15}$, determine:

a) the logarithmic decrement;
b) the damping ratio;
c) the damped angular frequency;
d) the undamped angular frequency;
e) the critical damping coefficient;
f) the damping coefficient;
g) the stiffness.

Solution

a) From Equation (6.63), the logarithmic decrement δ is:

$$\delta = \ln \frac{x_1}{x_2} = \ln 15 = 2.708$$

b) From Equation (6.64), the damping ratio ζ is:

$$\zeta = \frac{\delta}{\sqrt{(2\pi)^2 + \delta^2}} = \frac{2.708}{\sqrt{(2\pi)^2 + 2.708^2}} = 0.3958$$

c) From Equation (6.61), the damped angular frequency ω_d is calculated as:

$$\tau_d = \frac{2\pi}{\omega_d} = 1.2 \Rightarrow \omega_d = \frac{2\pi}{1.2} = 1.667\pi = 5.236 \, \text{rad/s}$$

d) From Equation (6.53), the undamped angular frequency ω_n is calculated as:

$$\omega_n = \frac{\omega_d}{\sqrt{1 - \zeta^2}} = \frac{1.667\pi}{\sqrt{1 - 0.3958^2}} = 5.7 \, \text{rad/s}$$

e) From Equation (6.36), the critical damping coefficient c_c is:

$$c_c = 2m\omega_n = 2 \times 300 \times 5.7 = 3420 \,\text{N.s/m}$$

f) From Equation (6.37), the damping coefficient c is:

$$c = \zeta c_c = 0.3958 \times 3420 = 1353.64 \,\text{N.s/m}$$

g) From Equation (6.34), the stiffness k is calculated as:

$$\omega_n = \sqrt{\frac{k}{m}} \Rightarrow k = m\omega_n^2 = 300 \times 5.7^2 = 9747 \,\text{N/m}$$

Example 6.8 Damped free vibration II

The damped SDOF system shown in Figure 6.25, has an undamped angular frequency of ω_n rad/s. If the mass m is released from an initial displacement of x_0 and zero initial velocity, calculate the displacement as a function of ω_n and x_0, the damping ratio ζ and the time t for the following cases:

a) an underdamped system;
b) a critically damped system;
c) an overdamped system.

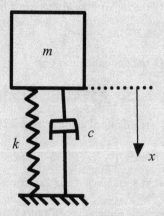

Figure 6.25: Representation of Question 6.8

Solution

a) For an underdamped system, the displacement time response is obtained using Equation (6.60) as:

$$x = \left[x_o \cos \omega_d t + \frac{\dot{x}_o + \zeta \omega_n x_o}{\omega_d} \sin \omega_d t \right] e^{-\zeta \omega_n t}$$

Applying the initial conditions, $t = 0$, $\dot{x}_o = 0$ gives:

$$x = \left[x_o \cos \omega_d t + \frac{\zeta \omega_n x_o}{\omega_d} \sin \omega_d t \right] e^{-\zeta \omega_n t}$$

Substituting $\omega_d = \sqrt{1 - \zeta^2} \omega_n$, from Equation (6.53), gives:

$$x = \left[x_o \cos \sqrt{1 - \zeta^2} \omega_n t + \frac{\zeta x_o}{\sqrt{1 - \zeta^2}} \sin \sqrt{1 - \zeta^2} \omega_n t \right] e^{-\zeta \omega_n t}$$

b) For a critically damped system, the displacement time response is obtained using Equation (6.51):

$$x = (x_o + (\dot{x}_o + \omega_n x_o)t)e^{-\omega_n t}$$

Applying initial conditions, $t = 0$, $\dot{x}_o = 0$ gives:

$$x = x_o(1 + \omega_n t)e^{-\omega_n t}$$

c) For an overdamped system, the displacement time response is obtained using Equation (6.45):

$$x = \frac{\dot{x}_o + x_o \omega_n(\zeta + \sqrt{\zeta^2 - 1})}{2\omega_n \sqrt{\zeta^2 - 1}} e^{(-\zeta + \sqrt{\zeta^2 - 1})\omega_n t}$$
$$- \frac{\dot{x}_o + x_o \omega_n(\zeta - \sqrt{\zeta^2 - 1})}{2\omega_n \sqrt{\zeta^2 - 1}} e^{(-\zeta - \sqrt{\zeta^2 - 1})\omega_n t}$$

Applying the initial conditions, $t = 0$, $\dot{x} = 0$ gives:

$$x = \frac{x_o(\zeta + \sqrt{\zeta^2 - 1})}{2\sqrt{\zeta^2 - 1}} e^{(-\zeta + \sqrt{\zeta^2 - 1})\omega_n t} - \frac{x_o(\zeta - \sqrt{\zeta^2 - 1})}{2\sqrt{\zeta^2 - 1}} e^{(-\zeta - \sqrt{\zeta^2 - 1})\omega_n t}$$

Example 6.9 Damped free vibration III

Find the damping ratio for the underdamped SDOF system in Figure 6.26, if its damped natural frequency is 1 Hz, and the ratio between its amplitude of vibration in the first cycle and that after two cycles is 1:8, i.e. $x_3 = \frac{x_1}{8}$, where x_3 and x_1 are as shown in Figure 6.26.

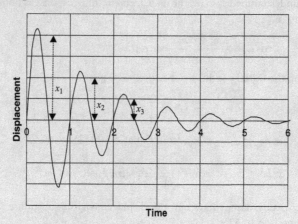

Figure 6.26: Representation of Example 6.9

Solution

From Equation (6.56), the ratio between the first and the third amplitudes is given by:

$$\frac{x_1}{x_3} = \frac{De^{-\zeta\omega_n t_1}\sin(\omega_d t_1 + \theta)}{De^{-\zeta\omega_n(t_1 + 2\tau_d)}\sin(\omega_d(t_1 + 2\tau_d) + \theta)}$$

This gives:

$$\frac{x_1}{x_3} = e^{2\zeta\omega_n\tau_d} \quad \text{and} \quad \ln\frac{x_1}{x_3} = 2\zeta\omega_n\tau_d = \ln 8$$

As $\delta = \zeta\omega_n\tau_d$, from Equation (6.63), $\delta = \frac{\ln 8}{2} = 1.03972$. From Equation (6.64), the damping ratio is calculated as:

$$\zeta = \frac{\delta}{\sqrt{(2\pi)^2 + \delta^2}} = \frac{1.03972}{\sqrt{(2\pi)^2 + 1.03972^2}} = 0.163$$

Example 6.10 Damped free vibration IV

A mass of 10 kg has an initial displacement of 0.1 m, is suspended from a spring of stiffness 40 N/m and has a viscous damper of damping coefficient 20 N.s/m as shown in Figure 6.27. If the mass is released from rest, determine its displacement after 1 s.

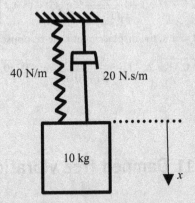

Figure 6.27: Representation of Example 6.10

Solution

From Equation (6.34), the undamped angular frequency ω_n is:

$$\omega_n = \sqrt{\frac{k}{m}} = \sqrt{\frac{40}{10}} = 2 \, \text{rad/s}$$

From Equation (6.36), the critical damping coefficient c_c is:

$$c_c = 2m\omega_n = 2 \times 10 \times 2 = 40 \, \text{N.s/m}$$

From Equation (6.37), the damping ratio is:

$$\zeta = \frac{c}{c_c} = \frac{20}{40} = 0.5$$

The system is underdamped ($\zeta < 1$) thus, using Equation (6.60), the displacement time response is obtained as:

$$x = \left[x_o \cos \omega_d t + \frac{\dot{x}_o + \zeta \omega_n x_o}{\omega_d} \sin \omega_d t \right] e^{-\zeta \omega_n t}$$

Applying initial conditions, $t = 0$, $\dot{x}_o = 0$ gives:

$$x = \left[x_o \cos \omega_d t + \frac{\zeta \omega_n x_o}{\omega_d} \sin \omega_d t \right] e^{-\zeta \omega_n t}$$

Substituting $\omega_d = \sqrt{1 - \zeta^2}\,\omega_n$ (Equation (6.53)) gives:

$$x = \left[x_o \cos\sqrt{1 - \zeta^2}\,\omega_n t + \frac{\zeta x_o}{\sqrt{1 - \zeta^2}}\sin\sqrt{1 - \zeta^2}\,\omega_n t \right] e^{-\zeta\omega_n t}$$

For $x_o = 0.1$ m and $t = 1$ s, the displacement time response is calculated as:

$$x = \left[0.1 \times \cos\sqrt{1 - 0.5^2} \times 2 \times 1 + \frac{0.5 \times 0.1}{\sqrt{1 - 0.5^2}}\sin\sqrt{1 - 0.5^2} \times 2 \times 1 \right] e^{-0.5 \times 2 \times 1}$$

$$= 0.015\,\text{m} = 15\,\text{mm}$$

Example 6.11 Damped free vibration V

A railway boxcar of a mass of 2.1×10^3 kg is travelling at a constant speed of 8 m/s when it hits a buffer (a spring and damper system) at the end of the track as shown in Figure 6.28. If the spring has a stiffness of 44 kN/m and the damping coefficient of the damper is 22 kN.s/m, determine the maximum displacement of the boxcar.

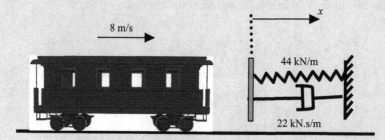

Figure 6.28: Representation of Example 6.11

Solution

From Equation (6.34), the undamped angular frequency ω_n is:

$$\omega_n = \sqrt{\frac{k}{m}} = \sqrt{\frac{44\,000}{2100}} = 4.577\,\text{rad/s}$$

From Equation (6.36), the critical damping coefficient c_c is:

$$c_c = 2m\omega_n = 2 \times 2100 \times 4.5772 = 19\,225\,\text{N.s/m}$$

From Equation (6.37), the damping ratio is:

$$\zeta = \frac{c}{c_c} = \frac{22\,000}{19\,225} = 1.144$$

The system is overdamped ($\zeta > 1$) thus, using Equation (6.45), the displacement time response is obtained as:

$$x = \frac{\dot{x}_o + x_o\omega_n(\zeta + \sqrt{\zeta^2 - 1})}{2\omega_n\sqrt{\zeta^2 - 1}}e^{(-\zeta + \sqrt{\zeta^2 - 1})\omega_n t}$$
$$- \frac{\dot{x}_o + x_o\omega_n(\zeta - \sqrt{\zeta^2 - 1})}{2\omega_n\sqrt{\zeta^2 - 1}}e^{(-\zeta - \sqrt{\zeta^2 - 1})\omega_n t}$$

For initial condition $t = 0$, $x = 0$, the displacement time response becomes:

$$x = \frac{\dot{x}_o}{2\omega_n\sqrt{\zeta^2 - 1}}e^{(-\zeta + \sqrt{\zeta^2 - 1})\omega_n t} - \frac{\dot{x}_o}{2\omega_n\sqrt{\zeta^2 - 1}}e^{(-\zeta - \sqrt{\zeta^2 - 1})\omega_n t}$$

$$x = \frac{8}{2 \times 4.577 \times \sqrt{1.144^2 - 1}}(e^{-1.144 + \sqrt{1.144^2 - 1} \times 4.577t} - e^{-1.144 - \sqrt{1.144^2 - 1} \times 4.577t})$$

This gives

$$x = 1.57284 \times (e^{-2.693t} - e^{-7.7792t})$$

By plotting x against t, as shown in Figure 6.29, the maximum displacement is 0.586 m and takes place at $t = 0.21$ s.

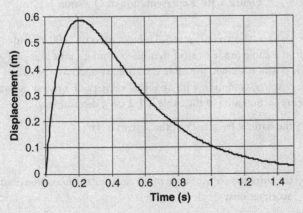

Figure 6.29: Displacement time response of Example 6.11

6.4 Tutorial Sheet

6.4.1 Undamped systems

Q6.1   The undamped SDOF system shown in Figure 6.30, comprises a mass of 50 kg and a spring of stiffness 10 kN/m. If the system starts its motion with an initial displacement of 0.4 m and an initial velocity of 1.2 m/s, determine:

a) the natural frequency of the system in Hz;

[2.25 Hz]

b) the static deflection δ_o in mm;

[49 mm]

c) the maximum velocity and acceleration of the system.

[5.78 m/s, 81.78 m/s^2]

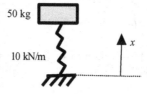

50 kg

10 kN/m

x

Figure 6.30: Representation of Question 6.1

Q6.2   A lift cage, shown in Figure 6.31, has a mass of 1200 kg and is suspended from a cable made of steel that has a Young's modulus of 200 GPa. At the position shown, the cable is stationary and has a length of 60 m. The cage starts its motion with an initial velocity of 3 m/s downwards. If the cross-section area of the cable is 1.2 cm^2, determine:

a) the natural frequency of the system in Hz;

[2.9 Hz]

b) the amplitude of vibration of the cage and the maximum and minimum accelerations;

[0.164 m, ±54.664 m/s^2]

c) the maximum tension force in the cable.

[77.37 kN]

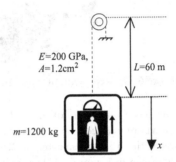

Figure 6.31: Representation of Question 6.2

Q6.3 Use a system with a single degree of freedom to determine the fundamental natural frequency of the aircraft wing shown in Figure 6.32. If the average cross-section of the wing can be idealized as a hollow box 300 mm × 2000 mm with 2 mm thickness, as shown, and the wing is considered to be fixed to the fuselage at one end and free at the other, use $E = 70$ GPa and $\rho = 2770$ kg/m^3.

[4.06 Hz]

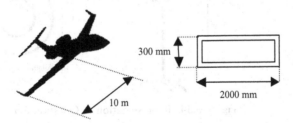

Figure 6.32: Representation of Question 6.3

Q6.4 A rotor blade of a helicopter, shown in Figure 6.33, has a constant cross-section of 1 cm × 20 cm and a length of 0.7 m. The material of the blade is aluminium, for which Young's modulus is 70 GPa and the density is 2800 kg/m^3. Use a system with a single degree of freedom to determine the fundamental frequency of the blade.

[16.73 Hz]

Rotor blade

Figure 6.33: Representation of Question 6.4

Q6.5   A water tank has a total mass of 320×10^3 kg, including water. It is supported by a reinforced concrete column of 95 m height and an annular cross-section with an inner diameter of 2.5 m and an outer diameter of 3.2 m as shown in Figure 6.34. Determine the natural frequency of the system for transverse vibration and the displacement response after 1 second due to an initial displacement of 0.2 m and zero initial velocity. Use an elastic modulus for reinforced concrete of 30 GPa and ignore the column's weight.

[0.164 Hz, 0.103 m]

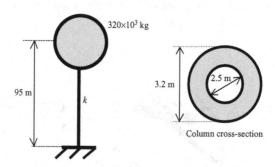

320×10^3 kg

95 m

k

3.2 m

2.5 m

Column cross-section

Figure 6.34: Representation of Question 6.5

Q6.6   An SDOF system has a natural period of 0.2 s. Determine its natural period if the spring stiffness:

a) increases by 30%;

[0.175 s]

b) decreases by 30%.

[0.239 s]

Q6.7   Determine the natural frequency of a car, modelled as an undamped SDOF system, as shown in Figure 6.35, that has a mass of 1800 kg and deflects due to its weight by 18 mm.

[3.72 Hz]

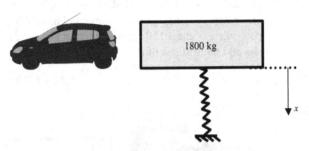

Figure 6.35: Representation of Question 6.7

Q6.8   A cantilever beam of 2 m length is supported by a simply supported beam of 3 m in length at its midpoint as shown in Figure 6.36. Both beams are made of steel with E-modulus of 200 GPa and have a cross-section of 15 cm width and 4 cm height. The system carries a mass of 100 kg at point C, the midpoint of the simply supported beam. Determine the natural frequency of the system.

[9.34 Hz]

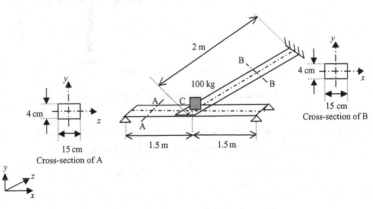

Figure 6.36: Representation of Question 6.8

Q6.9   An SDOF system consisting of a spring of stiffness k and a mass m has a natural frequency of 3 Hz. When a mass of 1.5 kg is added to the system, its frequency becomes 1.5 Hz. Determine the stiffness k and the mass m.

[177.65 N/m, 0.5 kg]

Q6.10   A car, modelled as an undamped SDOF system, has a natural frequency of 3.35 Hz without passengers, as shown in Figure 6.37(a) and 3.18 Hz with passengers of mass 150 kg, as shown in Figure 6.37(b). Determine the mass and the stiffness of the car.

[1.37×10^3 kg, 605.37 kN/m]

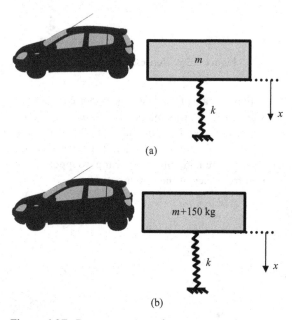

(a)

(b)

Figure 6.37: Representation of Question 6.10

Q6.11   A platform of mass 4.2×10^3 kg is supported by three springs, each of stiffness k as shown in Figure 6.38. Determine:

a) the spring's stiffness, k, so that the natural frequency of the platform equals 2.8 Hz;

[433.3 kN/m]

b) the natural frequency of the platform when a truck of mass 42×10^3 kg is loaded onto it.

[0.844 Hz]

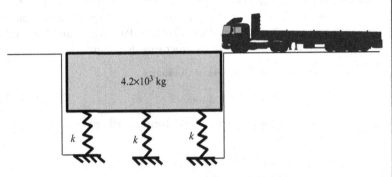

Figure 6.38: Representation of Question 6.11

Q6.12 The bell crank system ABC, shown in Figure 6.39, is supported at point A by a spring of stiffness k and at point B by a hinged support. When the crank is in a static position, the spring is stretched by 3 cm from its original un-deformed length in order to balance the bell mass of 2.5 kg at point C. If the mass of the crank is negligible, determine:

a) the spring stiffness k;

[3270 N/m]

b) the natural frequency of the system in Hz.

[1.44 Hz]

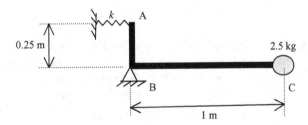

Figure 6.39: Representation of Question 6.12

Q6.13   A simple pendulum has a natural frequency of 1 Hz. Determine the length of the pendulum.

[0.249 m]

Q6.14   Figure 6.40 shows a rotor with moment of inertia $J = 500\,000$ kg.m^2, mounted on a shaft of torsional stiffness, a length of 2 m, a diameter of 50 cm and a shear modulus $G = 80$ GPa. If the shaft is of a negligible mass and is clamped at the other end, determine:

a) the torsional stiffness of the shaft;

[245.44 MN.m]

b) the natural frequency of free torsional vibration of the system in Hz.

[3.53 Hz]

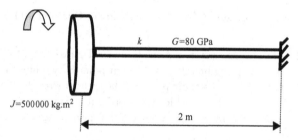

Figure 6.40: Representation of Question 6.14

Q6.15   Determine the natural frequency of torsional vibration of a 0.8 m diameter disc mounted on a steel shaft of 0.04 m diameter and 2 m length. The thickness of the disc is 0.1 m and both shaft and disc are made of steel for which the shear modulus is equal to 77 GPa and density is equal to 7800 kg/m^3.

[2.795 Hz]

Q6.16   A system with a single degree of freedom, shown in Figure 6.41, consists of a helical spring, which has a stiffness k and mass $m = 10$ kg. The spring requires a force of 100 N to produce an elongation of 10 mm. If an initial velocity of 1 m/s is applied, determine:

a) the natural frequency of the system in Hz;

[5.03 Hz]

b) the displacement time response and the amplitude of vibration;

$$[x = 0.0316 \sin 31.622t \text{ m}, \, 31.6 \text{ mm}]$$

c) the velocity of the system;

$$[\dot{x} = \cos 31.622t]$$

d) the maximum acceleration;

$$[31.62 \text{ m/s}^2]$$

e) the maximum force in the spring.

$$[316.2 \text{ N}]$$

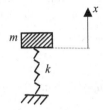

Figure 6.41: Representation of Question 6.16

Q6.17 For the SDOF system shown in Figure 6.42, if the mass m is 5 kg and the stiffness k is 1 kN/m, determine the natural frequency in Hz.

$$[2.25 \text{ Hz}]$$

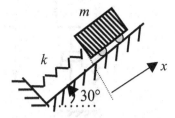

Figure 6.42: Representation of Question 6.17

Q6.18 A boy of mass 50 kg stands on a simply supported wooden board at its midpoint as shown in Figure 6.43. If the board deflects 2 cm under the boy's weight, determine the natural frequency of the system ignoring the board's weight.

$$[3.53 \text{ Hz}]$$

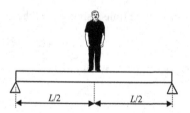

Figure 6.43: Representation of Question 6.18

6.4.2 Viscous damped systems

Q6.19   The damped SDOF system shown in Figure 6.44 has an undamped angular frequency of $\omega_n = 2\pi$ rad/s. If the mass m is released from an initial displacement of $x_0 = 1$ m and zero initial velocity, calculate the displacement at time $t = 2$ s for the following cases:

a)   an underdamped system with $\zeta = 0.02$ and $\zeta = 0.2$;

[0.78 m, 0.074 m]

b)   a critically damped system;

[4.73×10^{-5} m]

c)   an overdamped system with $\zeta = 2$ and $\zeta = 20$.

[0.037 m, 0.73 m]

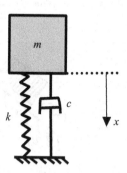

Figure 6.44: Representation of Question 6.19

Complete this table and plot the calculated displacements at 2 seconds against the given damping ratio, ζ, using a logarithmic scale.

	$\zeta = 0.02$	$\zeta = 0.2$	$\zeta = 1$	$\zeta = 2$	$\zeta = 20$
Displacement (m)					

Q6.20 If the SDOF system in Figure 6.44 is underdamped with a damped natural frequency of 1 Hz and the ratio between its amplitude of vibration in the first cycle and that after two cycles is 1:10, i.e. $x_3 = \frac{x_1}{10}$, where x_3 and x_1 are as shown in Figure 6.45. If the system has a mass $m = 25$ kg, determine:

a) the damping ratio ζ;

[0.1802]

b) the undamped natural frequency;

[1.02 Hz]

c) the stiffness k;

[1021 N/m]

d) the critical damping coefficient c_c;

[319.5 N.s/m]

e) the damping coefficient c.

[57.57 N.s/m]

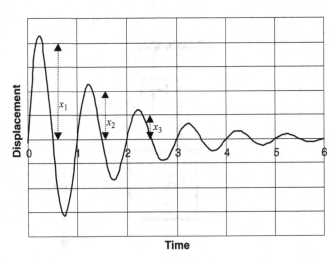

Figure 6.45: Representation of Question 6.20

Q6.21   The platform in Question 6.11 has two viscous dampers, each of damping
coefficient c, added to reduce the amplitude, as shown in Figure 6.46. If
the ratio of two successive positive amplitudes of vibration is 5, determine
the damping coefficient c.

[2918 N.s/m]

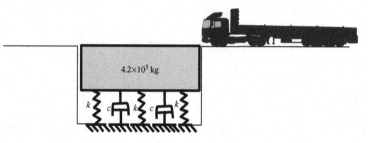

Figure 6.46: Representation of Question 6.21

Q6.22   The system shown in Figure 6.47 has a mass of 10 kg, angular frequency
of 2 rad/s, an initial displacement of 0.1 m and an initial velocity of 1 m/s.
If the damping coefficient of the viscous damper is 20 N.s/m, determine
the mass displacement after 1 s.

[225 mm]

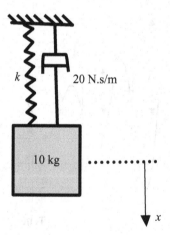

Figure 6.47: Representation of Question 6.22

Q6.23 A railway boxcar of a mass of 2.1×10^3 kg is travelling at a constant speed of 8 m/s when it hits a buffer (spring and damper system) at the end of the track, as shown in Figure 6.48. If the spring has a stiffness of 44 kN/m, determine the displacement of the boxcar when $t = 0.2$ s for the following viscous damping coefficients:

a) $c = 20$ kN.s/m

[0.63 m]

b) $c = 30$ kN.s/m

[0.47 m]

c) $c = 50$ kN.s/m

[0.3 m]

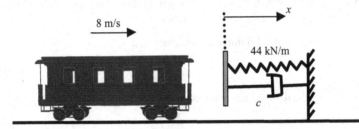

Figure 6.48: Representation of Question 6.23

Q6.24 A man of 80 kg is riding a bicycle of 10 kg as shown in Figure 6.49. If the man and the bicycle are modelled as an SDOF system with equivalent stiffness of 45 kN/m and equivalent viscous damping coefficient of 900 N.s/m, determine:

a) the damping ratio;

[0.22]

b) the damped natural frequency;

[3.47 Hz]

c) the displacement response after 0.3 s if the system has an initial displacement of 4 cm.

[9.2 mm]

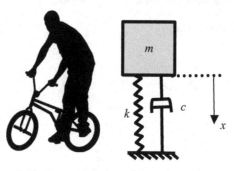

Figure 6.49: Representation of Question 6.24

Q6.25   For the spring–mass–damper system shown in Figure 6.50, with $m = 50$ kg and $k = 5000$ N/m, determine:

a) the critical damping coefficient (c_c);

[1000 N.s/m]

b) the damped natural frequency in Hz when $c = c_c/2$;

[1.38 Hz]

c) the logarithmic decrement.

[3.63]

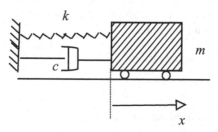

Figure 6.50: Representation of Question 6.25

Q6.26   For the damped SDOF system in Figure 6.50, if the mass m is 5 kg and the stiffness k is 1 kN/m, determine the viscous damping coefficient when the system is critically damped.

[141.4 N.s/m]

Q6.27   An underdamped SDOF system has a displacement response similar to that shown in Figure 6.45 with two successive amplitudes equal to $x_1 = 5$ mm and $x_2 = 4$ mm. If the system has a mass of 2 kg and an undamped natural frequency of 20 Hz, determine the viscous damping coefficient of the system.

[17.84 N.s/m]

Q6.28   For the damped SDOF system shown in Figure 6.51, if the mass m is 5 kg, the stiffness k is 1 kN/m and the viscous damping coefficient c is 25 N.s/m, determine the damping ratio.

[0.177]

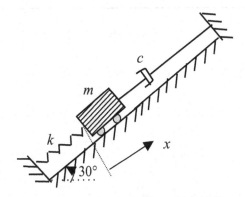

Figure 6.51: Representation of Question 6.28

Q6.29   A mass of 2 kg is suspended from a spring of stiffness 50 N/s and immersed in a liquid as shown in Figure 6.52 so that the equivalent viscous damping coefficient is 0.9 N.s/m. If the mass is released from rest and has an initial displacement downwards of 0.5 m, determine the displacement of the mass after 2 seconds.

[−275 mm]

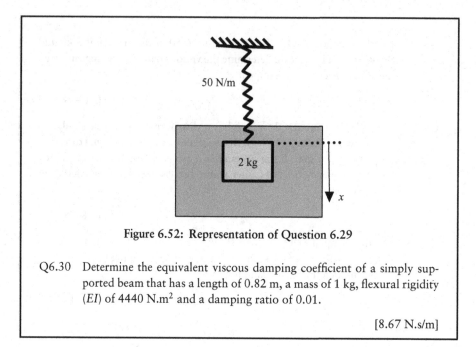

Figure 6.52: Representation of Question 6.29

Q6.30   Determine the equivalent viscous damping coefficient of a simply supported beam that has a length of 0.82 m, a mass of 1 kg, flexural rigidity
(*EI*) of 4440 N.m^2 and a damping ratio of 0.01.

[8.67 N.s/m]

CHAPTER 7

Forced Vibration of Systems with a Single Degree of Freedom

7.1 INTRODUCTION

Forced vibration is one of the most important classes of vibration problem, especially when the excitation force is continuously applied to the structure. The dynamic force could be externally applied, generated by an unbalanced rotating machine or due to motion of foundation, e.g. an earthquake. A dynamic force is a function of time and could have different relationships with time and take different shapes, e.g. periodic, non-periodic, harmonic, square, triangle, saw tooth, half sine, step, ramp, impulse or random. A force of a harmonic type, i.e. $f(t) = F_o \sin \omega t$, is shown in Figure 7.1(a), where ω is the forcing frequency (also called the excitation frequency) and a force of a random type is shown in Figure 7.1(b).

Using Fourier transformation, any random force can be represented by a summation of a series of harmonic forces:

$$f(t) = A_o + \sum (A_i \sin \omega_i t + B_i \cos \omega_i t) \tag{7.1}$$

where A_o, A_i and B_i are constants.

When a harmonic force is applied to a system with a single degree of freedom (SDOF), the vibration consists of two motions, one at the forcing frequency and the other at the natural frequency of the system. The vibration at the natural frequency dies out after some time due to damping, while the vibration at the forcing frequency continues. The vibration at the forcing frequency is called steady-state forced vibration and its amplitude depends on the ratio between the forcing frequency, the natural frequency of the system and the damping ratio of the system. When the forcing frequency is equal to the natural frequency of the system, resonance takes place and the amplitude of vibration builds up very quickly to a very high value, which is limited by the level of damping present in the system. Resonance phenomena should be avoided in the design of structures and mechanical systems in order to prevent failure and collapses of the system.

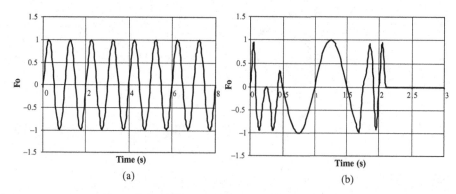

Figure 7.1: a) Harmonic force and b) random force

In this chapter, we consider the undamped and damped forced vibration mainly due to harmonic force. The response due to the general force function is briefly presented. The concept of vibration isolation is introduced and practical applications are presented.

7.2 UNDAMPED FORCED VIBRATION – HARMONIC FORCE

7.2.1 Deriving the equation of motion

For an undamped forced-vibration SDOF system, the damping coefficient c equals zero and the system is as shown in Figure 7.2(a). The free-body diagram of such a system is drawn in Figure 7.2(b), from which the equation of motion, derived from Newton's second law, is:

$$-kx + F_o \sin \omega t = m\ddot{x} \tag{7.2}$$

Rearranging Equation (7.2) gives:

$$m\ddot{x} + kx = F_o \sin \omega t \tag{7.3}$$

or

$$\ddot{x} + \omega_n^2 x = \frac{F_o}{m} \sin \omega t \tag{7.4}$$

where $\omega_n = \sqrt{\frac{k}{m}}$ is the angular frequency in rad/s.

7.2.2 Time response

The solution of the second-order differential equation, Equation (7.4), has two parts: the complementary solution and the particular solution. The complementary solution (also called the transient solution), denoted as x_c, is the solution of free vibration equation ($\ddot{x} + \omega_n^2 x = 0$); the particular solution (also called the steady-state solution), denoted as x_p, can be any solution that satisfies Equation (7.4). The total solution is, therefore, the summation of both solutions:

$$x = x_c + x_p \tag{7.5}$$

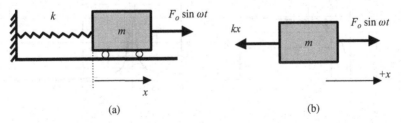

(a)   (b)

Figure 7.2: a) An undamped forced-vibration SDOF system and b) its free-body diagram

From Equation (6.16) in Chapter 6, the transient solution is:

$$x_c = A\cos\omega_n t + B\sin\omega_n t \tag{7.6}$$

and the steady-state solution that satisfies Equation (7.4) is:

$$x_p = \chi \sin\omega t \tag{7.7}$$

where χ is the maximum displacement (amplitude). Thus, the total displacement time response of the system is:

$$x = A\cos\omega_n t + B\sin\omega_n t + \chi \sin\omega t \tag{7.8}$$

Differentiating Equation (7.7) twice with respect to time and substituting into Equation (7.4) gives:

$$-\chi\omega^2 \sin\omega t + \chi\omega_n^2 \sin\omega t = \frac{F_o}{m}\sin\omega t$$

from which the steady-state amplitude χ is determined as:

$$\chi = \frac{F_o/m}{\omega_n^2 - \omega^2} = \frac{F_o/(m\omega_n^2)}{1 - \left(\frac{\omega}{\omega_n}\right)^2} \tag{7.9}$$

Replacing m by k/ω_n^2, χ becomes

$$\chi = \frac{F_o/k}{1 - \left(\frac{\omega}{\omega_n}\right)^2} \tag{7.10}$$

The variation of the steady-state amplitude χ as a function of $\frac{\omega}{\omega_n}$ is shown in Figure 7.3(a). When $0 < \frac{\omega}{\omega_n} < 1$, the steady-state amplitude χ is in phase with the external force; when $\frac{\omega}{\omega_n} > 1$, the steady-state amplitude χ is anti-phase (180° out of phase) with the external force. In Figure 7.3(b), the absolute value of χ is plotted against $\frac{\omega}{\omega_n}$. It can be seen that when $\frac{\omega}{\omega_n} = 1$, the amplitude $\chi \to \infty$ (resonance occurs).

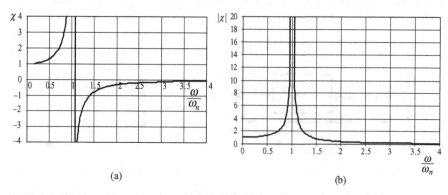

(a) (b)

Figure 7.3: a) Steady-state amplitude and b) absolute value of the steady-state amplitude

Example 7.1 Undamped forced vibration I

A fixed-fixed beam has a length of 3 m and carries an unbalanced motor of a mass of 250 kg in its middle as shown in Figure 7.4. The beam is made of steel, $E = 200$ GPa, and has a cross-section of 20 × 400 mm. Determine:

a) the amplitude of vibration if the force produced by the motor is equal to $f(t) = 300 \sin 40t$ N, in the case where the beam's weight is ignored and in the case where the beam's weight is taken into account (steel density 7800 kg/m³);

b) the total displacement time response of the beam, taking into account the beam's weight, if only an initial velocity of 173 mm/s is applied to the beam (initial displacement is zero).

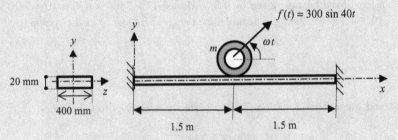

Figure 7.4: Representation of Example 7.1

Solution

a) The area moment of inertia of the beam about the z axis (see Appendix B) is:

$$I = \frac{(400 \times 10^{-3}) \times (20 \times 10^{-3})^3}{12} = 2.66667 \times 10^{-7} \, \text{m}^4$$

For a fixed-fixed beam subjected to a load F in the middle, the static deflection is calculated from Table 6.1 in Chapter 6 as:

$$u = \frac{FL^3}{192EI}$$

Thus the stiffness k can be calculated as:

$$k = \frac{F}{u} = \frac{192EI}{L^3} = \frac{192 \times 200 \times 10^9 \times 2.66667 \times 10^{-7}}{3^3} = 379\,259 \, \text{N/m}$$

Ignoring the beam's weight, the undamped angular frequency of the system is:

$$\omega_n = \sqrt{\frac{k}{m}} = \sqrt{\frac{379\,259}{250}} = 38.9491\,\text{rad/s}$$

The forcing frequency $\omega = 40$ rad/s thus, from Equation (7.10), the steady-state amplitude is given by:

$$\chi = \frac{F_o/k}{1 - \left(\frac{\omega}{\omega_n}\right)^2} = \frac{300/379\,259}{1 - \left(\frac{40}{38.9491}\right)^2} = -0.01446\,\text{m} = -15\,\text{mm}$$

Taking the beam's weight into account, from Table 6.1, the equivalent mass of a fixed-fixed beam is given by $0.37\rho AL = 0.37 \times 7800 \times 400 \times 10^{-3} \times 20 \times 10^{-3} \times 3 = 69.264$ kg. The undamped angular frequency of the system is:

$$\omega_n = \sqrt{\frac{k}{m}} = \sqrt{\frac{379\,259}{250 + 69.264}} = 34.4662\,\text{rad/s}$$

and, again from Equation (7.10), the steady-state amplitude becomes:

$$\chi = \frac{F_o/k}{1 - \left(\frac{\omega}{\omega_n}\right)^2} = \frac{300/379\,259}{1 - \left(\frac{40}{34.4662}\right)^2} = -0.00228\,\text{m} = -2.3\,\text{mm}$$

b) The steady-state response of the beam (see Figure 7.5(a)) is:

$$x_p = \chi \sin \omega t = -2.3 \sin 40t\,\text{mm} \qquad\qquad \text{(E7.1a)}$$

The transient solution for an undamped SDOF system (from Equation (6.20) in Chapter 6) is:

$$x_c = C \sin(\omega_n t + \psi)$$

where

$$C = \sqrt{x_o^2 + \left(\frac{\dot{x}_o}{\omega_n}\right)^2} = \sqrt{0 + \left(\frac{0.173}{34.4662}\right)^2} = 0.005\,\text{m}$$

and

$$\psi = \tan^{-1}\frac{x_o \omega_n}{\dot{x}_o} = \tan^{-1}\frac{0}{2} = 0$$

Thus x_c becomes (see Figure 7.5(b)):

$$x_c = 5 \sin 34.4662t \text{ mm} \qquad\qquad \text{(E7.1b)}$$

And the total response is obtained by adding Equations (E7.1a) and (E7.1b) as plotted in Figure 7.5(c):

$$x = x_c + x_p = 5 \sin 34.4662t - 2.3 \sin 40t \text{ mm}$$

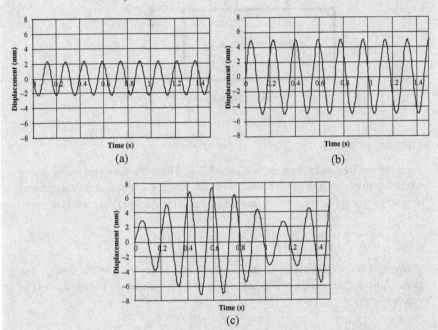

Figure 7.5: Time response of Example 7.1: a) steady state, b) transient solution and c) total

Example 7.2 Undamped forced vibration II

A mass m is supported by two springs of stiffness k_1 and k_2, which are resting on an elastic foundation as shown in Figure 7.6. The foundation undergoes harmonic motion given by $x_e = x_o \sin \omega t$, where x_o is the displacement amplitude and ω is the forcing frequency. Determine the equation of motion and the amplitude of the steady-state vibration in terms of m, k_1, k_2, x_o and ω.

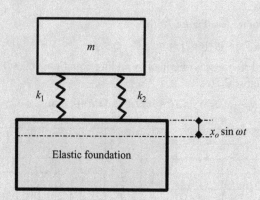

Figure 7.6: Representation of Example 7.2

Solution

Consider the free-body diagram in Figure 7.7(a). The force acting on the mass due to the harmonic motion of the foundation is $(k_1 + k_2) \times x_o \sin \omega t$ and thus the total dynamic force is $(k_1 + k_2) \times (x + x_o \sin \omega t)$. Applying Newton's second law gives:

$$mg - (k_1 + k_2) \times \delta_o - (k_1 + k_2) \times (x + x_o \sin \omega t) = m\ddot{x} \qquad \text{(E7.2a)}$$

where δ_o is the static deflection and $(k_1 + k_2) \times \delta_o$ is the force in the springs due to δ_o. Consider the static equilibrium in Figure 7.7(b) where the summation of the forces is zero:

$$mg - (k_1 + k_2) \times \delta_o = 0 \Rightarrow (k_1 + k_2) \times \delta_o = mg \qquad \text{(E7.2b)}$$

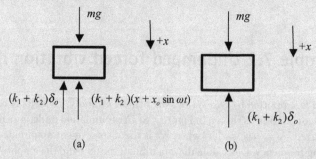

Figure 7.7: Example 7.2: a) free-body diagram and b) static equilibrium

Substituting Equation (E7.2b) into Equation (E7.2a) gives:

$$-(k_1 + k_2) \times (x + x_o \sin \omega t) = m\ddot{x}$$

Rearranging, the equation of motion is obtained as:

$$\ddot{x} + \frac{(k_1 + k_2)}{m}x = \frac{-x_o(k_1 + k_2)}{m} \sin \omega t$$

or

$$\ddot{x} + \omega_n^2 x = \frac{F_o}{m} \sin \omega t$$

where

$$\omega_n = \sqrt{\frac{(k_1 + k_2)}{m}} \quad \text{and} \quad F_o = -x_o(k_1 + k_2)$$

Thus, the equivalent force due to the motion of the foundation is:

$$F_o = -x_o(k_1 + k_2)$$

From Equation (7.10), the steady-state amplitude of vibration is obtained as:

$$\chi = \frac{F_o/k}{1 - \left(\frac{\omega}{\omega_n}\right)^2} = \frac{-x_o(k_1 + k_2)/(k_1 + k_2)}{1 - \left(\frac{\omega}{\omega_n}\right)^2}$$

This gives

$$\chi = \frac{-x_o}{1 - \left(\frac{\omega}{\omega_n}\right)^2}$$

Example 7.3 Undamped forced vibration III

An unbalanced centrifugal pump of mass 50 kg produces an out-of-balance force of 2 kN and is supported by four springs each of 10 kN/m as shown in Figure 7.8. If the pump operates at 800 rev/min, determine the steady-state amplitude of vibration.

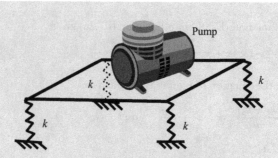

Figure 7.8: Representation of Example 7.3

Solution

Since the overall stiffness is $4k$ (see Appendix C), the undamped angular frequency of the system is calculated as:

$$\omega_n = \sqrt{\frac{4k}{m}} = \sqrt{\frac{4 \times 10000}{50}} = 28.28427 \, \text{rad/s}$$

The forcing frequency is given by:

$$\omega = 800 \times \frac{2\pi}{60} = 83.7758 \, \text{rad/s}$$

From Equation (7.10), using the force amplitude of 2000 N, the steady-state amplitude of vibration is calculated as:

$$\chi = \frac{F_o/k}{1 - \left(\frac{\omega}{\omega_n}\right)^2} = \frac{2000/(4 \times 10000)}{1 - \left(\frac{83.7758}{28.28427}\right)^2} = -6.43 \times 10^{-3} \, \text{m} = -6.43 \, \text{mm}$$

7.3 VISCOUS DAMPED FORCED VIBRATION – HARMONIC FORCE

7.3.1 Deriving the equation of motion

A viscous damped SDOF system subjected to a harmonic force, $f(t) = F_o \sin \omega t$ is shown in Figure 7.9(a), where c is the viscous damping coefficient. The free-body diagram of the system is shown in Figure 7.9(b).

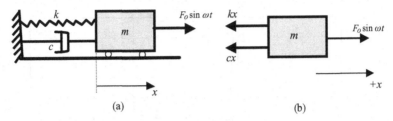

Figure 7.9: a) A damped forced vibration SDOF and b) its free-body diagram

From the free-body diagram, Newton's second law can be written as:

$$-kx - c\dot{x} + F_o \sin \omega t = m\ddot{x} \qquad (7.11)$$

Rearranging Equation (7.11) gives:

$$m\ddot{x} + c\dot{x} + kx = F_o \sin \omega t \qquad (7.12)$$

or

$$\ddot{x} + 2\zeta\omega_n\dot{x} + \omega_n^2 x = \frac{F_o}{m}\sin \omega t \qquad (7.13)$$

where

$\omega_n = \sqrt{\frac{k}{m}}$ is the angular frequency in rad/s and $\zeta = \frac{c}{2m\omega_n}$ is the damping ratio.

7.3.2 Time response

The solution of Equation (7.13) has two parts: the complementary solution and the particular solution. Similarly to the case of undamped forced vibration, the complementary or transient solution, x_c, is the solution of free vibration equation ($\ddot{x} + 2\zeta\omega_n\dot{x} + \omega_n^2 x = 0$) and the particular or steady-state solution, x_p, can be any solution that satisfies Equation (7.13). Again, the total solution is the summation of both solutions, i.e. $x = x_c + x_p$, where the transient solution, from Equation (6.41) in Chapter 6 is:

$$x_c = A_1 e^{(-\zeta + \sqrt{\zeta^2 - 1})\omega_n t} + A_2 e^{(-\zeta - \sqrt{\zeta^2 - 1})\omega_n t} \qquad (7.14)$$

and the steady-state solution that satisfies Equation (7.14) is:

$$x_p = \chi \sin(\omega t - \phi) \qquad (7.15)$$

Thus the total displacement time response of the system is:

$$x = A_1 e^{(-\zeta + \sqrt{\zeta^2 - 1})\omega_n t} + A_2 e^{(-\zeta - \sqrt{\zeta^2 - 1})\omega_n t} + \chi \sin(\omega t - \phi) \qquad (7.16)$$

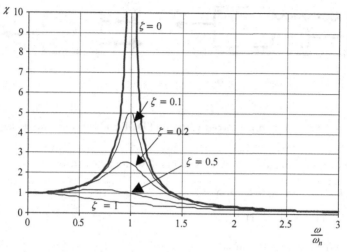

Figure 7.10: Steady-state amplitude χ against $\dfrac{\omega}{\omega_n}$ for different values of ζ

Differentiating Equation (7.15) twice with respect to time, substituting into Equation (7.13), and solving for χ and ϕ, yields:

$$\chi = \frac{F_o/k}{\sqrt{\left(1 - \left(\frac{\omega}{\omega_n}\right)^2\right)^2 + \left(\frac{2\zeta\omega}{\omega_n}\right)^2}} \tag{7.17}$$

and

$$\phi = \tan^{-1} \frac{2\zeta\,(\omega/\omega_n)}{1 - \left(\frac{\omega}{\omega_n}\right)^2} \tag{7.18}$$

In Figure 7.10, the variation of the steady-state amplitude χ is shown as a function of $\frac{\omega}{\omega_n}$ for different damping ratios ζ. It can be seen from Figure 7.10 that for undamped systems the maximum steady-state amplitude goes to infinity and occurs at $\frac{\omega}{\omega_n} = 1$, at which resonance takes place. By increasing the damping ratio ζ, the maximum steady-state amplitude at resonance is reduced and shifted to $\frac{\omega}{\omega_n} < 1$ (as can be also deduced from Equation (7.17)).

Example 7.4 Damped forced vibration I

A machine, shown in Figure 7.11, is supported by a spring of stiffness 120 kN/m and a viscous damper of damping coefficient 1.5 kN.s/m. The machine has a mass of 500 kg and is subject to a harmonic force of $250 \sin 12t$(N). Determine:

a) the solution of the steady-state vibration;
b) the total displacement time response of the system if only an initial displacement of -10 mm is applied.

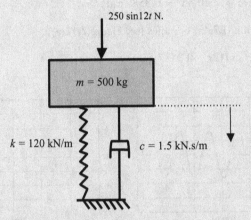

Figure 7.11: Representation of Example 7.4

Solution

a) From Equation (7.15), the solution of the steady-state vibration is:

$$x_p = \chi \sin(\omega t - \phi) \qquad \text{(E7.4a)}$$

The undamped angular frequency of the system is calculated as:

$$\omega_n = \sqrt{\frac{k}{m}} = \sqrt{\frac{120\,000}{500}} = 15.4919 \text{ rad/s}$$

and the damping ratio (from Equation (6.37)) is:

$$\zeta = \frac{c}{c_c} = \frac{c}{2\sqrt{km}} = \frac{1500}{2\sqrt{120\,000 \times 500}} = 0.0968$$

From Equation (7.17), the steady-state amplitude can be calculated as:

$$\chi = \frac{F_o/k}{\sqrt{\left(1 - \left(\frac{\omega}{\omega_n}\right)^2\right)^2 + \left(\frac{2\zeta\omega}{\omega_n}\right)^2}} = \frac{250/120\,000}{\sqrt{\left(1 - \left(\frac{12}{15.4919}\right)^2\right)^2 + \left(\frac{2 \times 0.0968 \times 12}{15.4919}\right)^2}}$$

$$= 0.004876 \text{ m} = 4.9 \text{ mm}$$

From Equation (7.18), the phase angle can be calculated as:

$$\phi = \tan^{-1} \frac{2\zeta\,(\omega/\omega_n)}{1 - \left(\frac{\omega}{\omega_n}\right)^2} = \tan^{-1} \frac{2 \times 0.0968 \times (12/15.4919)}{1 - \left(\frac{12}{15.4919}\right)^2}$$

$$\Rightarrow \phi = 20.55° = 0.3587\,\text{rad}$$

Thus, Equation (E7.4a) becomes (see Figure 7.12(a)):

$$x_p = 4.9\sin(12t - 0.3587)\,\text{mm} \tag{E7.4b}$$

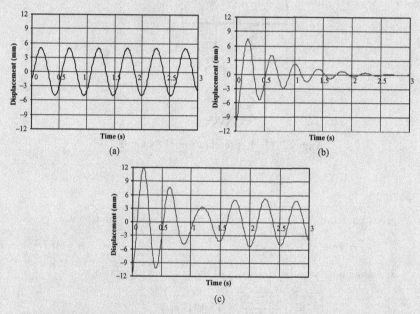

Figure 7.12: Displacement time response of Example 7.4: a) steady state, b) transient solution and c) total

b) The system is underdamped as $\zeta < 1$, thus from Equation (6.60) in Chapter 6, the transient solution of the displacement time response is given by:

$$x_c = \left[x_o\cos\omega_d t + \frac{\dot{x}_o + \zeta\omega_n x_o}{\omega_d}\sin\omega_d t\right]e^{-\zeta\omega_n t}$$

Applying initial conditions, $t = 0$, $x = x_o$ and $\dot{x} = 0$ gives:

$$x_c = [x_o\cos\omega_d t]\,e^{-\zeta\omega_n t} \tag{E7.4c}$$

where $\omega_n = 15.4919$ rad/s, $x_0 = -10$ mm and $\zeta = 0.0968$. From Equation (6.53), the damped natural frequency is:

$$\omega_d = \sqrt{1 - \zeta^2}\,\omega_n = \sqrt{1 - 0.0968^2} \times 15.4919 = 15.4191$$

and the transient solution, Equation (E7.4c), becomes (see Figure 7.12(b)):

$$x_c = -10\cos 15.4191t \times e^{-1.5t} \text{ mm} \qquad (E7.4d)$$

The total displacement time response is obtained by adding Equations (E7.4b) and (E7.4d), as plotted in Figure 7.12(c):

$$x = x_c + x_p = -10\cos 15.4191t \times e^{-1.5t} + 4.9\sin(12t - 0.3587) \text{ mm}$$

Example 7.5 Damped forced vibration II

An electric motor of mass 32 kg is supported by four springs each of stiffness 250 N/m and a damper of damping coefficient 40 N.s/m as shown in Figure 7.13. The motor has an out-of-balance mass m located at distance of 60 mm from its centre of rotation and is running with an angular velocity of 10 rad/s. If the amplitude of steady-state vibration is 10 mm, determine the out-of-balance mass m.

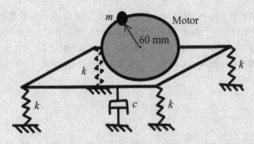

Figure 7.13: Representation of Example 7.5

Solution

From Appendix C, the overall stiffness is $4k$, and hence the undamped angular frequency of the system is calculated as:

$$\omega_n = \sqrt{\frac{4k}{m}} = \sqrt{\frac{4 \times 250}{32}} = 5.59 \text{ rad/s}$$

From Equation (6.37), the damping ratio is:

$$\zeta = \frac{c}{c_c} = \frac{c}{2\sqrt{(4k)m}} = \frac{40}{2\sqrt{4 \times 250 \times 32}} = 0.112$$

From Equation (7.17), the magnitude of the excitation force can be calculated as:

$$\chi = \frac{F_o/(4k)}{\sqrt{\left(1 - \left(\frac{\omega}{\omega_n}\right)^2\right)^2 + \left(\frac{2\zeta\omega}{\omega_n}\right)^2}} \Rightarrow 0.01 = \frac{F_o/(4 \times 250)}{\sqrt{\left(1 - \left(\frac{10}{5.59}\right)^2\right)^2 + \left(\frac{2 \times 0.112 \times 10}{5.59}\right)^2}}$$

$$\Rightarrow F_o = 22.36\,\text{N}$$

From Equation (5.1) in Chapter 5, the out-of-balance force produced by a rotating mass is given by:

$$F_o = mr\omega^2 \Rightarrow 22.36 = m \times 0.06 \times 10^2$$

From which the out-of-balance mass m is calculated as:

$$m = 3.73\,\text{kg}$$

Example 7.6 Damped forced vibration III

An SDOF system consists of a spring, a mass and a damper. When a harmonic force is applied to the system, the amplitude of steady-state vibration is 16 mm at resonance. If the damping ratio of the system is 0.1, determine the amplitude of steady-state vibration at a frequency equivalent to 70% of the resonance frequency.

Solution

At resonance $\frac{\omega}{\omega_n} = 1$, the steady-state amplitude is obtained from Equation (7.17) as:

$$\chi = \frac{F_o/k}{\sqrt{\left(1 - \left(\frac{\omega}{\omega_n}\right)^2\right)^2 + \left(\frac{2\zeta\omega}{\omega_n}\right)^2}} \Rightarrow 0.016 = \frac{F_o/k}{\sqrt{\left(1 - (1)^2\right)^2 + (2 \times \zeta \times 1)^2}}$$

from which the static displacement of the system under the action of F_o is:

$$F_o/k = 0.032\zeta = 0.032 \times 0.1 = 0.0032\,\text{m}$$

At a frequency of 70% of the resonance frequency, $\frac{\omega}{\omega_n} = 0.7$ and the steady-state amplitude is calculated as:

$$\chi = \frac{0.0032}{\sqrt{\left(1 - (0.7)^2\right)^2 + (2 \times 0.1 \times 0.7)^2}} = 6.05 \times 10^{-3}\,\text{m} = 6.1\,\text{mm}$$

Example 7.7 Damped forced vibration IV

A car is modelled as an SDOF system with a mass m in kg, an equivalent stiffness k in N/m and an equivalent damping coefficient c in N.s/m as shown in Figure 7.14. If the road is assumed to provide a base-motion vertical displacement equivalent to $y = Y \sin \omega t$, where y is in mm and t is the time in seconds, determine the amplitude of vibration of the car due to the road motion.

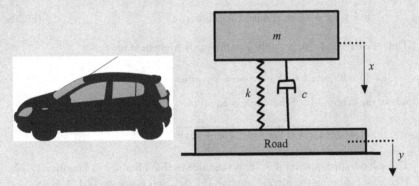

Figure 7.14: Representation of Example 7.7

Solution

The motion of the car can be analyzed as a damped SDOF system under harmonic motion of the base, where the harmonic displacement is $y = Y \sin \omega t$, where Y is the excitation amplitude and ω is the excitation frequency. Consider the free-body diagram of the car (in Figure 7.15); the equation of motion is obtained as:

$$-c(\dot{x} - \dot{y}) - k(x - y) = m\ddot{x}$$
$$m\ddot{x} + c(\dot{x} - \dot{y}) + k(x - y) = 0 \qquad\qquad (\text{E7.7a})$$

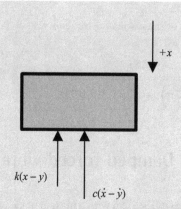

Figure 7.15: Free-body diagram for Example 7.7

Substituting $y = Y \sin \omega t$ and its derivative with respect to time into Equation (E7.7a) gives:

$$m\ddot{x} + c\dot{x} + kx = kY \sin \omega t + cY\omega \cos \omega t \tag{E7.7b}$$

Thus the harmonic force applied to the car is equivalent to:

$$F = kY \sin \omega t + cY\omega \cos \omega t = F_o \sin(\omega t - \varphi)$$

where the magnitude of this force is equal to:

$$F_o = \sqrt{(kY)^2 + (cY\omega)^2} = Y\sqrt{k^2 + (c\omega)^2} \tag{E7.7c}$$

The solution of Equation (E7.7b) is the same as that obtained in Equation (7.17). Hence, substituting the force, Equation (E7.7c), into Equation (7.17), gives

$$\chi = \frac{Y\sqrt{k^2 + (c\omega)^2}/k}{\sqrt{\left(1 - \left(\frac{\omega}{\omega_n}\right)^2\right)^2 + \left(\frac{2\zeta\omega}{\omega_n}\right)^2}}$$

which can be written as:

$$\chi = \frac{Y\sqrt{1 + \left(\frac{c\omega}{k}\right)^2}}{\sqrt{\left(1 - \left(\frac{\omega}{\omega_n}\right)^2\right)^2 + \left(\frac{2\zeta\omega}{\omega_n}\right)^2}}$$

7.4 GENERAL FORCED RESPONSE

The equation of motion of a damped SDOF system subjected to a general force, $f(t)$, can be obtained in a way similar to that used to derive Equation (7.13):

$$\ddot{x} + 2\zeta\omega_n\dot{x} + \omega_n^2 x = \frac{f(t)}{m} \qquad (7.19)$$

The solution of Equation (7.19) mainly depends on the type of the function $f(t)$.

7.4.1 Response to an impulse

For an impulse or impact, the force suddenly reaches a maximum value in a very short time as shown in Figure 7.16, then drops to zero. From the linear impulse–momentum relationship (Equation (3.31) in Chapter 3), the impulse force $\overline{F}$ can be written as:

$$\overline{F} = F_o\Delta t = m\dot{x}_o \qquad (7.20)$$

where F_o is the force amplitude as shown in Figure 7.16 and $\dot{x}_o$ is the initial velocity.

From Equation (7.20), the initial velocity is:

$$\dot{x}_o = \frac{F_o\Delta t}{m} \qquad (7.21)$$

For an undamped free vibration SDOF system, Equation (6.19) in Chapter 6 gives the displacement time response as:

$$x = x_o \cos\omega_n t + \frac{\dot{x}_o}{\omega_n} \sin\omega_n t$$

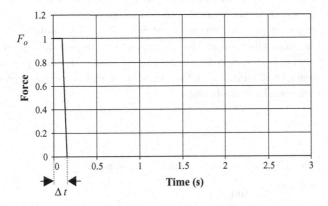

Figure 7.16: Impulse force as a function of time

In case of zero initial displacement $x_o = 0$ and initial velocity $\dot{x}_o$, the displacement time response reduces to:

$$x = \frac{\dot{x}_o}{\omega_n} \sin \omega_n t \qquad (7.22)$$

Substituting Equation (7.21) into Equation (7.22) gives:

$$x = \frac{F_o \Delta t}{m \omega_n} \sin \omega_n t$$

The same procedures can be applied to an underdamped SDOF system. From Equation (6.60), the displacement time response for an underdamped system is:

$$x = \left[x_o \cos \omega_d t + \frac{\dot{x}_o + \zeta \omega_n x_o}{\omega_d} \sin \omega_d t \right] e^{-\zeta \omega_n t}$$

Applying initial conditions, at $t = 0$, initial displacement $x = 0$ and initial velocity $\dot{x} = \dot{x}_o$, yields:

$$x = \frac{\dot{x}_o}{\omega_d} e^{-\zeta \omega_n t} \sin(\omega_d t) \qquad (7.23)$$

Substituting Equation (7.21) into Equation (7.23) gives:

$$x = \frac{F_o \Delta t}{m \omega_d} e^{-\zeta \omega_n t} \sin(\omega_d t) \qquad (7.24)$$

Example 7.8 General forced response I

A simply supported beam with a clear length of 0.82 m between the two supports and mass $m = 1$ kg, as shown in Figure 7.17, is excited at its midpoint with an impact hammer that produces an impulse of $\overline{F} = 1.5$ N.s. The fundamental natural frequency of lateral vibration and the corresponding damping ratio are measured as 142 Hz and 0.01, respectively, using an accelerometer, which is placed on top of the beam at its midpoint. Use an SDOF system to predict the displacement time response in the middle of the beam.

Solution

From Equation (7.24), the displacement time response is:

$$x = \frac{F_o \Delta t}{m \omega_d} e^{-\zeta \omega_n t} \sin(\omega_d t) \qquad (E7.8)$$

Figure 7.17: Representation of Example 7.8

The measured natural frequency is the damped frequency; the damped angular frequency is $\omega_d = 142 \times 2\pi = 892.21$ rad/s. From Table 6.1, the equivalent mass of a simply-supported beam is $m = 0.4857 \times 1 = 0.4857$ kg and the undamped angular frequency is calculated from Equation (6.53) as:

$$\omega_n = \frac{\omega_d}{\sqrt{1 - \zeta^2}} = \frac{892.21}{\sqrt{1 - 0.01^2}} = 892.255 \, \text{rad/s}$$

and the impulse force is:

$$\overline{F} = F_o \Delta t = 1.5 \, \text{N.s}$$

Thus, Equation (E7.8) becomes:

$$x = \frac{F_o \Delta t}{m\omega_d} e^{-\zeta \omega_n t} \sin(\omega_d t) = \frac{1.5}{0.4857 \times 892.21} e^{-0.01 \times 892.255t} \sin(892.21t)$$

from which the displacement time response is:

$$x = 0.00346 e^{-8.9225t} \sin(892.21t) \, \text{m}$$

or

$$x = 3.46 e^{-8.9225t} \sin(892.21t) \, \text{mm}$$

The displacement time response of the beam under the impact force is shown in Figure 7.18.

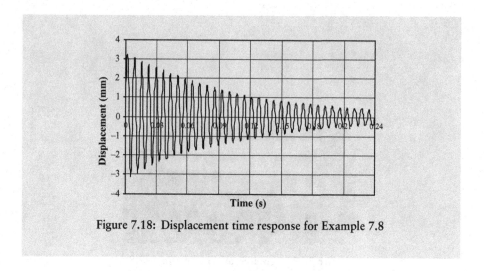

Figure 7.18: Displacement time response for Example 7.8

7.4.2 Response to a general forcing condition

If an arbitrary external force, as shown in Figure 7.19, is considered, the impulse function for a small time interval $\Delta\tau$ can be written as:

$$f(\tau) = F(\tau)\Delta\tau \tag{7.25}$$

Using the impulse equation, Equation (7.24), for an underdamped system, it is possible to express the displacement time response just after the time τ, thus at time $(t - \tau)$:

$$\Delta x(t) = F(\tau)\Delta\tau \times \frac{1}{m\omega_d} e^{-\zeta\omega_n(t-\tau)} \sin(\omega_d(t - \tau)) \tag{7.26}$$

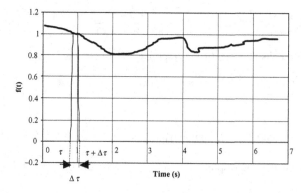

Figure 7.19: Arbitrary force function

Integrating equation (7.26) over time gives:

$$x(t) = \frac{1}{m\omega_d} \int_0^t F(\tau) \times e^{-\zeta\omega_n(t-\tau)} \sin(\omega_d(t-\tau))d\tau \tag{7.27}$$

The integration in Equation (7.27) mainly depends on the type of function $F(\tau)$. Note that the elapsed time since the impulse is $(t-\tau)$, thus the response of the system is due to impulse alone.

Example 7.9 General forced response II

If the simply-supported beam in Figure 7.17 is subjected to a sudden force $F_o = 200$ N after a time $t_o = 1$ s, as shown in Figure 7.20. Determine the displacement time response in the middle of the beam.

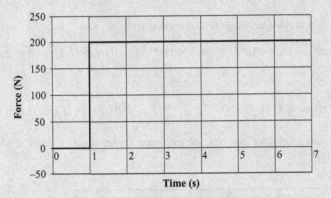

Figure 7.20: Representation of Example 7.9

Solution

Since the force is constant for $t > t_o$, i.e. $F(\tau) = F_o$, Equation (7.27) becomes:

$$x(t) = \frac{F_o}{m\omega_d} \int_{t_o}^t e^{-\zeta\omega_n(t-\tau)} \sin(\omega_d(t-\tau))d\tau$$

Integrating over time (integration by parts) gives:

$$x(t) = \frac{F_o}{k}\left(1 - \frac{1}{\sqrt{1-\zeta^2}}e^{-\zeta\omega_n(t-t_o)}\cos(\omega_d(t-t_o) - \varphi)\right) \tag{E7.9}$$

where

$$\varphi = \tan^{-1} \frac{\zeta}{\sqrt{1-\zeta^2}}$$

For the simply-supported beam in Figure E7.17, the mass is $m = 0.4857$ kg, the damped angular frequency is $\omega_d = 892.21$ rad/s, the stiffness is calculated as $k = \omega_n^2 \times m = 892.255^2 \times 0.4857 = 386675$ N/m, the damping ratio is $\zeta = 0.01$, the force amplitude is $F = 200$ N at $t_o = 1$ s and the phase angle is:

$$\varphi = \tan^{-1} \frac{\zeta}{\sqrt{1-\zeta^2}} = \tan^{-1} \frac{0.01}{\sqrt{1-0.01^2}} = 0.5729° = 0.01 \text{ rad}$$

Equation (E7.9) becomes:

$$x(t) = \frac{200}{386675} \left(1 - \frac{1}{\sqrt{1-0.01^2}} e^{-0.01 \times 892.225(t-1)} \cos(892.21(t-1) - 0.01) \right)$$

which gives the displacement time response as:

$$x(t) = 5.17 \times 10^{-4} \left(1 - 1.00005 e^{-8.9225(t-1)} \cos(892.21t - 892.22) \right) \text{ m}$$

or

$$x(t) = 0.52 \left(1 - 1.00005 e^{-8.9225(t-1)} \cos(892.21t - 892.22) \right) \text{ mm}$$

The displacement time response of the beam under the step force is shown in Figure 7.21.

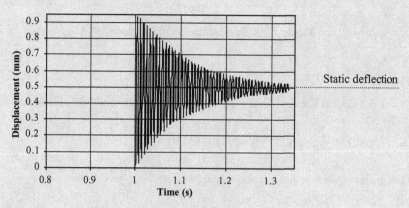

Figure 7.21: Representation of Example 7.9

7.5 VIBRATION ISOLATION

In order to reduce unwanted vibration and its effect on structures, vibration isolators are often used. Vibration isolators can be used either to isolate the source of vibration or to isolate the system of interest. This can be achieved by inserting highly damped materials between the source of vibration and the system of interest so that the stiffness and damping between them are altered. Figure 7.22 shows a vibration isolator model used to protect a foundation from the vibration produced by a machine. The unwanted vibration could be due to unbalanced forces, as in the case of rotating machines and engines, or impact forces, as in the case of stamping presses. The vibration isolator system in Figure 7.22 consists of a spring of stiffness k and a viscous damper of damping coefficient c.

The force transmitted from the machine to the foundation, denoted as F_t, through the spring and the damper is obtained as:

$$F_t = kx + c\dot{x} \tag{7.28}$$

If the force produced by the machine varies harmonically, i.e. $f = F_o \sin \omega t$, the equation of motion, Equation (7.12) is applicable and has the form:

$$m\ddot{x} + c\dot{x} + kx = F_o \sin \omega t \tag{7.29}$$

The steady-state solution of Equation (7.29) was obtained in Section 7.3.2 and is given by:

$$x_p = \chi \sin(\omega t - \phi) \tag{7.30}$$

where

$$\chi = \frac{F_o/k}{\sqrt{\left(1 - \left(\frac{\omega}{\omega_n}\right)^2\right)^2 + \left(\frac{2\zeta\omega}{\omega_n}\right)^2}} \tag{7.31}$$

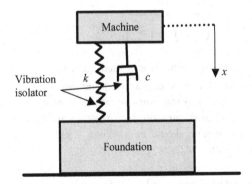

Figure 7.22: Vibration isolator system

and

$$\phi = \tan^{-1} \frac{2\zeta\,(\omega/\omega_n)}{1 - \left(\frac{\omega}{\omega_n}\right)^2} \tag{7.32}$$

Differentiating Equation (7.30) with respect to time and substituting Equation (7.30) and its derivative into Equation (7.28) gives:

$$F_t = k\chi\,\sin(\omega t - \phi) - c\chi\omega\cos(\omega t - \phi) \tag{7.33}$$

The magnitude of the force F_t is obtained as:

$$F_t = \sqrt{(k\chi)^2 + (c\chi\omega)^2} = \chi\sqrt{k^2 + (c\omega)^2} \tag{7.34}$$

Substituting Equation (7.31) into Equation (7.34), the transmitted force is given by:

$$F_t = \frac{F_o k\sqrt{k^2 + (c\omega)^2}}{\sqrt{\left(1 - \left(\frac{\omega}{\omega_n}\right)^2\right)^2 + \left(\frac{2\zeta\omega}{\omega_n}\right)^2}} \tag{7.35}$$

Rearranging Equation (7.35) gives:

$$\frac{F_t}{F_o} = \sqrt{\frac{1 + \left(\frac{c\omega}{k}\right)^2}{\left(1 - \left(\frac{\omega}{\omega_n}\right)^2\right)^2 + \left(\frac{2\zeta\omega}{\omega_n}\right)^2}} \tag{7.36}$$

The ratio between the transmitted force F_t and the excitation force F_o is called the transmissibility, or transmission ratio of the isolator and is denoted by T_r. Since $k = m\omega_n^2$ and $c = \zeta c_c = 2\zeta m\omega_n$, the ratio $\frac{c\omega}{k}$ becomes $\frac{c\omega}{k} = \frac{2\zeta m\omega_n\omega}{m\omega_n^2} = \frac{2\zeta\omega}{\omega_n}$ and the transmission ratio becomes:

$$T_r = \frac{F_t}{F_o} = \sqrt{\frac{1 + \left(\frac{2\zeta\omega}{\omega_n}\right)^2}{\left(1 - \left(\frac{\omega}{\omega_n}\right)^2\right)^2 + \left(\frac{2\zeta\omega}{\omega_n}\right)^2}} \tag{7.37}$$

In Figure 7.23, the transmission ratio T_r is plotted against $\frac{\omega}{\omega_n}$ for different damping ratios ζ. Figure 7.23 shows that all curves meet at one point where $\frac{\omega}{\omega_n} = \sqrt{2}$. At that point, the transmission ratio is equal to 1. A transmission ratio of 1 means that the force transmitted to the foundation is equal to the amplitude of the excitation force. In order to achieve isolation, the transmission ratio should be less than 1 and in such a case the frequency ratio $\frac{\omega}{\omega_n}$ should be larger than $\sqrt{2}$.

In case of a small amount of damping and for $\frac{\omega}{\omega_n}$ larger than $\sqrt{2}$, damping is less significant as can be seen from Figure 7.23, and therefore it can be ignored in Equation (7.37), which

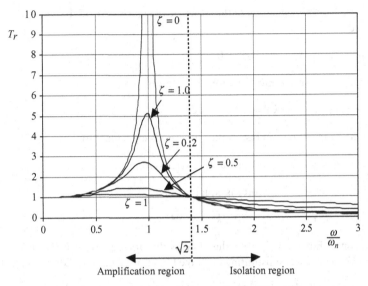

Figure 7.23: Transmission ratio T_r versus $\frac{\omega}{\omega_n}$

reduces to:

$$T_r = \frac{F_t}{F_o} = \frac{1}{\left(\frac{\omega}{\omega_n}\right)^2 - 1} \tag{7.38}$$

Another vibration isolator system, shown in Figure 7.24(a), is used to protect a structure from the vibration of the foundation, e.g. because of an earthquake. The foundation has a harmonic displacement $y = Y \sin \omega t$, where Y is the excitation amplitude.

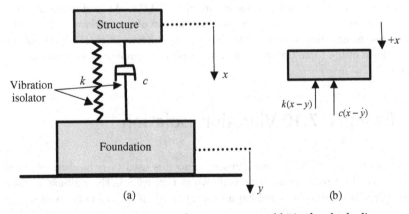

Figure 7.24: a) Vibration isolator for a structure and b) its free-body diagram

Consider the free-body diagram of the structure in Figure 7.24(b); the equation of motion is obtained as:

$$m\ddot{x} + c(\dot{x} - \dot{y}) + k(x - y) = 0 \qquad (7.39)$$

Substituting $y = Y \sin \omega t$ into Equation (7.39) gives:

$$m\ddot{x} + c\dot{x} + kx = kY \sin \omega t + cY\omega \cos \omega t \qquad (7.40)$$

Thus the harmonic force applied to the structure is equivalent to:

$$F = F_o \sin(\omega t - \varphi) = kY \sin \omega t + cY\omega \cos \omega t \qquad (7.41)$$

where the magnitude of this force is equal to:

$$F_o = \sqrt{(kY)^2 + (cY\omega)^2} = Y\sqrt{k^2 + (c\omega)^2} \qquad (7.42)$$

The solution of Equation (7.40) is the same as that for Equation (7.29) and is given by Equation (7.31). Hence, substituting the force Equation (7.42) into Equation (7.31), gives

$$\chi = \frac{Y\sqrt{k^2 + (c\omega)^2}/k}{\sqrt{\left(1 - \left(\frac{\omega}{\omega_n}\right)^2\right)^2 + \left(\frac{2\zeta\omega}{\omega_n}\right)^2}} \qquad (7.43)$$

Rearranging in a similar way to Equation (7.36), the displacement transmissibility, T_d, is obtained as:

$$T_d = \frac{\chi}{Y} = \sqrt{\frac{1 + \left(\frac{2\zeta\omega}{\omega_n}\right)^2}{\left(1 - \left(\frac{\omega}{\omega_n}\right)^2\right)^2 + \left(\frac{2\zeta\omega}{\omega_n}\right)^2}} \qquad (7.44)$$

It can be seen that the right-hand side in Equation (7.44) is identical to that in Equation (7.36). This means that the isolation and amplification of the displacement transmission in the case of harmonic motion of the base or foundation (Figure 7.24(a)) are identical to those of the force transmission in the case of harmonic motion of the machine (Figure 7.22).

Example 7.10 Vibration isolation I

An isolator of stiffness k and damping ratio ζ is used to reduce the vibration produced by a washing machine as shown in Figure 7.25. The washing machine operates at 1200 rev/min and has a mass of 60 kg. If the isolator provides 80%

isolation, determine the isolator stiffness k for the following cases:

a) ignore damping;
b) consider a damping ratio of $\zeta = 0.01$;
c) consider a damping ratio of $\zeta = 0.1$.

Figure 7.25: Representation of Example 7.10

Solution

The transmission ratio is $T_r = 1-$ isolation ratio $= 1-0.8 = 0.2$.

a) If damping is neglected, Equation (7.38) can be used as follows:

$$T_r = \frac{1}{\left(\frac{\omega}{\omega_n}\right)^2 - 1} \Rightarrow 0.2 = \frac{1}{\left(\frac{\omega}{\omega_n}\right)^2 - 1}$$

from which

$$\left(\frac{\omega}{\omega_n}\right) = 2.4495$$

Since the forcing frequency $\omega = 1200 \times \frac{2\pi}{60} = 125.6637$ rad/s, the undamped angular frequency is calculated as:

$$\left(\frac{125.6637}{\omega_n}\right) = 2.4495 \Rightarrow \omega_n = 51.3018 \text{ rad/s}$$

The undamped angular frequency is also given by:

$$\omega_n = 51.3018 = \sqrt{\frac{k}{m}} = \sqrt{\frac{k}{60}}$$

from which the stiffness is calculated as:

$$k = 157\,912\,\text{N} = 157.91\,\text{kN}$$

b) If damping is taken into account, the transmission ratio is obtained from Equation (7.37) as:

$$T_r = \sqrt{\frac{1 + \left(\frac{2\zeta\omega}{\omega_n}\right)^2}{\left(1 - \left(\frac{\omega}{\omega_n}\right)^2\right)^2 + \left(\frac{2\zeta\omega}{\omega_n}\right)^2}} \qquad (E7.10)$$

In order to determine the stiffness, $\frac{\omega}{\omega_n}$ should be calculated from the above equation. Rearranging Equation (E7.10) gives:

$$T_r^2 \left(1 - \left(\frac{\omega}{\omega_n}\right)^2\right)^2 + \left(\frac{2\zeta\omega}{\omega_n}\right)^2 = 1 + \left(\frac{2\zeta\omega}{\omega_n}\right)^2$$

Expanding this equation, gives:

$$T_r^2 \left(\frac{\omega}{\omega_n}\right)^4 - 2\left(\frac{\omega}{\omega_n}\right)^2 (T_r^2 - 2\zeta^2 T_r^2 + 2\zeta^2) + \left(T_r^2 - 1\right) = 0$$

Solving using the quadratic formula, leads to:

$$\left(\frac{\omega}{\omega_n}\right)^2 = \frac{-b \pm \sqrt{b^2 - 4 \times a \times c}}{2 \times a}$$

where

$$a = T_r^2$$
$$b = -2 \times (T_r^2 - 2\zeta^2 T_r^2 + 2\zeta^2)$$
$$c = T_r^2 - 1$$

For $T_r = 0.2$ and $\zeta = 0.01$, the constants a, b and c are calculated as:

$$a = 0.2^2 = 0.04$$
$$b = -2 \times (0.2^2 - 2 \times 0.01^2 \times 0.2^2 + 2 \times 0.01^2) = -0.080384$$
$$c = 0.2^2 - 1 = -0.96$$

Thus, the solution of $\frac{\omega}{\omega_n}$ is given by:

$$\left(\frac{\omega}{\omega_n}\right)^2 = \frac{0.080384 \pm \sqrt{0.080384^2 - 4 \times 0.04 \times (-0.96)}}{2 \times 0.04}$$

$$\Rightarrow \left(\frac{\omega}{\omega_n}\right) = 2.45067$$

Since the forcing frequency is $\omega = 125.6637$ rad/s, the undamped angular frequency is calculated as:

$$\left(\frac{125.6637}{\omega_n}\right) = 2.46067 \Rightarrow \omega_n = 51.277 \, \text{rad/s}$$

which is also given by:

$$\omega_n = 51.277 = \sqrt{\frac{k}{m}} = \sqrt{\frac{k}{60}}$$

From which the stiffness is calculated as:

$$k = 157762 \, \text{N} = 157.76 \, \text{kN}$$

c) For $T_r = 0.2$ and $\zeta = 0.1$, the constants a, b and c are calculated as:

$$a = 0.2^2 = 0.04$$

$$b = -2 \times (0.2^2 - 2 \times 0.1^2 \times 0.2^2 + 2 \times 0.1^2) = -0.1184$$

$$c = 0.2^2 - 1 = -0.96$$

Thus, the solution of $\frac{\omega}{\omega_n}$ is given by:

$$\left(\frac{\omega}{\omega_n}\right)^2 = \frac{0.1184 \pm \sqrt{0.1184^2 - 4 \times 0.04 \times (-0.96)}}{2 \times 0.04}$$

$$\Rightarrow \left(\frac{\omega}{\omega_n}\right) = 2.5686$$

Since the forcing frequency is $\omega = 125.6637$ rad/s, the undamped angular frequency is calculated as:

$$\left(\frac{125.6637}{\omega_n}\right) = 2.5686 \Rightarrow \omega_n = 48.923 = \text{rad/s}$$

which is also given by:

$$\omega_n = 48.923 = \sqrt{\frac{k}{m}} = \sqrt{\frac{k}{60}}$$

From which the stiffness is calculated as:

$$k = 143\,608 \, \text{N} = 143.61 \, \text{kN}$$

Comparing the answers obtained in these cases, it can be seen that the approximation of Equation (7.38) for small or negligible damping (case a) is acceptable for a damping ratio of $\zeta = 0.01$ (case b), but is unacceptable for a damping ratio of $\zeta = 0.1$ (case c).

Example 7.11 Vibration isolation II

A vibration isolator system is designed for a machine that has a rotating unbalanced mass of 1 kg at a distance of 0.25 m from the axis of rotation and operates at a speed of ω as shown in Figure 7.26. The isolator is a damped system with negligible damping and a stiffness of 400 kN/m and the machine has a mass of 100 kg. Determine the angular speed of the machine ω so that the force transmitted to the foundation is less than 1.2 kN.

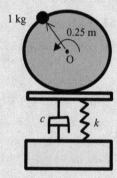

Figure 7.26: Representation of Example 7.11

Solution

From Equation (5.1) in Chapter 5, the amplitude of the force produced by the machine is:

$$F_o = mr\omega^2 = 1 \times 0.25 \times \omega^2 = 0.25\omega^2$$

Since the transmitted force $F_t = 1200$ N, the force transmission ratio is calculated as:

$$T_r = \frac{F_t}{F_o} = \frac{1200}{0.25\omega^2} = \frac{4800}{\omega^2}$$

The undamped angular frequency of the system is:

$$\omega_n = \sqrt{\frac{400\,000}{100}} = 63.2455 \, \text{rad/s}$$

Using Equation (7.38) for negligible damping gives:

$$T_r = \frac{4800}{\omega^2} = \frac{1}{\left(\frac{\omega}{63.2455}\right)^2 - 1}$$

from which the angular speed of the machine is calculated as:

$$\omega = 154.9185\,\text{rad/s} = 154.9185 \times \frac{60}{2\pi} = 1479\,\text{rev/min}$$

Example 7.12 Vibration isolation III

A vibration isolator with negligible damping is used to reduce the force produced by an engine and transmitted to a foundation, as shown in Figure 7.27. If the engine runs at 2500 rev/min and the isolator deflects by 8 mm due to the engine's weight, determine the transmission ratio and hence estimate the percentage reduction in the force transmitted to the foundation.

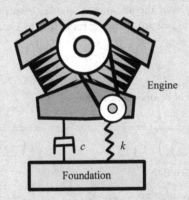

Figure 7.27: Representation of Example 7.12

Solution

Since the stiffness is given by $k = \frac{mg}{\delta}$, where mg is the engine's weight and δ is the static deflection of the vibrator due to the engine's weight, the undamped angular frequency of the system is given by:

$$\omega_n = \sqrt{\frac{k}{m}} = \sqrt{\frac{mg}{m\delta}} = \sqrt{\frac{g}{\delta}} = \sqrt{\frac{9.81}{8 \times 10^{-3}}} = 35.01785\,\text{rad/s}$$

The forcing frequency of the engine is:

$$\omega = 2500 \times \frac{2\pi}{60} = 261.8\,\text{rad/s}$$

From Equation (7.38) for negligible damping, the transmission ratio is:

$$T_r = \frac{1}{\left(\frac{261.8}{35.01785}\right)^2 - 1} = 0.0182$$

The percentage reduction (R) in the force transmitted to the foundation is:

$$R = (1 - T_r) \times 100\% = 98.2\%$$

Example 7.13 Vibration isolation IV

An unbalanced electric motor is mounted on an isolator. If the steady-state amplitude of vibration of the motor is 20 mm at resonance and 7 mm at a forcing frequency equal to 80% of the undamped angular frequency of the system, i.e. $\frac{\omega}{\omega_n} = 0.8$, determine the isolator damping ratio and the excitation amplitude.

Solution

Using Equation (7.31) at resonance, becomes:

$$\chi = \frac{F_o/k}{\sqrt{\left(1 - \left(\frac{\omega}{\omega_n}\right)^2\right)^2 + \left(\frac{2\zeta\omega}{\omega_n}\right)^2}} \Rightarrow 0.02 = \frac{F_o/k}{\sqrt{\left(1 - (1)^2\right)^2 + (2 \times \zeta \times 1)^2}}$$

$$\Rightarrow F_o/k = 0.04\zeta \tag{E7.13a}$$

When $\frac{\omega}{\omega_n} = 0.8$, Equation (7.31) gives:

$$0.007 = \frac{F_o/k}{\sqrt{\left(1 - (0.8)^2\right)^2 + (2 \times \zeta \times 0.8)^2}} \Rightarrow 4.9 \times 10^{-5}$$

$$\times (0.1296 + 2.56\zeta^2) = (F_o/k)^2 \tag{E7.13b}$$

Solving (E7.13a) and (E7.13b), yields:

$$\zeta = 0.065625 = 6.6\%$$

and

$$F_o/k = 2.625 \times 10^{-3} \text{ m}$$

Example 7.14 Vibration isolation V

An exhaust fan has a mass of 90 kg, operates at 1100 rev/min and is supported by an undamped vibration isolator that has a stiffness of 220 kN. If the harmonic force produced by the fan has a magnitude of 12 kN, determine:

a) the transmitted force through the isolator;
b) the excitation amplitude of the exhaust fan;
c) the steady-state amplitude of the exhaust fan.

Solution

The angular frequency of the isolator is:

$$\omega_n = \sqrt{\frac{k}{m}} = \sqrt{\frac{220\,000}{90}} = 49.4413\,\text{rad/s}$$

The operating frequency is:

$$\omega = 1100 \times \frac{2\pi}{60} = 115.1917\,\text{rad/s}$$

The transmission ratio, from Equation (7.38), is:

$$T_r = \frac{1}{\left(\frac{115.1917}{49.4413}\right)^2 - 1} = 0.2258$$

a) From Equation (7.37), the force transmission ratio is given by:

$$T_r = \frac{F_t}{F_o} \Rightarrow 0.2258 = \frac{F_t}{12000}$$

from which the transmitted force is calculated as:

$$F_t = 2709.856\,\text{N} = 2.71\,\text{kN}$$

b) The excitation amplitude is:

$$Y = \frac{F_o}{k} = \frac{12\,000}{220\,000} = 0.05454\,\text{m} = 54.5\,\text{mm}$$

c) Since the system is undamped, the steady-state amplitude is calculated as:

$$X = \frac{F_t}{k} = \frac{2709.856}{220\,000} = 0.0123\,\text{m} = 12.3\,\text{mm}$$

7.6 Tutorial Sheet

7.6.1 Undamped systems

Q7.1   A motor is mounted on an elastic beam, at its midpoint, as shown in Figure 7.28. It produces a vertical downward harmonic force of 1000 $\sin \omega t$ in Newtons, where ω is the forcing frequency in rad/s. The beam deflects 1 mm at its midpoint under the total mass of the system (the weight of the motor and the beam), which is estimated as 120 kg. Determine:

a) the stiffness of the beam;

[1177 kN/m]

b) the angular frequency of the system in rev/min;

[946 rev/min]

c) the maximum and minimum forcing frequencies in which the motor could operate such that the displacement on the beam should be within the range of ± 4 mm.

[839 rev/min and 1041 rev/min]

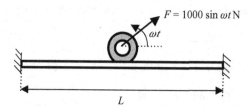

Figure 7.28: Representation of Question 7.1

Q7.2   For the undamped system, shown in Figure 7.29, the mass $m = 100$ kg is supported by two springs of stiffness $k_1 = 1.25$ kN/m and $k_2 = 1.5$ kN/m, which are resting on an elastic foundation. If the foundation undergoes harmonic motion of $2.5 \sin \omega t$ mm, where ω is the forcing frequency in rad/s, determine:

a) the natural frequency of the system in Hz;

[0.835 Hz]

b) the maximum dynamic force magnitude acting on the system;

[−6.875 N]

c) the amplitude of the steady-state vibration if the forcing frequency is 5 rad/s.

[−27.5 mm]

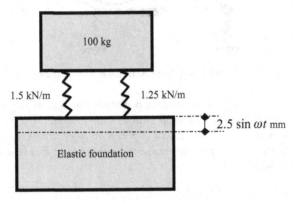

Figure 7.29: Representation of Question 7.2

Q7.3 A pump of 70 kg mass is mounted on a beam of 3 m length at its midpoint. Both ends of the beam are clamped. The beam is made of steel, which has a Young's modulus of 200 GPa and a density of 7800 kg/m^3, and has a cross section of 0.6 m width and 0.012 m thickness. The pump produces a harmonic force of $F = 200 \sin 50t$, where F is in newtons and t is the time in seconds. Determine the steady-state amplitude of vibration of the beam.

[−0.96 mm]

Q7.4 A spring has a stiffness of 4200 N/m and supports a mass m, which is subjected to a harmonic force of $F = 120 \sin 8\pi t$, where F is in newtons and t is the time in seconds. If the amplitude of the steady-state vibration is −0.03 m, determine the mass of the system m.

[12.98 kg]

Q7.5   An unbalanced centrifugal pump of mass 50 kg is supported by four springs each of 8000 N/m as shown in Figure 7.30. If the pump operates at 900 rev/min and the steady-state amplitude should not exceed 4 mm, determine the maximum allowable out-of-balance force.

[−1.65 kN]

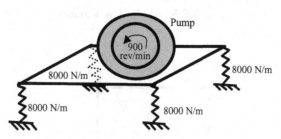

Figure 7.30: Representation of Question 7.5

Q7.6   An instrument of mass 25 kg is mounted on a plate that is supported by four springs each of 1000 N/m as shown in Figure 7.31. If the plate is subjected to a displacement of $x = 8 \sin 10\pi t$, where x is in mm and t is the time in seconds, determine the amplitude of steady-state vibration of the instrument.

[21.33 mm]

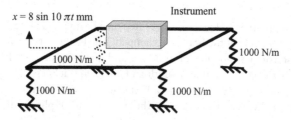

Figure 7.31: Representation of Question 7.6

Q7.7   A building is approximated to an SDOF system with an equivalent mass of 102×10^3 kg at the top of the building supported by two columns each of stiffness 3600 kN/m as shown in Figure 7.32. Determine the amplitude of vibration at the top of the building due to an earthquake that produces a displacement amplitude of 80 mm at a frequency of 7.5 rad/s.

[0.394 m]

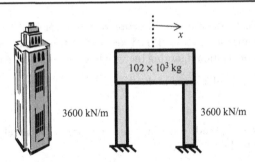

Figure 7.32: Representation of Question 7.7

Q7.8 A fixed-fixed steel beam, shown in Figure 7.33, has a length of 5 m, width of 0.5 m, thickness of 0.1 m and elastic modulus of 200 GPa. The beam carries an electric motor of $m = 75$ kg mass in its midpoint. If the motor is operating at $\omega = 1200$ rev/min speed and producing an out-of-balance force of $F_o = 5000$ N, determine:

a) the amplitude of the steady-state vibration if the mass of the beam is ignored;

[0.43 mm]

b) the amplitude of the steady-state vibration if the mass of the beam is taken into account (consider that the beam density is 7800 kg/m³ and the total mass of the system is the mass of the motor plus 0.37× the mass of the beam).

[22.4 mm]

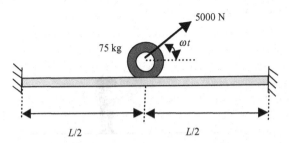

Figure 7.33: Representation of Question 7.8

Q7.9 A simply-supported beam carries at its midpoint an electric motor of mass 80 kg that has an eccentric flywheel (Figure 7.34). The beam deflects 15 mm under the action of the motor's weight. The flywheel has an unbalanced mass of 0.25 kg located at a distance of 0.2 m from the axis of rotation. Ignoring the weight of the beam, determine:

a) the angular speed of the flywheel at resonance;

[25.57 rad/s]

b) the amplitude of steady-state vibration if the angular speed of the flywheel is 25 rad/s.

[13.5 mm]

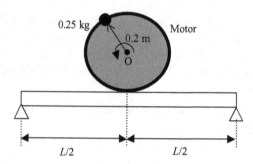

Figure 7.34: Representation of Question 7.9

Q7.10 A boy of mass 50 kg jumps in regular time in the middle of a simply supported wooden board (Figure 7.35) so that he touches the board every 0.5 s. If the board deflects 3 cm under the boy's weight in a static condition, determine the amplitude of steady-state vibration of the board. Ignore the board's weight.

[5.8 cm]

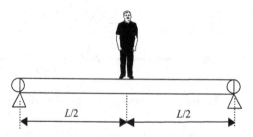

Figure 7.35: Representation of Question 7.10

7.6.2 Damped systems

Q7.11 A harmonic force of $F = 100 \sin 25t$, where F is in newtons and t is the time in seconds, is acting on a machine of mass 200 kg, which is supported by a spring and a damper as shown in Figure 7.36. If the spring has a stiffness of 125 kN/m and the damping ratio is $\zeta = 0.15$, determine:

a) the undamped and damped natural frequencies of the system in Hz;

[3.98 Hz, 3.93 Hz]

b) the amplitude of the steady-state vibration.

[2.67 mm]

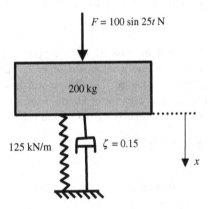

Figure 7.36: Representation of Question 7.11

Q7.12 For an underdamped forced vibration SDOF, the ratio between the steady-state amplitude of vibration (χ) and the excitation amplitude (F_o/k) is given by:

$$\frac{\chi}{F_o/k} = \frac{1}{\sqrt{\left(1 - \left(\frac{\omega}{\omega_n}\right)^2\right)^2 + \left(\frac{2\zeta\omega}{\omega_n}\right)^2}}$$

where m is the mass of the body, k is the stiffness of the spring, ω is the forcing frequency, ω_n the undamped natural frequency and ζ the damping ratio. Calculate the ratio $\frac{\chi}{F_o/k}$ for the values of ζ and $\frac{\omega}{\omega_n}$ given in the table.

$$\left(\tfrac{\omega}{\omega_n}\right) = 0.85 \quad \left(\tfrac{\omega}{\omega_n}\right) = 0.9 \quad \left(\tfrac{\omega}{\omega_n}\right) = 0.95 \quad \left(\tfrac{\omega}{\omega_n}\right) = 1 \quad \left(\tfrac{\omega}{\omega_n}\right) = 1.05$$

$\zeta = 0.1$
$\zeta = 0.2$
$\zeta = 0.3$

What are the effects of ζ and $\tfrac{\omega}{\omega_n}$ on the amplitude of the steady-state vibration?

Q7.13 A single degree of freedom consists of a spring, a mass and a damper. When a harmonic force is applied to the system, the amplitude of steady-state vibration is 15 mm at resonance and 9 mm at a frequency of 80% of the resonance frequency. Determine the damping ratio of the system.

[0.123]

Q7.14 An electric motor of mass 32 kg is supported by four springs each of stiffness 250 N/m and a viscous damper of damping coefficient 42.93 N.s/m as shown in Figure 7.37. The motor has an out-of-balance mass of 5 kg located at a distance of 60 mm from its centre of rotation and is running with an angular velocity of 9 rad/s. Determine:

a) the damping ratio;

[0.12]

b) the out-of-balance force;

[24.3 N]

c) the amplitude of steady-state vibration.

[14.8 mm]

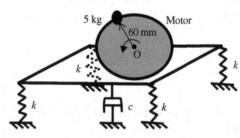

Figure 7.37: Representation of Question 7.14

Q7.15 A car is modelled as an SDOF system with a mass of 1000 kg, an equivalent
stiffness of 400 kN/m, and an equivalent damping coefficient of 20 kN.s/m,
as shown in Figure 7.38. If the road is assumed to provide a base-motion
vertical displacement equivalent to $y = 10 \sin 6t$, where y is in mm and t
is the time in seconds, determine:

a) the natural frequency of the car;

[3.18 Hz]

b) the damping ratio;

[0.5]

c) the amplitude of vibration of the car due to the road motion.

[10.9 mm]

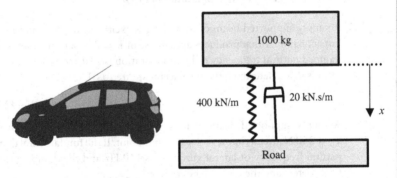

Figure 7.38: Representation of Question 7.15

Q7.16 An unbalanced machine of 100 kg is mounted on a spring of stiffness
780 kN/m and a damper of damping coefficient 490 N.s/m as shown in
Figure 7.39. The machine produces an out-of-balance force of 370 N due
to a mass of 1 kg located at a distance r from the axis of rotation. If the
machine runs at 300 rad/s, determine:

a) the natural frequency of the system;

[14.06 Hz]

b) the damping ratio;

[0.0277]

c) the amplitude of steady-state vibration;

[0.045 mm]

d) the distance r where the out-of-balance mass is located.

[4.11 mm]

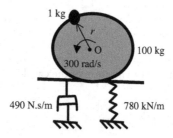

Figure 7.39: Representation of Question 7.16

Q7.17 A simply supported beam of mass 1 kg is excited at its midpoint with an impact hammer that produces an impulse of $\overline{F} = 2$ N.s. If the fundamental damped natural frequency of lateral vibration is 140 Hz and the damping ratio is 0.01, determine the displacement after 0.01 s.

[2.45 mm]

Q7.18 A simply supported beam of mass 1 kg is subjected to a sudden force $F_o = 250$ N after a time $t_o = 1$ s at its midpoint. If the fundamental damped natural frequency of lateral vibration is 140 Hz and the damping ratio is 0.01, determine the displacement after 1.02 s.

[0.48 mm]

Q7.19 Show that, for a critically damped system, the impulse displacement time response function is given by:

$$x = \frac{\overline{F}}{m} t.e^{-\omega_n t}$$

where $\overline{F}$ is the impulse force.

Q7.20 Calculate the response of an underdamped SDOF system, where $m = 2$ kg, $c = 4$ N.s/m and $k = 32$ N/m, under an impulse of $\overline{F} = 4$ N.s

$$[x = 0.5164e^{-t} \sin 3.873t \text{ m}]$$

7.6.3 Vibration isolators

Q7.21 A vibration isolator of stiffness k and negligible damping is used to reduce the vibration produced by a washing machine. The washing machine operates at 1100 rev/min and has a mass of 65 kg. If the isolator provides 75% isolation, determine the isolator stiffness k.

$$[172.51 \text{ kN/m}]$$

Q7.22 An instrument of mass 10 kg is isolated from a device that produces vibration at a frequency of 30 Hz using a vibration isolator system consisting of a spring of stiffness k. If the isolator provides 80% isolation, determine:

a) the isolator stiffness k;

$$[59.22 \text{ kN/m}]$$

b) the static deflection of the isolator.

$$[1.66 \text{ mm}]$$

Q7.23 A vibration isolator system is designed for a machine that has a rotating unbalanced mass of m at a distance of 0.2 m from the axis of rotation and operates at a speed of 80 rad/s as shown in Figure 7.40. The vibration isolator has negligible damping and a stiffness of 450 kN/m, and the machine has a mass of 100 kg. Determine the minimum unbalanced mass m so that the force transmitted to the foundation is less than 2.2 kN

$$[0.73 \text{ kg}]$$

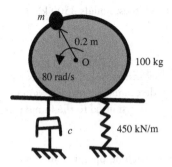

Figure 7.40: Representation of Question 7.23

Q7.24 A washing machine produces an unbalanced force of 199 N and operates at 1100 rev/min. If the machine is mounted on a vibration isolator with negligible damping that deflects 4 mm under the static load of the machine, determine:

a) the transmission ratio;

[0.227]

b) the force transmitted to the foundation.

[45.1 N]

Q7.25 An unbalanced electric motor is mounted on a vibration isolator. If the steady-state amplitude of vibration of the motor is 20 mm at resonance and 5 mm away from resonance ($\frac{\omega}{\omega_n} \approx 0$), determine:

a) the isolator damping ratio;

[0.125]

b) the unbalanced force produced by the motor if the isolator stiffness is 1000 N/m.

[5 N]

Q7.26 A machine of a mass 1100 kg produces an unbalanced force of 22 kN and operates at a speed of 550 rev/min. If the machine is mounted on an undamped vibration isolator, determine the transmission ratio and the force transmitted to the foundation for the following cases:

a) the isolator stiffness equals 1500 kN/m

[0.698, 15.36 kN]

b) the isolator stiffness equals 3500 kN/m

[23.49, 516.77 kN]

Q7.27 An engine running at 2500 rev/min is mounted on a foundation through an undamped vibration isolator. If the isolator deflects 12 mm under the engine weight, determine the transmission ratio and the percentage reduction in the force transmitted to the foundation.

[0.012, 98.8%]

Q7.28 A dishwashing machine operates at 350 rev/min and is mounted on an isolator with negligible damping. If the mass of the machine is 70 kg, determine the isolator stiffness and its maximum deflection in order to achieve 65% isolation.

[24.38 kN, 28 mm]

Q7.29 An engine of mass 600 kg (see Figure 7.41) is mounted on a rigid foundation through a vibration isolator of negligible damping and produces a force of $F = 12\,000 \sin 26t$, where F is in newtons and t is the time in seconds. If the maximum allowable force that could be transmitted to the foundation is 4000 N, determine the isolator stiffness.

[101.4 kN/m]

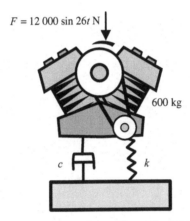

Figure 7.41: Representation of Question 7.29

Q7.30 An exhaust fan has a mass of 90 kg, operates at 1100 rev/min and is supported by an undamped vibration isolator. If the harmonic force produced by the fan has a magnitude of 12 kN and the transmitted force through the isolator is 2400 N, determine:

a) the isolator stiffness;

[199.07 kN]

b) the static deflection of the exhaust fan;

[60.3 mm]

c) the steady-state amplitude of the exhaust fan.

[12.1 mm]

CHAPTER 8

Vibration of Systems with Two Degrees of Freedom

8.1 INTRODUCTION

Chapters 6 and 7 have considered the vibration of systems with one degree of freedom. In this chapter, the derivation is extended to systems with two degrees of freedom. In a system with two degrees of freedom, two independent motions are used to define the vibration of the system. It follows that two co-ordinates and two equations of motion are required to describe the motion. These equations are, in general, coupled differential equations, i.e. each equation involves both degrees of freedom. The number of natural frequencies is equal to the number of degrees of freedom in the system; for a system with two degrees of freedom, there are two natural frequencies. For each natural frequency, there is a normal mode of vibration, also called the principal mode, the natural mode or the mode shape of vibration. A normal mode for a system with two degrees of freedom is the ratio between the two amplitudes at a specific natural frequency. The overall vibration of the system has a contribution from each normal mode. The amount of contribution of each mode depends on how and where the structure or the system is excited. If a harmonic force is applied to the system at a forcing frequency close or equal to any of its natural frequencies, resonance takes place. An example of a system with two degrees of freedom is shown in Figure 8.1, in which a vehicle suspension and tyre assembly is modelled as a system with two masses and two springs. The axle mass is m_1, the tyre stiffness is k_1, the trailer mass is m_2 and the suspension stiffness is k_2.

Another example of a system with two degrees of freedom is shown in Figure 8.2, where a bicycle, a motorcycle or a vehicle could be approximated as a rigid bar supported by front and rear springs having stiffness k_f and k_r, respectively. The two degrees of freedom of the system can be either the vertical displacements at the front and rear springs, y_f and y_r, or the vertical displacement (y_C) and rotation (θ_C) at the centre of gravity of the rigid bar (point CG).

Since the derivation of the equations of motion for a system with two degrees of freedom makes use of matrix manipulation, it serves as a good introduction to systems with multiple degrees of freedom. These equations can easily be extended to systems with multiple degrees of freedom.

In the following sections, firstly the equations of motion for undamped systems are derived using Newton's second law. Then, the solution of these equations for undamped free

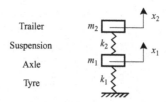

Figure 8.1: A system with two masses and two springs

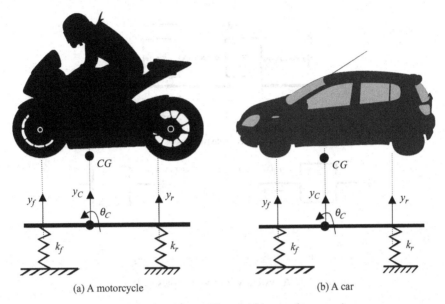

(a) A motorcycle (b) A car

Figure 8.2: A system with a rigid bar and two springs

vibration is presented and the two natural frequencies of the system are determined. Next, the forced vibration of undamped systems is discussed and the principle of vibration absorbers is introduced. Finally, a brief introduction to solving free and forced viscous damped systems with two degrees of freedom is presented.

8.2 DERIVING THE EQUATIONS OF MOTION

Consider the two masses m_1 and m_2 in Figure 8.3(a), which are attached to two springs having stiffness constants k_1 and k_2. As the masses can move only in the x direction, two co-ordinates are required to define the motion of the system, which represents the horizontal displacements of the two masses m_1 and m_2. These two displacements are denoted as x_1 and x_2, for masses m_1 and m_2, respectively. The external forces f_1 and f_2, which are functions of time, are applied to the masses m_1 and m_2, respectively. The free-body diagrams for both masses are shown in Figure 8.3(b), in which all forces acting on the masses by the springs, as well as the external forces, are drawn.

Using Newton's second law and assuming that there is neither friction nor damping in the system, the following equations of motion for m_1 and m_2 are obtained:

$$\sum F_{x1} = m_1 \ddot{x}_1 \qquad (8.1)$$

$$\sum F_{x2} = m_2 \ddot{x}_2 \qquad (8.2)$$

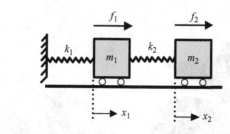

(a) A system with two masses

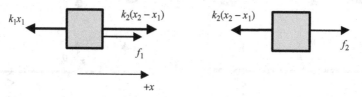

(b) Free-body diagrams for masses m_1 and m_2

Figure 8.3: A system with two degrees of freedom

Applying Equations (8.1) and (8.2) to the free-body diagrams in Figure 8.3(b), gives:

$$-k_1 x_1 + k_2(x_2 - x_1) + f_1 = m_1 \ddot{x}_1 \tag{8.3}$$

$$-k_2(x_2 - x_1) + f_2 = m_2 \ddot{x}_2 \tag{8.4}$$

If the two masses are suspended from two springs as shown in Figure 8.4(a), the weights of the masses $m_1 g$ and $m_2 g$ and the static deflections δ_1 and δ_2 for masses m_1 and m_2, respectively, should be taken into account when writing Newton's second law. From the free-body diagrams in Figure 8.4(b), the two Newton's second law equations are written as:

$$-k_1(x_1 + \delta_1) + k_2(x_2 + \delta_2 - x_1 - \delta_1) + m_1 g + f_1 = m_1 \ddot{x}_1 \tag{8.5}$$

$$-k_2(x_2 + \delta_2 - x_1 - \delta_1) + m_2 g + f_2 = m_2 \ddot{x}_2 \tag{8.6}$$

However, from the static equilibrium in Figure 8.4(c), the static deflections δ_1 and δ_2 are related to the springs' stiffness k_1 and k_2, and the weights $m_1 g$ and $m_2 g$ as:

$$-k_1 \delta_1 + k_2(\delta_2 - \delta_1) + m_1 g = 0 \tag{8.7}$$

$$-k_2(\delta_2 - \delta_1) + m_2 g = 0 \tag{8.8}$$

Substituting Equation (8.7) into (8.5) and Equation (8.8) into (8.6), the terms containing δ_1, δ_2, $m_1 g$ and $m_2 g$ are eliminated from Equations (8.5) and (8.6) and expressions identical to Equations (8.3) and (8.4) are obtained. Thus, the weights of the masses do not contribute in the equations of motion of the system.

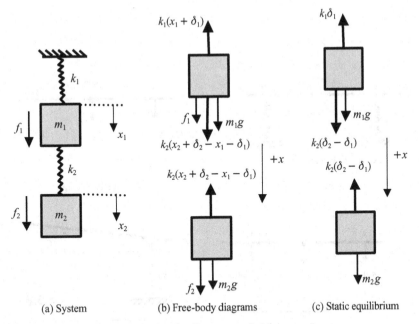

(a) System   (b) Free-body diagrams   (c) Static equilibrium

Figure 8.4: Two bodies suspended from springs

Re-arranging the terms in Equations (8.3) and (8.4), by moving the internal and inertia forces to the left-hand side and the external forces to the right-hand side, gives:

$$m_1\ddot{x}_1 + (k_1 + k_2)x_1 - k_2x_2 = f_1 \tag{8.9}$$

$$m_2\ddot{x}_2 - k_2x_1 + k_2x_2 = f_2 \tag{8.10}$$

In matrix form, they become:

$$\begin{bmatrix} m_1 & 0 \\ 0 & m_2 \end{bmatrix} \begin{bmatrix} \ddot{x}_1 \\ \ddot{x}_2 \end{bmatrix} + \begin{bmatrix} k_1 + k_2 & -k_2 \\ -k_2 & k_2 \end{bmatrix} \begin{bmatrix} x_1 \\ x_2 \end{bmatrix} = \begin{bmatrix} f_1 \\ f_2 \end{bmatrix} \tag{8.11}$$

The rigid bar supported by two springs, shown in Figure 8.5(a), can translate and rotate only in the x-y plane and is constrained to move only in the x direction. The rigid bar has a mass m and a mass moment of inertia I_C. The two springs have stiffness constants k_1 and k_2. The centre of gravity of the rigid bar is located at point C, which is at a distance L_1 from point A and a distance L_2 from point B. The two degrees of freedom that are considered herein are the vertical translation of the centre of gravity y_C and the bar rotation at the centre of gravity θ_C, measured anti-clockwise. Consider the free-body diagram of the rigid bar in Figure 8.5(b); the forces acting on the rigid bar are those

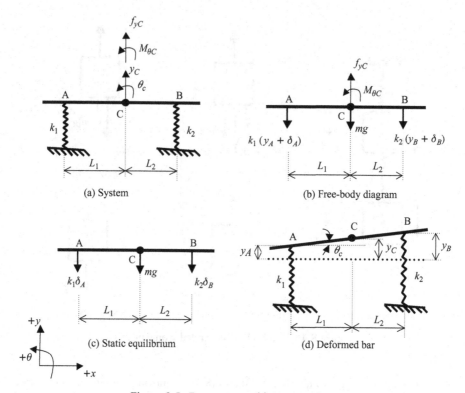

Figure 8.5: Bar supported by two springs

exerted by the springs and the applied external vertical force f_{yC} and bending moment $M_{\theta C}$ at the centre of gravity C.

Two equations of motion can be written, which are Newton's second law in the y direction and the moment equation equivalent to Newton's second law:

$$\sum F_y = m\ddot{y}_C \tag{8.12}$$
$$\sum M_C = I_C \ddot{\theta}_C \tag{8.13}$$

By summing all forces in the y direction in Figure 8.5(b), Equation (8.12) becomes:

$$-k_1(y_A + \delta_A) - k_2(y_B + \delta_B) - mg + f_{yC} = m\ddot{y}_C \tag{8.14}$$

Taking moment about point C, Equation (8.13) becomes:

$$k_1(y_A + \delta_A)L_1 - k_2(y_B + \delta_B)L_2 + M_{\theta C} = I_C \ddot{\theta}_C \tag{8.15}$$

where δ_A and δ_B are the static deflections at points A and B, respectively, due to the bar's weight, mg. However, the static equilibrium equations can be obtained by equating the right-hand side in Equations (8.12) and (8.13) to zero, since there is no motion:

$$\sum F_y = 0 \tag{8.16}$$

$$\sum M_C = 0 \tag{8.17}$$

Referring to the free-body diagram in Figure 8.5(c), summing the forces in the y direction, Equation (8.16) becomes:

$$-k_1 \delta_A - k_2 \delta_B - mg = 0 \tag{8.18}$$

Taking moment about C, Equation (8.17) becomes:

$$k_1 \delta_A L_1 - k_2 \delta_B L_2 = 0 \tag{8.19}$$

Substituting Equation (8.18) into (8.14) and Equation (8.19) into (8.15), the following two equations of motion are obtained:

$$-k_1 y_A - k_2 y_B + f_{yC} = m \ddot{y}_C \tag{8.20}$$
$$k_1 y_A L_1 - k_2 y_B L_2 + M_{\theta C} = I_C \ddot{\theta}_C \tag{8.21}$$

Again, it can be seen from Equations (8.20) and (8.21) that the bar's weight does not contribute in the equations of motion.

Consider Figure 8.5(d). As the bar is rigid, the angle θ_C is given by $\tan \theta_C = \frac{y_C - y_A}{L_1}$, from which y_A is related to the degrees of freedom y_C and θ_C as:

$$y_A = y_C - L_1 \tan \theta_C \tag{8.22}$$

Similarly, from Figure 8.5(d), the angle θ_C is given by $\tan \theta_C = \frac{y_B - y_C}{L_2}$, from which y_B is related to the degrees of freedom y_C and θ_C as:

$$y_B = y_C + L_2 \tan \theta_C \tag{8.23}$$

For small vibration $\tan \theta_C \approx \theta_C$, thus, Equations (8.22) and (8.23) become:

$$y_A = y_C - L_1 \theta_C \tag{8.24}$$
$$y_B = y_C + L_2 \theta_C \tag{8.25}$$

Substituting Equations (8.24) and (8.25) into Equations (8.20) and (8.21) and rearranging terms, yields:

$$m \ddot{y}_c + (k_1 + k_2) y_C - (k_1 L_1 - k_2 L_2) \theta_C = f_{yC} \tag{8.26}$$
$$I_C \ddot{\theta}_C - (k_1 L_1 - k_2 L_2) y_C + (k_1 L_1^2 + k_2 L_2^2) \theta_C = M_{\theta C} \tag{8.27}$$

And in matrix form:

$$\begin{bmatrix} m & 0 \\ 0 & I_C \end{bmatrix} \begin{bmatrix} \ddot{y}_c \\ \ddot{\theta}_C \end{bmatrix} + \begin{bmatrix} k_1 + k_2 & k_2 L_2 - k_1 L_1 \\ k_2 L_2 - k_1 L_1 & k_1 L_1^2 + k_2 L_2^2 \end{bmatrix} \begin{bmatrix} y_C \\ \theta_C \end{bmatrix} = \begin{bmatrix} f_{yC} \\ M_{\theta C} \end{bmatrix} \tag{8.28}$$

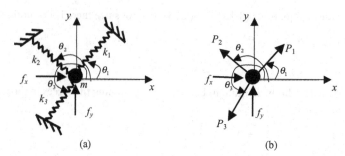

Figure 8.6: Spring-suspended mass: (a) the system and (b) its free-body diagram

Figure 8.6(a) shows another example of a system with two degrees of freedom: a system with a spring-suspended mass. The mass (m) is suspended from three springs and is constrained to move in the x-y plane. Therefore the motion can be defined by the x and y displacements. The three springs have stiffness constants k_1, k_2 and k_3, and the external forces f_x and f_y are acting in the x and y directions, respectively. The forces exerted on the mass by the springs (P_1, P_2 and P_3) are shown in Figure 8.6(b) and are related to (u, v) by:

$$P_i = -k_i(x\cos\theta_i + y\sin\theta_i) \tag{8.29}$$

where $i = 1, 2$ or 3. From the free-body diagram in Figure 8.6(b), Newton's second law in the x and y directions are then written as:

$$\sum_{i=1}^{3} P_i \cos\theta_i + f_x = m\ddot{x}$$
$$\sum_{i=1}^{3} P_i \sin\theta_i + f_y = m\ddot{y} \tag{8.30}$$

Substituting equation (8.29) into (8.30) and rearranging terms yields:

$$m\ddot{x} + \sum_{i=1}^{3} k_i(x\cos^2\theta_i + y\cos\theta_i\sin\theta_i) = f_x$$
$$m\ddot{y} + \sum_{i=1}^{3} k_i(x\sin\theta_i\cos\theta_i + y\sin^2\theta_i) = f_y \tag{8.31}$$

Rearranging in matrix form gives:

$$\begin{bmatrix} m & 0 \\ 0 & m \end{bmatrix}\begin{bmatrix} \ddot{x} \\ \ddot{y} \end{bmatrix} + \sum_{i=1}^{3} k_i \begin{bmatrix} \cos^2\theta_i & \sin\theta_i\cos\theta_i \\ \sin\theta_i\cos\theta_i & \sin^2\theta_i \end{bmatrix}\begin{bmatrix} x \\ y \end{bmatrix} = \begin{bmatrix} f_x \\ f_y \end{bmatrix} \tag{8.32}$$

From the above examples, it is possible to write the equations of motion of any system with two degrees of freedom in a general matrix form. The general matrix form of Equations (8.11), (8.28) and (8.32) is:

$$M\ddot{D} + SD = F \tag{8.33}$$

where M is the mass matrix, S the stiffness matrix, D the degree of freedom vector, $\ddot{D}$ the acceleration vector and F the force vector. The expanded form of Equation (8.33) is:

$$\begin{bmatrix} M_{11} & 0 \\ 0 & M_{22} \end{bmatrix} \begin{bmatrix} \ddot{D}_1 \\ \ddot{D}_2 \end{bmatrix} + \begin{bmatrix} S_{11} & S_{12} \\ S_{21} & S_{22} \end{bmatrix} \begin{bmatrix} D_1 \\ D_2 \end{bmatrix} = \begin{bmatrix} F_1 \\ F_2 \end{bmatrix} \tag{8.34}$$

8.3 UNDAMPED FREE VIBRATION

In the case of free vibration, no force is applied to the system and Equation (8.33) becomes:

$$M\ddot{D} + SD = 0 \tag{8.35}$$

In expanded form, that is:

$$\begin{bmatrix} M_{11} & 0 \\ 0 & M_{22} \end{bmatrix} \begin{bmatrix} \ddot{D}_1 \\ \ddot{D}_2 \end{bmatrix} + \begin{bmatrix} S_{11} & S_{12} \\ S_{21} & S_{22} \end{bmatrix} \begin{bmatrix} D_1 \\ D_2 \end{bmatrix} = \begin{bmatrix} 0 \\ 0 \end{bmatrix} \tag{8.36}$$

The displacement time response or transient solutions of these two differential equations are (see Equation (6.20) in Chapter 6):

$$D_1 = D_{m1} \sin(\omega_n t + \psi) \tag{8.37}$$

$$D_2 = D_{m2} \sin(\omega_n t + \psi) \tag{8.38}$$

where D_{m1} and D_{m2} are the maximum values (or amplitudes) of D_1 and D_2, respectively, ψ is the phase angle between initial displacement and initial velocity and ω_n represents the angular frequency of the system. Differentiating Equations (8.37) and (8.38) twice, the accelerations are obtained as:

$$\ddot{D}_1 = -\omega_n^2 D_{m1} \sin(\omega_n t + \psi) \tag{8.39}$$

$$\ddot{D}_2 = -\omega_n^2 D_{m2} \sin(\omega_n t + \psi) \tag{8.40}$$

Substituting Equations (8.37) to (8.40) into (8.36), multiplying and adding the matrices gives:

$$\begin{bmatrix} S_{11} - \omega_n^2 M_{11} & S_{12} \\ S_{21} & S_{22} - \omega_n^2 M_{22} \end{bmatrix} \begin{bmatrix} D_{m1} \\ D_{m2} \end{bmatrix} = \begin{bmatrix} 0 \\ 0 \end{bmatrix} \tag{8.41}$$

For non-trivial solutions, the determinant of the matrix in Equation (8.41) should be equal to zero:

$$\begin{bmatrix} S_{11} - \omega_n^2 M_{11} & S_{12} \\ S_{21} & S_{22} - \omega_n^2 M_{22} \end{bmatrix} = 0$$

from which

$$M_{11} M_{22} \omega_n^4 - (M_{11} S_{22} + M_{22} S_{11}) \omega_n^2 + S_{11} S_{22} - S_{12}^2 = 0 \tag{8.42}$$

Note that since the stiffness matrix is symmetric, $S_{21} = S_{12}$.

Solving Equation (8.42) using the quadratic formula leads to:

$$\omega_{1,2}^2 = \frac{-b \pm \sqrt{b^2 - 4ac}}{2a} \tag{8.43}$$

where

$$a = M_{11} M_{22}, \quad b = -(M_{11} S_{22} + M_{22} S_{11}), \quad c = S_{11} S_{22} - S_{12}^2$$

The term $b^2 - 4ac$ is equal to:

$$b^2 - 4ac = (M_{11} S_{22} + M_{22} S_{11})^2 - 4 M_{11} M_{22} \left(S_{11} S_{22} - S_{12}^2 \right)$$
$$= M_{11}^2 S_{22}^2 + M_{22}^2 S_{11}^2 - 2 M_{11} M_{22} S_{11} S_{22} + 4 M_{11} M_{22} S_{12}^2$$

which can be written as:

$$b^2 - 4ac = (M_{11} S_{22} - M_{22} S_{12})^2 + 4 M_{11} M_{22} S_{12}^2 \tag{8.44}$$

It can be seen from Equation (8.44) that both terms on the right-hand side ($(M_{11} S_{22} - M_{22} S_{12})^2$ and $4 M_{11} M_{22} S_{12}^2$ are always positive so that $b^2 - 4ac$ is positive and ω_1^2 and ω_2^2 are real. As the product of the diagonal of the stiffness matrix $S_{11} S_{22}$ is always larger than the square of the off-diagonal S_{12}^2, the constant $c = S_{11} S_{22} - S_{12}^2$ is always positive. Thus, the square-root in the quadratic equation $\sqrt{b^2 - 4ac}$ is smaller than b, which leads to positive ω_1^2 and ω_2^2. That proves that ω_1^2 and ω_2^2 should always be real positive numbers.

If ω_1^2 and ω_2^2 are known, the amplitude ratios, r_1 and r_2, can be obtained using Equation (8.41) as:

$$r_1 = \frac{D_{m1}}{D_{m2}} = \frac{-S_{12}}{S_{11} - \omega_1^2 M_{11}} = \frac{S_{22} - \omega_1^2 M_{22}}{-S_{21}} \tag{8.45}$$

$$r_2 = \frac{D_{m1}}{D_{m2}} = \frac{-S_{12}}{S_{11} - \omega_2^2 M_{11}} = \frac{S_{22} - \omega_2^2 M_{22}}{-S_{21}} \tag{8.46}$$

The amplitude of the mode shapes can be scaled as desired, to the first or the second degree of freedom, using the ratios r_1 and r_2. The amplitude ratios r_1 and r_2 are in fact the eigenvectors of the eigenvalues ω_1 and ω_2, respectively.

Example 8.1 Free vibration I

For the system with two degrees of freedom shown in Figure 8.7, the two masses are 20 kg and 10 kg. If the constants of the springs are equal to 15 kN/m and 25 kN/m, calculate the natural frequencies and the mode shape amplitude ratios.

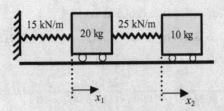

Figure 8.7: Representation of Example 8.1

Solution

Comparing the general form of the mass matrix in Equation (8.34) to that in Equation (8.11):

$$M = \begin{bmatrix} M_{11} & 0 \\ 0 & M_{22} \end{bmatrix} = \begin{bmatrix} m_1 & 0 \\ 0 & m_2 \end{bmatrix}$$

from which

$$M_{11} = m_1 = 20\,\text{kg}, \quad M_{22} = m_2 = 10\,\text{kg}$$

Similarly, comparing the general form of the stiffness matrix in Equation (8.34) to that in Equation (8.11):

$$S = \begin{bmatrix} S_{11} & S_{12} \\ S_{21} & S_{22} \end{bmatrix} = \begin{bmatrix} k_1 + k_2 & -k_2 \\ -k_2 & k_2 \end{bmatrix}$$

from which

$$S_{11} = k_1 + k_2 = 40\,000\,\text{N/m}, \quad S_{12} = S_{21} = -k_2 = -25\,000\,\text{N/m} \quad \text{and}$$
$$S_{22} = k_2 = 25\,000\,\text{N/m}$$

From Equation (8.43), the constants required for the quadratic formula are calculated as:

$$a = M_{11} M_{22} = 200, \quad b = -(M_{11} S_{22} + M_{22} S_{11}) = -9 \times 10^5,$$
$$c = S_{11} S_{22} - S_{12}^2 = 3.75 \times 10^8$$

Thus, the angular frequencies are:

$$\omega_{1,2}^2 = \frac{9 \times 10^5 \pm \sqrt{(9 \times 10^5)^2 - 4 \times 200 \times 3.75 \times 10^8}}{2 \times 200},$$
$$\omega_1 = 21.5556\,\text{rad/s}, \quad \omega_2 = 63.5245\,\text{rad/s}$$

And the natural frequencies are calculated as:

$$f_1 = \omega_1/2\pi = 3.43\,\text{Hz} \quad \text{and} \quad f_2 = \omega_2/2\pi = 10.11\,\text{Hz}$$

From Equations (8.45) and (8.46), the amplitude ratios are:

$$r_1 = \frac{-S_{12}}{S_{11} - \omega_1^2 M_{11}} = \frac{25\,000}{40\,000 - (21.5556)^2 \times 20} = 0.814 \quad \text{(in-phase)}$$

$$r_2 = \frac{-S_{12}}{S_{11} - \omega_2^2 M_{11}} = \frac{25\,000}{40\,000 - (63.5245)^2 \times 20} = -0.614 \quad \text{(anti-phase)}$$

Figure 8.8 shows the mode shapes scaled to the second degree of freedom.

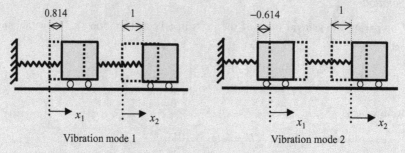

Figure 8.8: Mode shapes for Example 8.1

Example 8.2 Free vibration II

A racing car has a mass of 230 kg and dimensions as shown in Figure 8.9. The stiffness of the front and rear wheel suspensions are both equal to 17 600 N/m and the mass moment of inertia of the car about its centre of gravity is 405 kg.m². Determine the frequencies and mode shape amplitude ratios of the car.

Solution

Comparing the general form of the mass matrix in Equation (8.34) to that in Equation (8.28):

$$M = \begin{bmatrix} M_{11} & 0 \\ 0 & M_{22} \end{bmatrix} = \begin{bmatrix} m & 0 \\ 0 & I_C \end{bmatrix}$$

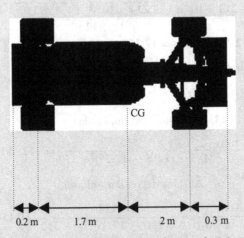

Figure 8.9: Representation of Example 8.2

from which:

$$M_{11} = m_1 = 230\,\text{kg}, \quad M_{22} = I_C = 405\,\text{kg.m}^2$$

Similarly, comparing the general form of the stiffness matrix in Equation (8.34) to that in Equation (8.28), with $k_1 = k_2 = 17\,600$ N/m, $L_1 = 1.7$ m and $L_2 = 2$ m:

$$S = \begin{bmatrix} S_{11} & S_{12} \\ S_{21} & S_{22} \end{bmatrix} = \begin{bmatrix} k_1 + k_2 & k_2 L_2 - k_1 L_1 \\ k_2 L_2 - k_1 L_1 & k_1 L_1^2 + k_2 L_2^2 \end{bmatrix}$$

from which:

$$S_{11} = k_1 + k_2 = 35\,200\,\text{N/m}$$
$$S_{12} = S_{21} = k_2 L_2 - k_1 L_1 = 5280\,\text{N}$$
$$S_{22} = k_1 L_1^2 + k_2 L_2^2 = 121\,264\,\text{N.m}$$

From Equation (8.43), the constants required for the quadratic formula are calculated as:

$$a = M_{11} M_{22} = 93150, \quad b = -(M_{11} S_{22} + M_{22} S_{11}) = -4.214 \times 10^7,$$
$$c = S_{11} S_{22} - S_{12}^2 = 4.24 \times 10^9$$

Thus, the angular frequencies are:

$$\omega_{1,2}^2 = \frac{4.214 \times 10^7 \pm \sqrt{(4.214 \times 10^7)^2 - 4 \times 93\,150 \times 4.24 \times 10^9}}{2 \times 93\,150},$$

$$\omega_1 = 12.29 \, \text{rad/s}, \, \omega_2 = 17.3589 \, \text{rad/s}$$

and the natural frequencies are calculated as:

$$f_1 = \frac{\omega_1}{2\pi} = 1.96 \, \text{Hz}, \quad f_2 = \frac{\omega_2}{2\pi} = 2.76 \, \text{Hz}$$

From Equations (8.45) and (8.46), the amplitude ratios are:

$$r_1 = \frac{-S_{12}}{S_{11} - \omega_1^2 M_{11}} = \frac{-5280}{35\,200 - (12.29)^2 \times 230} = -11.4818 \, \text{m/rad} \times 10^3$$

$$\times \frac{2\pi}{360} = -200 \, \text{mm/degree (anti-phase)},$$

$$r_2 = \frac{-S_{12}}{S_{11} - \omega_2^2 M_{11}} = \frac{-5280}{35\,200 - (17.3589)^2 \times 230} = 0.1548 \, \text{m/rad} \times 10^3$$

$$\times \frac{2\pi}{360} = 2.7 \, \text{mm/degree (in-phase)}.$$

The two vibration mode shapes are shown in Figure 8.10.

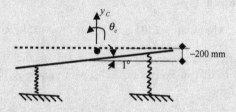

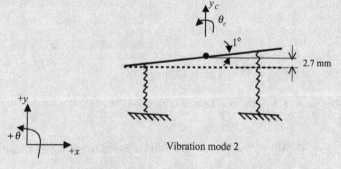

Figure 8.10: Mode shapes for Example 8.2

Example 8.3 Free vibration III

In the system shown in Figure 8.11, two masses m_1 and m_2 are attached by a spring of stiffness k. This system is known as a semidefinite or unconstrained system. It could represent two railway boxcars coupled with a spring. Determine the natural frequencies of the system.

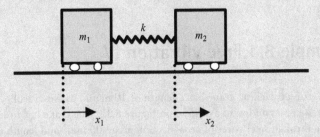

Figure 8.11: Representation of Example 8.3

Solution

From Equations (8.34) and (8.11), the masses of the system are:

$$M_{11} = m_1, \quad M_{22} = m_2$$

Since the stiffness k_1 equals 0, the stiffness matrix becomes:

$$S = \begin{bmatrix} S_{11} & S_{12} \\ S_{21} & S_{22} \end{bmatrix} = \begin{bmatrix} k & -k \\ -k & k \end{bmatrix}$$

from which:

$$S_{11} = k, \quad S_{12} = S_{21} = -k, \quad S_{22} = k$$

From Equation (8.43), the constants required for the quadratic formula are calculated as:

$$a = M_{11} M_{22} = m_1 m_2$$
$$b = -(M_{11} S_{22} + M_{22} S_{11}) = -m_1 k - m_2 k = -k(m_1 + m_2)$$
$$c = S_{11} S_{22} - S_{12}^2 = k^2 - k^2 = 0$$

Thus, the angular frequencies are calculated as:

$$\omega_{1,2}^2 = \frac{k(m_1 + m_2) \pm \sqrt{(k(m_1 + m_2))^2 - 4 \times m_1 m_2 \times 0}}{2 \times m_1 m_2}$$

from which:

$$\omega_1 = 0, \quad \omega_2 = \sqrt{\frac{k(m_1 + m_2)}{m_1 m_2}}$$

The first angular frequency, which is zero, refers to the rigid body mode. Since the system is unconstrained it can move as a whole (rigid body) without any relative motion.

Example 8.4 Free vibration IV

The girder of a travelling crane has a length of 10 m and carries a trolley of mass 3.5×10^3 kg at its midpoint as shown in Figure 8.12. The crane has to lift a load of 800 kg through steel wires of cross-section area 900 mm^2 and length 6 m. The girder is simply supported and has flexural rigidity EI of 18×10^9 N.m^2. Use model of a system with two degrees of freedom to determine the natural frequencies of the system. Ignore the mass of the girder and use the Young's modulus of steel (200 GPa).

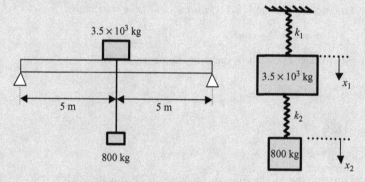

Figure 8.12: Representation of Example 8.4

Solution

From Table 6.1 in Chapter 6, the stiffness of a simply supported girder is calculated as:

$$k_1 = \frac{48EI}{L^3} = \frac{48 \times 18 \times 10^9}{10^3} = 8.64 \times 10^8 \,\text{N/m}$$

For the wire, the stiffness is calculated as:

$$k_2 = \frac{EA}{L} = \frac{200 \times 10^9 \times 900 \times 10^{-6}}{6} = 3 \times 10^7 \, \text{N/m}$$

From Equations (8.34) and (8.11), the masses and the stiffness of the system are given by:

$$M_{11} = m_1 = 3500 \, \text{kg}, \quad M_{22} = m_2 = 800 \, \text{kg}$$

$$S_{11} = k_1 + k_2 = 8.94 \times 10^8 \, \text{N/m}, \quad S_{12} = S_{21} = -k_2 = -3 \times 10^7 \, \text{N/m},$$

$$S_{22} = k_2 = 3 \times 10^7 \, \text{N/m}$$

From Equation (8.43), the constants required for the quadratic formula are calculated as:

$$a = M_{11} M_{22} = 2.8 \times 10^6, \quad b = -(M_{11} S_{22} + M_{22} S_{11}) = -8.202 \times 10^{11},$$

$$c = S_{11} S_{22} - S_{12}^2 = 2.592 \times 10^{16}$$

Thus, the angular frequencies are obtained as:

$$\omega_{1,2}^2 = \frac{8.202 \times 10^{11} \pm \sqrt{(8.202 \times 10^{11})^2 - 4 \times 2.8 \times 10^6 \times 2.5925 \times 10^{16}}}{2 \times 2.8 \times 10^6},$$

$$\omega_1 = 189.828 \, \text{rad/s}, \quad \omega_2 = 506.8467 \, \text{rad/s}$$

and the natural frequencies are calculated as:

$$f_1 = \omega_1/2\pi = 30.21 \, \text{Hz}, \quad f_2 = \omega_2/2\pi = 80.67 \, \text{Hz}$$

Example 8.5 Free vibration V

A two-storey building is modelled as a system with two degrees of freedom as shown in Figure 8.13. Each floor has a mass m and the stiffness constant k is given by $\frac{EI}{h}$, where EI is the column's flexural rigidity and h is the floor's height. Determine the natural frequencies of the building in terms of m and k.

Solution

The stiffness of each level is given by:

$$k_1 = k_2 = 24k + 24k = 48k$$

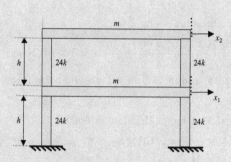

Figure 8.13: Representation of Example 8.5

From Equations (8.34) and (8.11), the mass and the stiffness matrix elements are given by:

$$M_{11} = m, \quad M_{22} = m$$
$$S_{11} = k_1 + k_2 = 96k, \quad S_{12} = S_{21} = -k_2 = -48k, \quad S_{22} = k_2 = 48k$$

From Equation (8.43), the constants required for the quadratic formula are calculated as:

$$a = M_{11} M_{22} = m^2, \quad b = -(M_{11} S_{22} + M_{22} S_{11}) = -144mk,$$
$$c = S_{11} S_{22} - S_{12}^2 = 2304k^2$$

Thus, the angular frequencies are:

$$\omega_{1,2}^2 = \frac{144mk \pm \sqrt{(144mk)^2 - 4 \times m^2 \times 2304 \times k^2}}{2 \times m^2},$$
$$\omega_1 = 4.2818\sqrt{\frac{k}{m}}\, \text{rad/s}, \quad \omega_2 = 11.21\sqrt{\frac{k}{m}}\, \text{rad/s}$$

and the natural frequencies are calculated as:

$$f_1 = \omega_1/2\pi = 0.68\sqrt{\frac{k}{m}}\, \text{Hz}, \quad f_2 = \omega_2/2\pi = 1.784\sqrt{\frac{k}{m}}\, \text{Hz}$$

Example 8.6 Free vibration VI

A mass m is suspended from three springs of stiffness k as shown in Figure 8.14. If the angles that the springs make with the horizontal axis are $\theta_1 = 45°$, $\theta_2 = 135°$

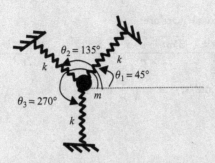

Figure 8.14: Representation of Example 8.6

and $\theta_3 = 270°$, determine the angular frequencies of the system in terms of k and m.

Solution

From Equation (8.32), the stiffness matrix elements are calculated as:

$$k_{11} = k \times \cos^2 45 + k \times \cos^2 135 + k \times \cos^2 270 = k$$

$$k_{22} = k \times \sin^2 45 + k \times \sin^2 135 + k \times \sin^2 270 = 2k$$

$$k_{12} = k_{21} = k \times \cos 45 \times \sin 45 + k \times \cos 135 \times \sin 135$$
$$+ \ k \times \cos 270 \times \sin 270 = 0$$

Comparing the general form of the equations of motion in matrix form, Equation (8.34), equations of motion for the spring-suspended mass, Equation (8.32), the mass and stiffness matrix elements are calculated as:

$$M_{11} = m, \quad M_{22} = m$$

$$S_{11} = k, \quad S_{12} = S_{21} = -0, \quad S_{22} = 2k$$

From Equation (8.43), the constants required for the quadratic formula are calculated as:

$$a = M_{11} M_{22} = m^2, \quad b = -(M_{11} S_{22} + M_{22} S_{11}) = -3km,$$

$$c = S_{11} S_{22} - S_{12}^2 = 2k^2$$

Thus, the angular frequencies are:

$$\omega_{1,2}^2 = \frac{3km \pm \sqrt{(3km)^2 - 4 \times m^2 \times (2k^2)}}{2 \times m^2} = \frac{3km \pm km}{2m^2}$$

$$\omega_1 = \sqrt{\frac{k}{m}} \text{ rad/s}, \quad \omega_2 = \sqrt{\frac{2k}{m}} \text{ rad/s}$$

Example 8.7 Free vibration VII

a) For the system with two degrees of freedom shown in Figure 8.15(a), write the equations of motion in matrix form.
b) Show that the ratio of the second natural frequency to the first natural frequency (f_2/f_1) is 5.93.
c) If the spring joining the two masses is removed, as shown in Figure 8.15(b), calculate the natural frequencies of the two systems.

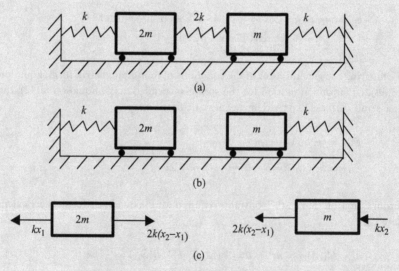

Figure 8.15: Representation of Question 8.7

Solution

a) To derive the equations of motion, the free-body diagrams are shown in Figure 8.15(c). Using Newton's second law, the equations of motion for m and $2m$ are obtained as:

$$-kx_1 + 2k(x_2 - x_1) = 2m\ddot{x}_1$$
$$-2k(x_2 - x_1) - kx_2 = m\ddot{x}_2$$

Rearranging terms gives:

$$2m\ddot{x}_1 + 3kx_1 - 2k_2x_2 = 0$$
$$m\ddot{x}_2 - 2kx_1 + 3kx_2 = 0$$

In matrix form, they are:

$$\begin{bmatrix} 2m & 0 \\ 0 & m \end{bmatrix} \begin{bmatrix} \ddot{x}_1 \\ \ddot{x}_2 \end{bmatrix} + \begin{bmatrix} 3k & -2k \\ -2k & 3k \end{bmatrix} \begin{bmatrix} x_1 \\ x_2 \end{bmatrix} = \begin{bmatrix} 0 \\ 0 \end{bmatrix}$$

b) Comparing the general equations of motion in matrix form, Equation (8.34), to the above equations of motion, the mass and stiffness matrix elements are calculated as:

$$M_{11} = 2m, \quad M_{22} = m, \quad S_{11} = 3k, \quad S_{12} = S_{21} = -2k, \quad S_{22} = 3k$$

From Equation (8.43), the constants required for the quadratic formula are calculated as:

$$a = M_{11}M_{22} = 2m^2, \quad b = -(M_{11}S_{22} + M_{22}S_{11})$$
$$= -(2m \times 3k + m \times 3k) = -9mk,$$
$$c = S_{11}S_{22} - S_{12}^2 = 3k \times 3k - (2k)^2 = 5k^2$$

Thus, the two angular frequencies of the system are:

$$\omega_{1,2}^2 = \frac{9mk \pm \sqrt{(9mk)^2 - 4 \times 2m^2 \times 5k^2}}{2 \times 2m^2} = \frac{k}{m}\left(\frac{9 \pm 6.4}{4}\right),$$
$$\omega_1 = 0.649\sqrt{\frac{k}{m}} \, \text{rad/s}, \, \omega_2 = 3.85\sqrt{\frac{k}{m}} \text{rad/s}$$

And the ratio between them is:

$$\omega_2/\omega_1 = 3.85/0.649 = 5.93$$

c) The system shown in Figure 8.15(b) comprises two separate systems with a single degree of freedom. The natural frequencies of the two systems are:

$$\omega_1 = \sqrt{\frac{k}{2m}} = 0.707\sqrt{\frac{k}{m}} \text{ rad/s} \Rightarrow f_1 = \frac{\omega_1}{2\pi} = 0.1125\sqrt{\frac{k}{m}} \text{ Hz}$$

$$\omega_2 = \sqrt{\frac{k}{m}} = \sqrt{\frac{k}{m}} \text{ rad/s} \Rightarrow f_2 = \frac{\omega_2}{2\pi} = 0.159\sqrt{\frac{k}{m}} \text{ Hz}$$

Example 8.8 Free vibration VIII

The uniform beam of 6 kg mass, shown in Figure 8.16, is supported at its ends by two springs A and B, each having the same stiffness k. If a 5 kg mass is placed at its centre, the period of the fundamental mode is 1.5 s. Determine the stiffness of each spring and the frequency of the second mode of vibration in Hz.

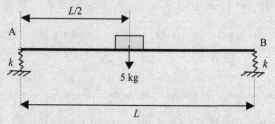

Figure 8.16: Representation of Question 8.8

Solution

The mass of the rod is $m = 6$ kg; the distance from each spring to the centre of gravity of the rod is $L_1 = L_2 = L/2$ and the stiffness of the springs is $k_1 = k_2 = k$. The total mass of the system is $6 + 5 = 11$ kg, while the mass moment of inertia of the rod (from Table 4.1 in Chapter 4) is $I_C = 6L^2/12 = L^2/2$ kg.m².

Comparing the general equations of motion in matrix form, Equations (8.34) to the equations of motion of the rigid bar/two-spring's system, Equation (8.28), the mass and stiffness matrix elements are calculated as:

$$M_{11} = m_1 = 11 \text{ kg}, \quad M_{22} = I_C = L^2/2 \text{ kg.m}^2$$

$$S_{11} = k_1 + k_2 = 2k, \quad S_{12} = S_{21} = k_2 L_2 - k_1 L_1 = 0,$$

$$S_{22} = k_1 L_1^2 + k_2 L_2^2 = kL^2/2$$

From Equation (8.43), the constants required for the quadratic formula are calculated as:

$$a = M_{11}M_{22} = 11 \times L^2/2, \quad b = -(M_{11}S_{22} + M_{22}S_{11})$$
$$= -(11kL^2/2 + kL^2) = -13kL^2/2$$
$$c = S_{11}S_{22} - S_{12}^2 = k^2L^2$$

Thus, the two angular frequencies of the system are:

$$\omega_{1,2}^2 = \frac{13kL^2/2 \pm \sqrt{(13kL^2/2)^2 - 4 \times 11L^2/2 \times k^2L^2}}{2 \times 11 \times L^2/2}$$
$$= \frac{13kL^2/2 \pm 9kL^2/2}{11 \times L^2} = \frac{13k \pm 9k}{22} \tag{E8.8a}$$

Since the period of the fundamental mode is 1.5 s, the first mode angular frequency is:

$$\omega_1 = \frac{2\pi}{1.5} = 0.67 \times 2\pi \text{ rad/s}$$

Substituting the first angular frequency in Equation (E8.8a), gives:

$$(0.67 \times 2\pi)^2 = \frac{13k - 9k}{22}$$

from which the stiffness is calculated as $k = 97.47\,\text{N/m}$.

The second natural frequency is calculated using Equation (E8.8a) as:

$$\omega_2^2 = \frac{13k + 9k}{22} = 97.47 \Rightarrow \omega_2 = 9.87\,\text{rad/s} \Rightarrow f_2 = 1.57\,\text{Hz}$$

8.4 TORSIONAL VIBRATION

The torsional vibration of a system with two degrees of freedom can be demonstrated by considering two disks mounted on a shaft, where the two degrees of freedom are the two angular displacements θ_1 and θ_2, as shown in Figure 8.17(a). The two segments of the shaft have stiffness k_1 and k_2, shear moduli G_1 and G_2, polar moment of inertia I_{P1} and I_{P2} and lengths L_1 and L_2. The mass moment of inertia of each disk is J_1 and J_2 and two external couples T_1 and T_2 are applied at disks 1 and 2, respectively.

Considering the free-body diagram in Figure 8.17(b) and applying the moment equation equivalent to Newton's second law, i.e. $\sum T = J\ddot{\theta}$ for each disk, gives:

$$-k_1\theta_1 + k_2(\theta_2 - \theta_1) + T_1 = J_1\ddot{\theta}_1 \tag{8.47}$$

$$-k_2(\theta_2 - \theta_1) + T_2 = J_2\ddot{\theta}_2 \tag{8.48}$$

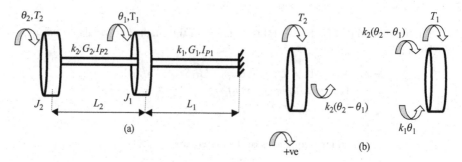

Figure 8.17: a) Torsional system with two degrees of freedom and b) its free-body diagram

where $\ddot{\theta}_1$ and $\ddot{\theta}_2$ are the torsional angular acceleration of disks 1 and 2, respectively.

Re-arranging Equations (8.47) and (8.48) leads to:

$$J_1\ddot{\theta}_1 + (k_1 + k_2)\theta_1 - k_2\theta_2 = T_1 \tag{8.49}$$

$$J_2\ddot{\theta}_2 - k_2\theta_1 + k_2\theta_2 = T_2 \tag{8.50}$$

where the torsional stiffness k_1 and k_2 are given by:

$$k_1 = \frac{G_1 I_{p1}}{L_1} \quad \text{and} \quad k_2 = \frac{G_2 I_{p2}}{L_2} \tag{8.51}$$

Equations (8.49) and (8.50) can be written in matrix form as:

$$\begin{bmatrix} J_1 & 0 \\ 0 & J_2 \end{bmatrix} \begin{bmatrix} \ddot{\theta}_1 \\ \ddot{\theta}_2 \end{bmatrix} + \begin{bmatrix} k_1 + k_2 & -k_2 \\ -k_2 & k_2 \end{bmatrix} \begin{bmatrix} \theta_1 \\ \theta_2 \end{bmatrix} = \begin{bmatrix} T_1 \\ T_2 \end{bmatrix} \tag{8.52}$$

For free vibration:

$$\begin{bmatrix} J_1 & 0 \\ 0 & J_2 \end{bmatrix} \begin{bmatrix} \ddot{\theta}_1 \\ \ddot{\theta}_2 \end{bmatrix} + \begin{bmatrix} k_1 + k_2 & -k_2 \\ -k_2 & k_2 \end{bmatrix} \begin{bmatrix} \theta_1 \\ \theta_2 \end{bmatrix} = \begin{bmatrix} 0 \\ 0 \end{bmatrix} \tag{8.53}$$

Equation (8.53) can be solved in the same way as followed in Section 8.3.

Example 8.9 Free torsional vibration

Two disks of masses 1 kg and 2 kg, having radii of 5 cm and 10 cm, respectively, are mounted on a solid steel shaft of diameter 15 mm and length 1 m as shown in Figure 8.18. Determine the torsional natural frequencies of the system and the corresponding amplitude ratios. Use steel shear modulus 77 GPa and ignore the mass of the shaft.

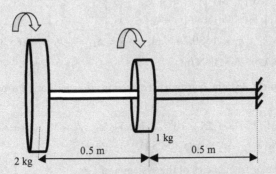

Figure 8.18: Representation of Example 8.9

Solution

The polar moment of inertia of the shaft is (see Appendix B):

$$I_p = \frac{\pi D^4}{32} = \frac{\pi \times 0.015^4}{32} = 4.97 \times 10^{-9}\,\text{m}^4$$

From Table 4.1 and Equation (4.32), the mass moments of inertia of the disks are:

$$J_1 = \frac{m_1 R_1^2}{2} = \frac{1 \times 0.05^2}{2} = 1.25 \times 10^{-3}\,\text{kg.m}^2$$

$$J_2 = \frac{m_2 R_2^2}{2} = \frac{2 \times 0.1^2}{2} = 0.01\,\text{kg.m}^2$$

The torsional stiffness of the shaft is constant and equal to (Equation (8.51)):

$$k_1 = k_2 = \frac{GI_p}{L} = \frac{77 \times 10^9 \times 4.97 \times 10^{-9}}{0.5} = 765.38\,\text{N.m}$$

Comparing the general form of the mass matrix, Equation (8.34), to Equation (8.52) leads to:

$$M = \begin{bmatrix} M_{11} & 0 \\ 0 & M_{22} \end{bmatrix} = \begin{bmatrix} J_1 & 0 \\ 0 & J_2 \end{bmatrix}$$

$$M_{11} = J_1 = 1.25 \times 10^{-3}\,\text{kg.m}^2, \quad M_{22} = J_2 = 0.01\,\text{kg.m}^2$$

Similarly, comparing the general form of the stiffness matrix in Equation (8.34) to Equation (8.52) leads to:

$$S = \begin{bmatrix} S_{11} & S_{12} \\ S_{21} & S_{22} \end{bmatrix} = \begin{bmatrix} k_1 + k_2 & -k_2 \\ -k_2 & k_2 \end{bmatrix}$$

$$S_{11} = k_1 + k_2 = 1530.76 \text{ N.m}, \quad S_{12} = S_{21} = -k_2 = -765.38 \text{ N.m},$$

$$S_{22} = k_2 = 765.38 \text{ N/m}$$

From Equation (8.43), the constants required for the quadratic formula are calculated as:

$$a = M_{11} M_{22} = 1.25 \times 10^{-5}, \quad b = -(M_{11} S_{22} + M_{22} S_{11}) = -16.264,$$

$$c = S_{11} S_{22} - S_{12}^2 = 585\,806.54$$

Thus, the angular frequencies of the system are:

$$\omega_{1,2}^2 = \frac{16.264 \pm \sqrt{16.264^2 - 4 \times 1.25 \times 10^{-5} \times 585\,806.54}}{2 \times 1.25 \times 10^{-5}},$$

$$\omega_1 = 192.55 \text{ rad/s}, \quad \omega_2 = 1124.2975 \text{ rad/s}$$

And the natural frequencies of the system are calculated as:

$$f_1 = \omega_1/2\pi = 30.65 \text{ Hz}, \quad f_2 = \omega_2/2\pi = 178.94 \text{ Hz}$$

From Equations (8.45) and (8.46), the amplitude ratios are:

$$r_1 = \frac{-S_{12}}{S_{11} - \omega_1^2 M_{11}} = \frac{765.38}{1530.76 - (192.55)^2 \times 1.25 \times 10^{-3}}$$

$$= 0.5156 \quad \text{(in-phase)}$$

$$r_2 = \frac{-S_{12}}{S_{11} - \omega_2^2 M_{11}} = \frac{765.38}{1530.76 - (1124.2975)^2 \times 1.25 \times 10^{-3}}$$

$$= -15.52 \quad \text{(anti-phase)}$$

8.5 UNDAMPED FORCED VIBRATIONS

As mentioned in Section 7.1 in Chapter 7, a dynamics force can be of different types, e.g. harmonic or random. In general, any random force can be approximated using series of harmonic forces. This section is devoted to the response of a system with two degrees of freedom under harmonic forces. Consider again the general equations of motion in matrix form, Equation (8.34), if the system is subjected to sinusoidal forcing functions, F_1 and F_2, given by:

$$F_1 = F_{o1} \sin \omega t \tag{8.54}$$

$$F_2 = F_{o2} \sin \omega t \tag{8.55}$$

where ω is the forcing or excitation angular frequency, which is assumed to be the same for both forces, and F_{o1} and F_{o2} are the magnitudes of the forcing functions F_1 and F_2, respectively. Equation (8.33) becomes:

$$M\ddot{D} + SD = F_o \sin \omega t \qquad (8.56)$$

and, in expanded form:

$$\begin{bmatrix} M_{11} & 0 \\ 0 & M_{22} \end{bmatrix} \begin{bmatrix} \ddot{D}_1 \\ \ddot{D}_2 \end{bmatrix} + \begin{bmatrix} S_{11} & S_{12} \\ S_{21} & S_{22} \end{bmatrix} \begin{bmatrix} D_1 \\ D_2 \end{bmatrix} = \begin{bmatrix} F_{o1} \\ F_{o2} \end{bmatrix} \sin \omega t \qquad (8.57)$$

The particular or steady-state solutions of these two differential equations are (see the solution of Equation (7.4) in Chapter 7):

$$D_1 = A_1 \sin \omega t \qquad (8.58)$$

$$D_2 = A_2 \sin \omega t \qquad (8.59)$$

where A_1 and A_2 are the amplitudes of the steady-state responses of D_1 and D_2, respectively. Differentiating Equations (8.58) and (8.59) twice with respect to time, the accelerations $\ddot{D}_1$ and $\ddot{D}_2$ are obtained as:

$$\ddot{D}_1 = -\omega^2 A_1 \sin \omega t \qquad (8.60)$$

$$\ddot{D}_2 = -\omega^2 A_2 \sin \omega t \qquad (8.61)$$

Substituting Equations (8.58) to (8.61) into Equation (8.57), and multiplying and adding the matrices, gives:

$$\begin{bmatrix} S_{11} - \omega^2 M_{11} & S_{12} \\ S_{21} & S_{22} - \omega^2 M_{22} \end{bmatrix} \begin{bmatrix} A_1 \\ A_2 \end{bmatrix} = \begin{bmatrix} F_{o1} \\ F_{o2} \end{bmatrix} \qquad (8.62)$$

Solving for $[A]$, gives:

$$\begin{bmatrix} A_1 \\ A_2 \end{bmatrix} = \frac{1}{C} \begin{bmatrix} S_{22} - \omega^2 M_{22} & -S_{12} \\ -S_{21} & S_{11} - \omega^2 M_{11} \end{bmatrix} \begin{bmatrix} F_{o1} \\ F_{o2} \end{bmatrix} \qquad (8.63)$$

where C is equal to:

$$C = (S_{11} - \omega^2 M_{11})(S_{22} - \omega^2 M_{22}) - S_{12}^2$$

$$= M_{11} M_{22} \omega^4 - (M_{11} S_{22} + M_{22} S_{11})\omega^2 + S_{11} S_{22} - S_{12}^2$$

The amplitudes of steady-state responses are then:

$$A_1 = \frac{(S_{22} - \omega^2 M_{22}) F_{o1} - S_{12} F_{o2}}{M_{11} M_{22} \omega^4 - (M_{11} S_{22} + M_{22} S_{11})\omega^2 + S_{11} S_{22} - S_{12}^2} \qquad (8.64)$$

$$A_2 = \frac{(S_{11} - \omega^2 M_{11}) F_{o2} - S_{21} F_{o1}}{M_{11} M_{22} \omega^4 - (M_{11} S_{22} + M_{22} S_{11})\omega^2 + S_{11} S_{22} - S_{12}^2} \qquad (8.65)$$

Comparing the denominator C in Equations (8.64) and (8.65) to Equation (8.42), it can be seen that they are identical. The root of Equation (8.42) is the two angular frequencies of the system ω_1 and ω_2. Therefore, the denominator C can be replaced by $M_{11} M_{22}(\omega^2 - \omega_1^2)(\omega^2 - \omega_2^2)$ and Equations (8.64) and (8.65) become:

$$A_1 = \frac{(S_{22} - \omega^2 M_{22})F_{o1} - S_{12} F_{o2}}{M_{11} M_{22}(\omega^2 - \omega_1^2)(\omega^2 - \omega_2^2)} \tag{8.66}$$

$$A_2 = \frac{(S_{11} - \omega^2 M_{11})F_{o2} - S_{21} F_{o1}}{M_{11} M_{22}(\omega^2 - \omega_1^2)(\omega^2 - \omega_2^2)} \tag{8.67}$$

This means the denominator is equal to zero if $\omega = \omega_1$ or $\omega = \omega_2$. In such a case, the amplitudes A_1 and A_2 are infinite, as will be demonstrated in Example 8.11.

Example 8.10 Forced vibration I

The bar shown in Figure 8.19(a) has a length of 2 m and a mass of 10 kg. The bar is supported by two springs at A and B, which have a stiffness of $k_A = 600$ N/m and $k_B = 900$ N/m.

a) Calculate the natural frequencies of the bar.
b) Determine the amplitude ratios (in mm/degree).
c) If a harmonic force of magnitude $P = 20$ N and forcing frequency of 10 rad/s is applied at G as shown in Figure 8.19(b), determine the absolute values of the two amplitudes.

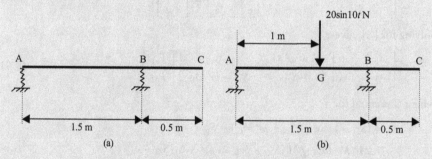

Figure 8.19: Representation of Question 8.10

Solution

a) The mass of the bar is $m = 10$ kg, the distance from spring A to the centre of gravity of the bar is $L_1 = 1$ m, the distance from spring B to the centre of gravity of the bar $L_2 = 0.5$ m, the stiffness of spring A is $k_1 = 600$ N/m and the stiffness of spring B is $k_2 = 900$ N/m.

To calculate the natural frequencies, use Equation (8.43), as before, and compare the general equations of motion in matrix form, Equation (8.34), to Equation (8.28), the mass and stiffness matrix elements are calculated as:

$$M_{11} = 10\,\text{kg}, \quad M_{22} = I_C = mL^2/12 = 3.33\,\text{kg.m}^2$$

$$S_{11} = 600 + 900 = 1500\,\text{N/m}, \quad S_{12} = S_{21} = 900 \times 0.5 - 600 \times 1$$

$$= -150\,\text{N/m},$$

$$S_{22} = 600 \times 1^2 + 900 \times 0.5^2 = 825\,\text{N/m}$$

From Equation (8.43), the constants required for the quadratic formula are calculated as:

$$a = M_{11}M_{22} = 33.333, \quad b = -(M_{11}S_{22} + M_{22}S_{11})$$

$$= -(10 \times 825 + 3.33 \times 1500) = -13\,250,$$

$$c = S_{11}S_{22} - S_{12}^2 = 1500 \times 825 - (-150)^2 = 1\,215\,000$$

The angular frequencies are obtained as:

$$\omega_{1,2}^2 = \frac{13\,250 \pm \sqrt{(13\,250)^2 - 4 \times 33.333 \times 1\,215\,000}}{2 \times 33.333}$$

$$= \frac{13\,250 \pm 3682.73}{66.667}, \quad \omega_1 = 11.979\,\text{rad/s}, \quad \omega_2 = 15.937\,\text{rad/s}$$

and the natural frequencies are calculated as:

$$f_1 = \frac{\omega_1}{2\pi} = 1.91\,\text{Hz}, \quad f_2 = \frac{\omega_2}{2\pi} = 2.54\,\text{Hz}$$

b) The amplitude ratios (from Equations (8.45) and (8.46)) are:

$$r_1 = \frac{-S_{12}}{S_{11} - \omega_1^2 M_{11}} = \frac{150}{1500 - (11.979)^2 \times 10} = 2.306\,\text{m/rad} \times 10^3$$

$$\times \frac{2\pi}{360} = 40.25\,\text{mm/}^\circ \quad \text{(in-phase)}$$

$$r_2 = \frac{-S_{12}}{S_{11} - \omega_2^2 M_{11}} = \frac{150}{1500 - (15.937)^2 \times 10} = -0.1442\,\text{m/rad}$$

$$\times 10^3 \times \frac{2\pi}{360} = -2.52\,\text{mm/}^\circ \quad \text{(anti-phase)}$$

c) To calculate the two steady-state amplitudes, apply Equations (8.64) and (8.65) and use $F_{o1} = 20$ N, $F_{o2} = 0$ and $\omega = 10$ rad/s to give:

$$y_C = A_1 = \frac{(S_{22} - \omega^2 M_{22})F_{o1} - S_{12}F_{o2}}{M_{11}M_{22}\omega^4 - (M_{11}S_{22} + M_{22}S_{11})\omega^2 + S_{11}S_{22} - S_{12}^2}$$

$$= \frac{(825 - (10)^2 \times 3.333) \times 20}{33.333 \times (10)^4 - 13\,250 \times (10)^2 + 1\,215\,000}$$

$$= \frac{9833.333}{223\,333.33} = 0.044\,\text{m} = 44\,\text{mm}$$

$$\theta_C = A_2 = \frac{(S_{11} - \omega^2 M_{11})F_{o2} - S_{21}F_{o1}}{M_{11}M_{22}\omega^4 - (M_{11}S_{22} + M_{22}S_{11})\omega^2 + S_{11}S_{22} - S_{12}^2}$$

$$= \frac{150 \times 20}{33.333 \times (10)^4 - 13\,250 \times (10)^2 + 1\,215\,000}$$

$$= \frac{3000}{223\,333.33} = 0.01343\,\text{rad} = 0.77°$$

Example 8.11 Forced vibration II

The system with two degrees of freedom from Example 8.1 (shown in Figure 8.20) is subjected to harmonic loading of magnitudes $F_{o1} = 100$ N and $F_{o2} = 0$. Draw the response spectrum (the curve of steady-state amplitudes versus ω).

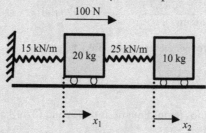

Figure 8.20: Representation of Example 8.11

Solution

From Example 8.1, the following parameters are calculated:

$$a = M_{11}M_{22} = 200, \quad b = -(M_{11}S_{22} + M_{22}S_{11}) = -9 \times 10^5,$$

$$c = S_{11}S_{22} - S_{12}^2 = 3.75 \times 10^8$$

Using Equations (8.64) and (8.65), and for $F_{o1} = 100$ N, $F_{o2} = 0$, the steady-state amplitudes are given by:

$$A_1 = \frac{(25\,000 - 10\omega^2) \times 100}{200\omega^4 - 9 \times 10^5 \omega^2 + 3.75 \times 10^8} = \frac{2.5 \times 10^6 - 1000\omega^2}{200\omega^4 - 9 \times 10^5 \omega^2 + 3.75 \times 10^8}$$

$$A_2 = \frac{-(25\,000) \times 100}{200\omega^4 - 9 \times 10^5 \omega^2 + 3.75 \times 10^8} = \frac{-2.5 \times 10^6}{200\omega^4 - 9 \times 10^5 \omega^2 + 3.75 \times 10^8}$$

The amplitudes A_1 and A_2 are plotted as a function of the forcing frequency ω in Figure 8.21. The absolute values for the amplitudes, $|A_1|$ and $|A_2|$, are plotted in Figure 8.22, from which it can be seen that both amplitudes go to infinity when ω approaches the natural frequencies of the system ω_1 or ω_2 (from Example 8.1, $\omega_1 = 21.5556$ rad/s, $\omega_2 = 63.5245$ rad/s). Also notice that the amplitude A_1 is equal to zero when $\omega = 50$ rad/s. This value corresponds to $\sqrt{\frac{S_{22}}{M_{22}}}$ as can be deduced from Equation (8.64). This phenomenon can be used as a vibration absorber for mass M_{11}. The vibration of M_{11} can be suppressed at a specific forcing frequency by adding to the system an appropriate mass M_{22} and stiffness S_{22}.

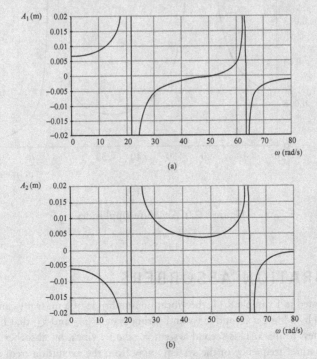

(a)

(b)

Figure 8.21: Amplitudes as a function of the forcing frequency

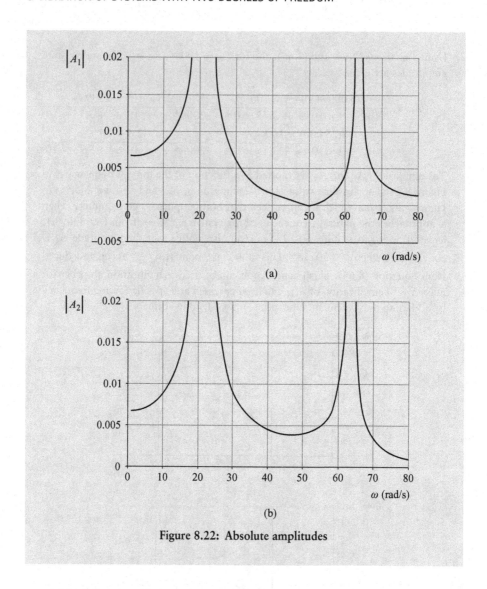

Figure 8.22: Absolute amplitudes

8.6 VIBRATION ABSORBERS

As demonstrated in Example 8.11, the vibration amplitude of a mass m_1 can be reduced or eliminated by adding a second mass and spring system (m_2 and k_2) that has a specific ratio of stiffness to mass. This second system is called a vibration absorber and it helps to shift the natural frequency of the system away from the excitation frequency so that resonance and excessive vibration do not take place. Figure 8.23 shows a machine model before and after adding a vibration absorber.

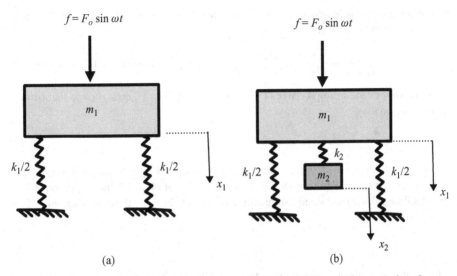

Figure 8.23: A machine model a) before and b) after adding a vibration absorber

The original system, Figure 8.23(a), has a single degree of freedom and only one natural frequency; the modified system after adding the vibration absorber, Figure 8.23(b), has two degrees of freedom and two natural frequencies. The system with a single degree of freedom was dealt with in Chapter 7: the amplitude of the mass m_1 in Figure 8.23(a) can be obtained from Equation (7.10) for an undamped system. The solution of the system with two degrees of freedom was derived in Section 8.5 and the two amplitudes can be obtained from Equations (8.66) and (8.67) as:

$$A_1 = \frac{(S_{22} - \omega^2 M_{22})F_{o1} - S_{12}F_{o2}}{(S_{11} - \omega^2 M_{11})(S_{22} - \omega^2 M_{22}) - S_{12}^2} \tag{8.68}$$

$$A_2 = \frac{(S_{11} - \omega^2 M_{11})F_{o2} - S_{21}F_{o1}}{(S_{11} - \omega^2 M_{11})(S_{22} - \omega^2 M_{22}) - S_{12}^2} \tag{8.69}$$

Using the force amplitudes, $F_{o1} = F_o$ and $F_{o2} = 0$, and the mass and stiffness elements $M_{11} = m_1$, $M_{22} = m_2$, $S_{11} = k_1 + k_2$, $S_{12} = S_{21} = -k_2$, and $S_{22} = k_2$ (as explained in Example 8.1) gives:

$$A_1 = \frac{(k_2 - \omega^2 m_2)F_o}{(k_1 + k_2 - \omega^2 m_1)(k_2 - \omega^2 m_2) - k_2^2} \tag{8.70}$$

$$A_2 = \frac{k_2 F_o}{(k_1 + k_2 - \omega^2 m_1)(k_2 - \omega^2 m_2) - k_2^2} \tag{8.71}$$

Vibration absorption is achieved when the amplitude of mass m_1 is zero, therefore, equating Equation (8.70) to zero, gives:

$$A_1 = \frac{(k_2 - \omega^2 m_2)F_o}{(k_1 + k_2 - \omega^2 m_1)(k_2 - \omega^2 m_2) - k_2^2} = 0 \Rightarrow (k_2 - \omega^2 m_2) = 0 \tag{8.72}$$

from which the relationship between the excitation frequency ω, mass m_2 and the stiffness k_2 is obtained as:

$$\omega^2 = \frac{k_2}{m_2} \tag{8.73}$$

An amplitude of mass m_1 equal to zero can be achieved by selecting a mass m_2 and stiffness k_2 that satisfy Equation (8.73). Substituting Equation (8.73) into Equation (8.71), the absolute value of the steady-state amplitude A_2 of the second mass, when $A_1 = 0$, is obtained as:

$$|A_2| = \frac{F_o}{k_2} \tag{8.74}$$

which imposes a second condition in that the stiffness k_2 must be selected so that the amplitude A_2 does not exceed a certain value.

Figure 8.24 uses the data in Example 8.11 to demonstrate the usefulness of the vibration absorber in shifting the natural frequency of the original system away from the exciting frequency. The original excitation and natural frequencies of the system, without the vibration absorber, is 50 rad/s, which is identical to the suppersion frequency for m_1 in

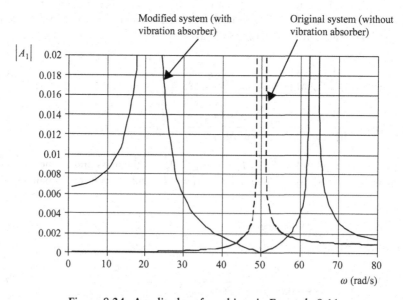

Figure 8.24: Amplitudes of machines in Example 8.11

Example 8.11. After adding the vibration absorber, this frequency is shifted to 21.55 rad/s, which is the fundamental frequency of the modified system. However, an additional frequency of 63.52 rad/s appears.

Example 8.12 Vibration absorbers I

A motor is mounted on a floor and operates at 650 rev/min. The floor and the motor have an equivalent mass of 1×10^3 kg and are supported by columns. The natural frequency of the system has been calculated and it was found that it is equal to the operating speed; excessive vibration is expected. A vibration absorber has to be designed by attaching a mass and spring to the floor that has a stiffness of 1000 kN/m. Determine the mass m_2 and the two natural frequencies of the system shown in Figure 8.25.

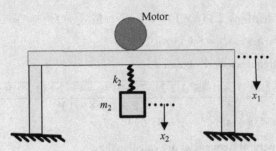

Figure 8.25: Representation of Example 8.12

Solution

Before adding the vibration absorber, the system can be modelled as one with a single degree of freedom with an angular frequency of:

$$\omega_1 = 650 \times \frac{2\pi}{60} = 68.0678 \, \text{rad/s}$$

which is also the operating speed of the motor (the exciting angular frequency). For a system with a single degree of freedom, the angular frequency is given from Equation (6.7), by:

$$\omega_1 = \sqrt{\frac{k_1}{m_1}} \Rightarrow 68.0678 = \sqrt{\frac{k_1}{1000}}$$

from which the stiffness of the floor is calculated as:

$$k_1 = 4.633 \times 10^6 \, \text{N/m}$$

Using Equation (8.73), the mass of the vibration absorber is given by:

$$\omega^2 = \frac{k_2}{m_2} \Rightarrow 68.0678^2 = \frac{1\,000\,000}{m_2} \Rightarrow m_2 = 215.83\,\text{kg}$$

To calculate the two natural frequencies of the modified system, comparing the general equations of motion, Equation (8.34), to the equations of motion of the system with two masses and two springs, Equation (8.11), the mass and stiffness matrix elements are calculated as:

$$M_{11} = m_1 = 1000\,\text{kg}, \quad M_{22} = m_2 = 215.83\,\text{kg}$$

$$S_{11} = k_1 + k_2 = 5.633 \times 10^6\,\text{N/m}, \quad S_{12} = S_{21} = -k_2 = -10^6\,\text{N/m},$$

$$S_{22} = k_2 = 10^6\,\text{N/m}$$

From Equation (8.43), the constants required for the quadratic formula are calculated as:

$$a = M_{11}M_{22} = 2.158 \times 10^5, \quad b = -(M_{11}S_{22} + M_{22}S_{11}) = -2.2157 \times 10^9,$$

$$c = S_{11}S_{22} - S_{12}^2 = 4.633 \times 10^{12}$$

The angular frequencies are obtained as:

$$\omega_{1,2}^2 = \frac{2.2157 \times 10^9 \pm \sqrt{(2.2157 \times 10^9)^2 - 4 \times 2.158 \times 10^5 \times 4.633 \times 10^{12}}}{2 \times 2.158 \times 10^5},$$

$$\omega_1 = 54.064\,\text{rad/s}, \quad \omega_2 = 85.6997\,\text{rad/s}$$

And the two natural frequencies are calculated as:

$$f_1 = \omega_1/2\pi = 8.6\,\text{Hz}, \quad f_2 = \omega_2/2\pi = 13.64\,\text{Hz}$$

Example 8.13 Vibration absorbers II

An unbalanced motor is mounted on a fixed-fixed steel beam at its midpoint as shown in Figure 8.26. The beam has a length of 3 m, is made of steel with $E = 200$ GPa and density 7800 kg/m^3, and has a cross-section of 20 × 400 mm. The motor has a mass of 250 kg and produces a force equal to $f(t) = 300 \sin 11\pi t\,(\text{N})$, which introduces excessive vibration to the beam. In order to eliminate the beam vibration, a vibration absorber of mass m_2 and stiffness k_2 is attached to the bottom of the beam. If the amplitude of the vibration absorber mass should not exceed 2 cm, determine:

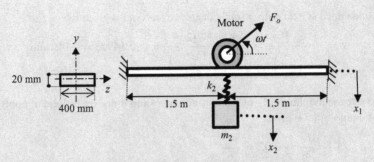

Figure 8.26: Representation of Example 8.13

a) the amplitude of the beam vibration before attaching the vibration absorber;
b) the stiffness and the mass of the vibration absorber.

Solution

a) The area moment of inertia of the beam about the z axis is (see Appendix B):

$$I = \frac{(400 \times 10^{-3}) \times (20 \times 10^{-3})^3}{12} = 2.66667 \times 10^{-7}\,\text{m}^4$$

For a fixed-fixed beam subjected to a load F in the middle, the static deflection is (see Table 6.1):

$$u = \frac{FL^3}{192EI}$$

Thus the stiffness k can be calculated as:

$$k = \frac{F}{u} = \frac{192EI}{L^3} = \frac{192 \times 200 \times 10^9 \times 2.66667 \times 10^{-7}}{3^3} = 379\,259\,\text{N/m}$$

Also, from Table 6.1, the equivalent mass is:

$$0.37\rho AL = 0.37 \times 7800 \times 400 \times 10^{-3} \times 20 \times 10^{-3} \times 3 = 69.264\,\text{kg}$$

Therefore, the angular frequency of the beam before attaching the vibration absorber (which is modelled as a system with a single degree of freedom system) is:

$$\omega_n = \sqrt{\frac{k}{m}} = \sqrt{\frac{379259}{250 + 69.264}} = 34.4662\,\text{rad/s}$$

The excitation force magnitude is $F_o = 300$ N and the excitation frequency is $\omega = 11\pi$ rad/s. From Equation (7.10), the steady-state amplitude of an undamped system with a single degree of freedom is given by:

$$X = \frac{F_o/k}{1 - \left(\frac{\omega}{\omega_n}\right)^2} = \frac{300/379\,259}{1 - \left(\frac{11\pi}{34.4662}\right)^2} = -0.149\,\text{m} = -149\,\text{mm}$$

b) Since the amplitude A_2 of the second mass should not exceed 2 cm, applying Equation (8.74) gives:

$$|A_2| = \frac{F_o}{k_2} \Rightarrow 0.02 = \frac{300}{k_2}$$

from which the stiffness of the vibration absorber is calculated as:

$$k_2 = 15\,000\,\text{N/m} = 15\,\text{kN/m}$$

And from Equation (8.73), the suppressed excitation frequency is:

$$\omega^2 = \frac{k_2}{m_2} \Rightarrow 34^2 = \frac{15\,000}{m_2}$$

From which the mass of the vibration absorber is obtained as:

$$m_2 = 12.975 = 13\,\text{kg}$$

Example 8.14 Vibration absorbers III

A bridge is modelled as a system with a single degree of freedom, as shown in Figure 8.27, that has a mass of $16\,000 \times 10^3$ kg and stiffness of $k = 2\,100\,000$ kN/m. Under the action of a vehicle that produces a harmonic force of 9000 N, the bridge undergoes excessive vibration. In order to suppress the vibration at its fundamental natural frequency, it is decided to use a vibration absorber. Design an undamped vibration absorber so that its amplitude does not exceed 2.5 cm.

Solution

The bridge's fundamental angular frequency is calculated from Equation (6.7) in Chapter 6 for a single degree of freedom system as :

$$\omega_n = \sqrt{\frac{k}{m}} = \sqrt{\frac{21 \times 10^8}{16 \times 10^6}} = 11.4564\,\text{rad/s}$$

And its natural frequency in Hz is:

$$f_n = \frac{\omega_n}{2\pi} = 1.823\,\text{Hz}$$

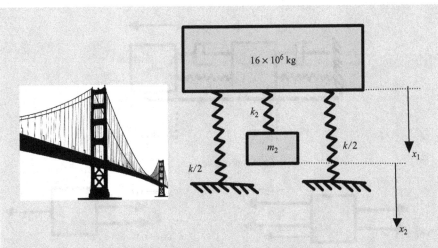

Figure 8.27: Representation of Example 8.14

The amplitude A_2 of the second mass should not exceed 2.5 cm, thus applying Equation (8.74) gives:

$$|A_2| = \frac{F_o}{k_2} \Rightarrow 0.025 = \frac{9000}{k_2}$$

from which the stiffness of the vibration absorber is calculated as:

$$k_2 = 360\,000\,\text{N/m} = 360\,\text{kN/m}$$

The suppressed excitation frequency is equal to the fundamental frequency of the bridge, i.e. 1.823 Hz. From Equation (8.73), the suppressed excitation frequency is:

$$\omega^2 = \frac{k_2}{m_2} \Rightarrow (1.823 \times 2\pi)^2 = \frac{360\,000}{m_2}$$

from which the mass of the vibration absorber is obtained as:

$$m_2 = 2753\,\text{kg}$$

8.7 VISCOUS DAMPING

A system with two degrees of freedom with two viscous dampers that have viscous damping coefficients c_1 and c_2 is shown in Figure 8.28(a). Two external forces, f_1 and f_2 are applied to the masses. The two equations of motion can be derived from the free-body diagrams in Figure 8.28(b) as:

$$-k_1 x_1 + k_2(x_2 - x_1) - c_1 \dot{x}_1 + c_2(\dot{x}_2 - \dot{x}_1) + f_1 = m_1 \ddot{x}_1$$
$$-k_2(x_2 - x_1) - c_2(\dot{x}_2 - \dot{x}_1) + f_2 = m_2 \ddot{x}_2$$

(8.75)

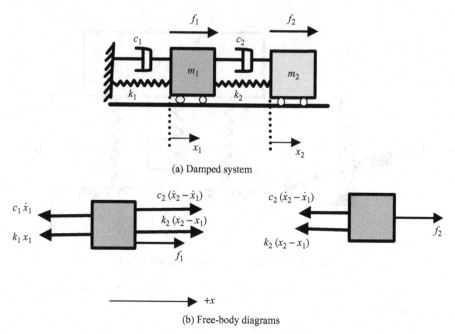

(a) Damped system

(b) Free-body diagrams

Figure 8.28: Damped system with two degrees of freedom

Rearranging terms gives:

$$m_1\ddot{x}_1 + (k_1 + k_2)x_1 - k_2 x_2 + (c_1 + c_2)\dot{x}_1 - c_2\dot{x}_2 = f_1$$
$$m_2\ddot{x}_2 - k_2 x_1 + k_2 x_2 - c_2\dot{x}_1 + c_2\dot{x}_2 = f_2$$

(8.76)

and, in matrix form:

$$\begin{bmatrix} m_1 & 0 \\ 0 & m_2 \end{bmatrix} \begin{bmatrix} \ddot{x}_1 \\ \ddot{x}_2 \end{bmatrix} + \begin{bmatrix} c_1 + c_2 & -c_2 \\ -c_2 & c_2 \end{bmatrix} \begin{bmatrix} \dot{x}_1 \\ \dot{x}_2 \end{bmatrix} + \begin{bmatrix} k_1 + k_2 & -k_2 \\ -k_2 & k_2 \end{bmatrix} \begin{bmatrix} x_1 \\ x_2 \end{bmatrix} = \begin{bmatrix} f_1 \\ f_2 \end{bmatrix}$$

(8.77)

8.7.1 Free vibration of damped system with two degrees of freedom

For free vibration, Equation (8.77) becomes:

$$\begin{bmatrix} m_1 & 0 \\ 0 & m_2 \end{bmatrix} \begin{bmatrix} \ddot{x}_1 \\ \ddot{x}_2 \end{bmatrix} + \begin{bmatrix} c_1 + c_2 & -c_2 \\ -c_2 & c_2 \end{bmatrix} \begin{bmatrix} \dot{x}_1 \\ \dot{x}_2 \end{bmatrix} + \begin{bmatrix} k_1 + k_2 & -k_2 \\ -k_2 & k_2 \end{bmatrix} \begin{bmatrix} x_1 \\ x_2 \end{bmatrix} = \begin{bmatrix} 0 \\ 0 \end{bmatrix}$$

(8.78)

The general form of equation (8.78) is:

$$M\ddot{D} + C\dot{D} + SD = 0$$

(8.79)

where the vectors D and $\ddot{D}$ are the displacement vector and the acceleration vector (see Equation (8.34)), respectively. $\dot{D}$ is the velocity vector and is given by:

$$\dot{D} = \begin{bmatrix} \dot{x}_1 \\ \dot{x}_2 \end{bmatrix}$$

The damping matrix C is given by:

$$C = \begin{bmatrix} C_{11} & C_{12} \\ C_{21} & C_{22} \end{bmatrix} = \begin{bmatrix} c_1 + c_2 & -c_2 \\ -c_2 & c_2 \end{bmatrix}$$

The solution of these differential equations is more complicated than for the undamped case. The general solution of Equation (8.78) is (see Equation (6.38) in Chapter 6):

$$x_1 = A_1 e^{\beta t},$$
$$x_2 = A_2 e^{\beta t} \tag{8.80}$$

Substituting Equation (8.80) and its derivatives into (8.78) gives:

$$\begin{bmatrix} M_{11}\beta^2 + C_{11}\beta + S_{11} & C_{12}\beta + S_{12} \\ C_{21}\beta + S_{21} & M_{22}\beta^2 + C_{22}\beta + S_{22} \end{bmatrix} \begin{bmatrix} A_1 \\ A_2 \end{bmatrix} = \begin{bmatrix} 0 \\ 0 \end{bmatrix} \tag{8.81}$$

For non-trivial solutions, the determinant of the matrix in Equation (8.81) should be equal to zero:

$$\begin{vmatrix} M_{11}\beta^2 + C_{11}\beta + S_{11} & C_{12}\beta + S_{12} \\ C_{21}\beta + S_{21} & M_{22}\beta^2 + C_{22}\beta + S_{22} \end{vmatrix} = 0$$

from which

$$M_{11}M_{22}\beta^4 + (M_{11}C_{22} + M_{22}C_{11})\beta^3 + (M_{11}S_{22} + C_{22}C_{11} + M_{22}S_{11} - C_{12}^2)\beta^2$$
$$+ (C_{11}S_{22} + S_{11}C_{22} - 2C_{12}S_{12})\beta + S_{11}S_{22} - S_{12}^2 = 0 \tag{8.82}$$

Equation (8.82) should be solved numerically by extracting the roots of β. The general form of the solution is complex and is given by:

$$\beta_{11} = -\eta_1 + i\omega_{d1}, \qquad \beta_{12} = -\eta_1 - i\omega_{d1}$$
$$\beta_{21} = -\eta_2 + i\omega_{d2}, \qquad \beta_{22} = -\eta_2 - i\omega_{d2}$$

where η_1 and η_2 are positive numbers representing damping and ω_{d1} and ω_{d2} are the damped angular frequencies. The amplitude ratios are obtained by substituting the roots into equation (8.81):

$$r_{ij} = \frac{-C_{12}\beta_{ij} - S_{12}}{M_{11}\beta_{ij}^2 + C_{11}\beta_{ij} + S_{11}} = \frac{M_{22}\beta_{ij}^2 + C_{22}\beta_{ij} + S_{22}}{-C_{21}\beta_{ij} - S_{21}}$$

where i, j are equal to 1 or 2. The amplitude ratios are complex conjugate pairs and have the form:

$$r_{11} = a + ib, \qquad r_{12} = a - ib$$
$$r_{21} = c + id, \qquad r_{22} = c - id$$

$$(8.83)$$

8.7.2 Forced vibration of damped system with two degrees of freedom

If the system in Figure 8.28 is subjected to harmonic excitation of the form:

$$f_1 = F_{o1} e^{i\omega t},$$
$$f_2 = F_{o2} e^{i\omega t}$$

$$(8.84)$$

where ω is the forcing angular frequency and F_{o1} and F_{o2} are the force magnitudes, Equation (8.79) becomes:

$$M\ddot{D} + C\dot{D} + SD = F_o e^{i\omega t}$$

and, in matrix form:

$$\begin{bmatrix} m_1 & 0 \\ 0 & m_2 \end{bmatrix} \begin{bmatrix} \ddot{x}_1 \\ \ddot{x}_2 \end{bmatrix} + \begin{bmatrix} c_1 + c_2 & -c_2 \\ -c_2 & c_2 \end{bmatrix} \begin{bmatrix} \dot{x}_1 \\ \dot{x}_2 \end{bmatrix} + \begin{bmatrix} k_1 + k_2 & -k_2 \\ -k_2 & k_2 \end{bmatrix} \begin{bmatrix} x_1 \\ x_2 \end{bmatrix} = \begin{bmatrix} F_{o1} \\ F_{o2} \end{bmatrix} e^{i\omega t}$$

$$(8.85)$$

The general solution of these differential equations has a complex form and is given by:

$$x_1 = A_1 e^{i\omega t}$$
$$x_2 = A_2 e^{i\omega t}$$

$$(8.86)$$

Substituting Equation (8.86) and its derivatives into (8.85) gives:

$$\begin{bmatrix} -M_{11}\beta^2 + iC_{11}\beta + S_{11} & iC_{12}\beta + S_{12} \\ iC_{21}\beta + S_{21} & -M_{22}\beta^2 + iC_{22}\beta + S_{22} \end{bmatrix} \begin{bmatrix} A_1 \\ A_2 \end{bmatrix} = \begin{bmatrix} F_{o1} \\ F_{o2} \end{bmatrix}$$

$$(8.87)$$

Solving for $[A]$ yields:

$$\begin{bmatrix} A_1 \\ A_2 \end{bmatrix} = \frac{1}{C} \begin{bmatrix} -M_{22}\beta^2 + iC_{22}\beta + S_{22} & iC_{12}\beta + S_{12} \\ iC_{21}\beta + S_{21} & -M_{11}\beta^2 + iC_{11}\beta + S_{11} \end{bmatrix} \begin{bmatrix} F_{o1} \\ F_{o2} \end{bmatrix}$$

$$(8.88)$$

where

$$C = (-M_{11}\beta^2 + iC_{11}\beta + S_{11})(-M_{22}\beta^2 + iC_{22}\beta + S_{22}) - (iC_{21}\beta + S_{21})^2 \qquad (8.89)$$

which has the complex form as in Equation (8.82).

8.8 Tutorial Sheet

8.8.1 Free vibration systems

Q8.1 For the system with two degrees of freedom shown in Figure 8.29:

a) Write the equations of motion in matrix form.

$$\left[\begin{bmatrix} m & 0 \\ 0 & 2m \end{bmatrix}\begin{bmatrix} \ddot{x}_1 \\ \ddot{x}_2 \end{bmatrix} + \begin{bmatrix} 3k & -2k \\ -2k & 5k \end{bmatrix}\begin{bmatrix} x_1 \\ x_2 \end{bmatrix} = \begin{bmatrix} F \\ F \end{bmatrix}\right]$$

b) Determine the natural frequencies of the system in terms of k and m.

$$\left[1.146\sqrt{\tfrac{k}{m}}\,\text{rad/s}, \quad 2.045\sqrt{\tfrac{k}{m}}\,\text{rad/s}\right]$$

c) Determine the mode shape ratios.

$$[1.185, \; -1.692]$$

d) Calculate the natural frequencies in Hz and draw the mode shapes for $m = 5$ kg and $k = 2$ kN/m.

$$[3.65 \text{ Hz}, 6.51 \text{ Hz}]$$

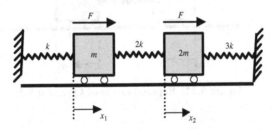

Figure 8.29: Representation of Question 8.1

Q8.2 A rigid bar AB of length 2 m and mass 10 kg is supported by two springs of stiffness constant k and $2k$ at point A and B, respectively, as shown in Figure 8.30. The two springs are located at an equal distance from the centre of gravity of the bar C. If $k = 1$ kN/m, determine:

a) the natural frequencies of the system in Hz;

$$[2.53 \text{ Hz}, 4.89 \text{ Hz}]$$

b) the amplitude ratios and sketch the two mode shapes.

$$[-37.77 \text{ mm/degree}, 2.7 \text{ mm/degree}]$$

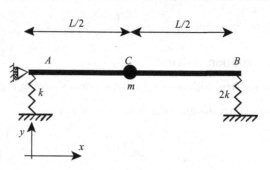

Figure 8.30: Representation of Question 8.2

Q8.3 A semidefinite system, shown in Figure 8.31, has two masses 10 kg and 20 kg, which are coupled using a spring of stiffness 25 kN/m. Determine the two natural frequencies of the system.

[0, 9.75 Hz]

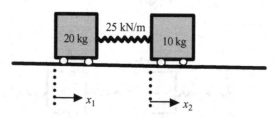

Figure 8.31: Representation of Question 8.3

Q8.4 A weather balloon lifts a mass m through elastic ropes that have an equivalent stiffness k and can be approximated to a system with two degrees of freedom, as shown in Figure 8.32. If the mass of the balloon is 1% of the lifted mass, determine:

a) the two natural frequencies of the system in terms of k and m;

$$\left[0, 1.6\sqrt{\tfrac{k}{m}}\ \text{Hz}\right]$$

b) the two amplitude ratios.

[1, −0.01]

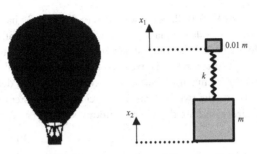

Figure 8.32: Representation of Question 8.4

Q8.5   Determine the natural frequencies and the amplitude ratios of the two railway boxcars, shown in Figure 8.33, that have mass of 12×10^3 kg each and are connected to one another by a spring of stiffness 250 kN/m

[0, 1.03 Hz, 1, −1]

Figure 8.33: Representation of Question 8.5

Q8.6   For the system with two degrees of freedom shown in Figure 8.34, determine the natural frequencies of vibration.

[0.51 Hz, 1.44 Hz]

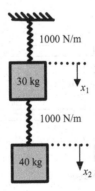

Figure 8.34: Representation of Question 8.6

Q8.7   The girder of a travelling crane shown in Figure 8.35 has a length of 10 m and carries a trolley of mass 3.5×10^3 kg at its midpoint. The crane has to lift a load of 800 kg through steel wires of cross-section area 900 mm^2 and 6 m length. If the girder is fixed at both ends and has flexural rigidity EI of 18×10^9 N.m^2, use a model of a system with two degrees of freedom to determine the two natural frequencies of the system. Ignore the mass of the girder and use the Young's modulus of steel as 200 GPa.

[30.67 Hz, 154.71 Hz]

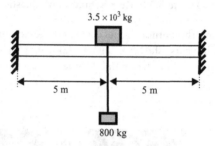

Figure 8.35: Representation of Question 8.7

Q8.8   For the two-storey building shown in Figure 8.36, determine:

a) the natural frequencies in terms of m and k;

$$\left[0.78 \sqrt{\tfrac{k}{m}} \, \text{Hz}, \, 1.56 \sqrt{\tfrac{k}{m}} \, \text{Hz} \right]$$

b) the natural frequencies if $m = 500 \times 10^3$ kg and $k = 20\,000$ kN/m.

[4.93 Hz, 9.87 Hz]

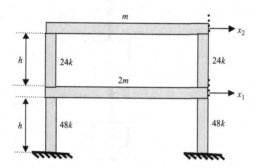

Figure 8.36: Representation of Question 8.8

Q8.9   A machine of mass 190×10^3 kg is supported by a foundation of mass 240×10^3 kg through an elastic pad of stiffness $k_e = 74\,000$ kN/m, as shown in Figure 8.37. If the soil stiffness is equal to $k_s = 145\,000$ kN/m, determine the fundamental frequency of the system.

[2.34 Hz]

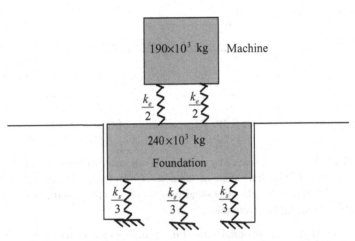

Figure 8.37: Representation of Question 8.9

Q8.10 A mass of 10 kg is suspended by three springs of stiffness 1000 N/m each, as shown in Figure 8.38. If the angles that the springs make with the horizontal axis are $\theta_1 = 30°$, $\theta_2 = 120°$ and $\theta_3 = 240°$, determine the two natural frequencies of the system.

[1.59 Hz, 2.25 Hz]

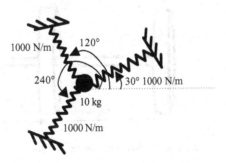

Figure 8.38: Representation of Question 8.10

Q8.11 The uniform beam of 10 kg mass shown in Figure 8.39 is supported at its ends by two springs A and B, each having stiffness of 200 N/m. If a mass of 6 kg is placed at the beam's midpoint, determine the two natural frequencies of the system in Hz.

[0.796 Hz, 1.74 Hz]

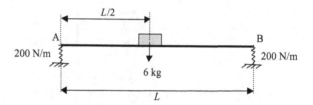

Figure 8.39: Representation of Question 8.11

Q8.12 Two disks of mass 2 kg each and having equal radius of 10 cm are mounted on a solid steel shaft of diameter 15 mm and length 1 m as shown in Figure 8.40. The steel shear modulus is 77 GPa and the mass of the shaft is negligible. Determine:

a) the two torsional natural frequencies of the system;

[28.42 Hz, 69.62 Hz]

b) the two amplitude ratios.

[0.5, −2]

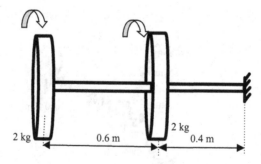

Figure 8.40: Representation of Question 8.12

Q8.13 For the torsional system with two degrees of freedom shown in Figure 8.41, determine the two torsional natural frequencies as functions of the mass moment of inertia J and the torsional stiffness k.

$$\left[0.1125\sqrt{\tfrac{k}{J}}\,\text{Hz},\quad 0.225\sqrt{\tfrac{k}{m}}\,\text{Hz}\right]$$

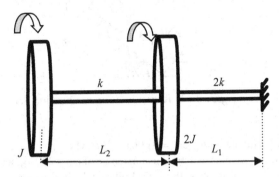

Figure 8.41: Representation of Question 8.13

8.8.2 Undamped forced vibration and vibration absorbers

Q8.14 In Question 8.2, if a vertical harmonic force of $15\sin\omega t$ (in N) is applied at the centre of gravity of the rigid bar (point C), determine:

a) the forcing angular frequency in rad/s at which the displacement at C is zero;

[30 rad/s]

b) the corresponding amplitude of θ_C.

[0.86°]

Q8.15 For the system with two degrees of freedom shown in Figure 8.42, determine the fundamental frequency and the steady-state amplitude of the 10 kg mass.

[4.43 Hz, 50.2 mm]

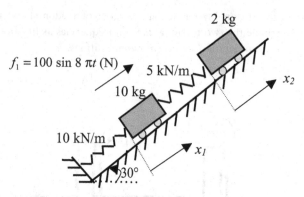

Figure 8.42: Representation of Question 8.15

Q8.16 A harmonic force of $f = 1000 \sin 24t$, where f is in newtons and t is the time in seconds, is acting on a machine of mass 200 kg, which is supported by two springs of stiffness 60 kN/m each, as shown in Figure 8.43(a). In order to reduce the machine's vibration amplitude, a mass m_2 is attached to the machine through a spring of stiffness 5 kN/m as shown in Figure 8.43(b). Determine the maximum displacement of the machine in mm for the following cases:

a) before adding the mass m_2;

[208.3 mm]

b) if $m_2 = 7$ kg;

[−62.4 mm]

c) if $m_2 = 10$ kg.

[23.4 mm]

d) Determine the mass m_2 so that the machine's displacement is zero.

[8.68 kg]

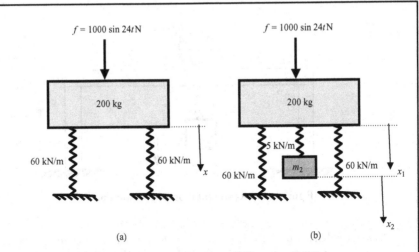

$f = 1000 \sin 24t\,\mathrm{N}$

$f = 1000 \sin 24t\,\mathrm{N}$

200 kg

200 kg

60 kN/m

60 kN/m

5 kN/m

x

60 kN/m

m_2

60 kN/m

x_1

x_2

(a)

(b)

Figure 8.43: Representation of Question 8.16

Q8.17 A motor is mounted on a floor and operates at 650 rev/min. The floor is supported by columns that have an overall stiffness of 5000 kN/m. The natural frequency of the system was calculated as 10.82 Hz and excessive vibration was expected. If a vibration absorber has to be designed by attaching to the floor a mass of 200 kg and a spring, as shown in Figure 8.44, determine:

a) the mass of the floor and the motor;

[1082 kg]

b) the stiffness of vibration absorber;

[926.6 kN]

c) the two frequencies of the modified system (floor, motor and vibration absorber).

[8.7 Hz, 13.4 Hz]

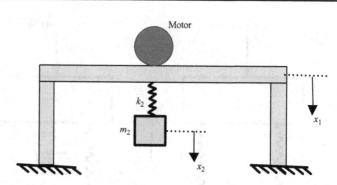

Figure 8.44: Representation of Question 8.17

Q8.18 An unbalanced motor is mounted at the free end of a cantilever beam, as shown in Figure 8.45. The beam has a length of 1 m, is made of steel with $E = 200$ GPa and density 7800 kg/m³ and has a cross-section of 20 × 400 mm. The motor has a mass of 150 kg and produces a force that is equal to $F = 200 \sin 31t$(N), where F is in newtons and t is the time in seconds, which introduces excessive vibration to the beam. In order to eliminate the vibration, an absorber is attached to the bottom of the beam at the free end. If the amplitude of the vibration absorber mass should not exceed 2.5 cm, determine:

a) the stiffness of the beam;

[160 kN/m]

b) the amplitude of the beam's vibration before attaching the absorber;

[114 mm]

c) the stiffness and the mass of the vibration absorber;

[8 kN/m, 8.32 kg]

d) the natural frequencies of the system after adding the vibration absorber.

[4.4 Hz, 5.5 Hz]

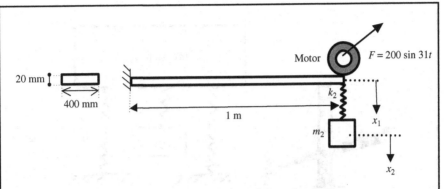

Figure 8.45: Representation of Question 8.18

Q8.19 A motor is mounted on a beam that has a natural frequency of 15 Hz and a mass of 10 kg. If a vibration absorber of mass 6 kg is attached to the beam in order to suppress vibration at 9 Hz, determine the stiffness of the vibration absorber.

[19.2 kN/m]

Q8.20 A bridge is modelled as a system with a single degree of freedom, as shown in Figure 8.46. The bridge has a mass of $16\,000 \times 10^3$ kg and a natural frequency of 2 Hz. Under the action of a vehicle that produces a harmonic force of 9500 N, the bridge undergoes excessive vibration. If it is decided to use an undamped vibration absorber of mass 3×10^3 kg in order to suppress the bridge's vibration at its fundamental natural frequency, determine:

a) the stiffness of the vibration absorber;

[473.71 kN]

b) the maximum amplitude of vibration of the absorber.

[2 cm]

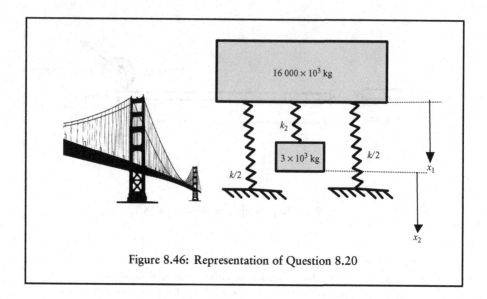

Figure 8.46: Representation of Question 8.20

CHAPTER 9

Vibration of Continuous Systems

9.1 INTRODUCTION

In a continuous system, stiffness, mass and damping are not considered at specific points (as they are in the case of discrete systems); rather, they are distributed over the whole structure. Chapters 6, 7 and 8 considered discrete systems in which the mass, the spring and the damper were concentrated at discrete points. In a continuous system, there are no discrete masses, springs or dampers.

A continuous system can be regarded as a system that has an infinite number of particles, each of which possesses a mass, a stiffness constant and a damping coefficient. Therefore, continuous systems have an infinite number of degrees of freedom. It follows that, in a continuous system, there is an infinite number of natural frequencies. Each of these natural frequencies has a normal mode or mode shape, which describes the shape of deformation at that natural frequency. The total vibration response of a continuous system consists of the contribution of all modes. The individual contribution of each mode in the total response depends on the position and the excitation frequency of the applied dynamic force. For a discrete system, the equations of motion are ordinary differential equations, which are straightforward and can be dealt with easily. If multiple degrees of freedom are considered, matrix manipulation (similar to that described in Chapter 8) can be performed by computers. In contrast, for a continuous system, the equations of motions are partial differential equations and are more complex.

For complex structures, it might be not possible to derive or to solve the required partial differential equation. In fact, many structures can be modelled using both concepts: they can be modelled as discrete systems by dividing the structure into masses, dampers and springs and as continuous systems by deriving a partial differential equation for the whole structure. However, the accuracy of the solution obtained using a continuous system approach is, in general, higher than that obtained using a discrete system approach.

The decision about which approach to use for a certain structure depends on the importance of the analysis, the accuracy required, and how quickly the results are required. This chapter considers continuous systems for simple structures such as strings, bars and beams. The equation of motion for each system is derived and solved for free vibration so that natural frequencies and mode shapes of the system are determined for different boundary conditions. The solution of the displacement time response due to some types of dynamic force applied to the string and the bar is also presented. Furthermore, the concept of whirling of shafts due to unbalanced rotating masses is introduced and discussed.

9.2 LATERAL VIBRATION OF A CABLE OR STRING

9.2.1 Deriving the equation of motion

In order to write the equation of motion of a continuous system, the forces acting on a small segment are analyzed and Newton's second law is applied. Figure 9.1(a) shows a

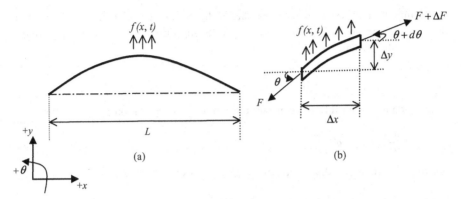

Figure 9.1: a) A vibrating cable and b) the forces acting on an infinitesimal segment of it

cable or a string of length L and mass per unit length ρ'. The cable is stretched with a tension force F and is subjected to transverse force $f(x, t)$ per unit length. The vertical co-ordinate y, which represents the transverse displacement of any point of the cable, is a function of the horizontal co-ordinate x, the distance along the cable, and time t, i.e. $y(x, t)$. A free-body diagram showing the forces acting on a cable segment of an infinitesimal length Δx is shown in Figure 9.1(b).

From Newton's second law, the sum of the forces in the y direction is given by:

$$\sum F_y = m\ddot{y} \tag{9.1}$$

where $m = \rho' \Delta x$ is the mass of the segment and $\ddot{y} = \frac{\partial^2 y}{\partial t^2}$ is the acceleration in the y direction. Thus, from Figure 9.1, Equation (9.1) is expanded as:

$$(F + \Delta F)\sin(\theta + d\theta) + f\,\Delta x - F\sin\theta = \rho'\,\Delta x\frac{\partial^2 y}{\partial t^2} \tag{9.2}$$

where θ is the angle between the deformed cable and the x axis. For an infinitesimal length Δx and small angle θ, the following approximations can be made:

$$\Delta F = \frac{\partial F}{\partial x}\Delta x, \quad \sin\theta \approx \tan\theta = \frac{\Delta y}{\Delta x} = \frac{\partial y}{\partial x},$$

$$\sin(\theta + d\theta) \approx \tan(\theta + d\theta) \approx \frac{\partial y}{\partial x} + \frac{\partial^2 y}{\partial x^2}\Delta x \tag{9.3}$$

Substituting Equation (9.3) into Equation (9.2) yields

$$\left(F + \frac{\partial F}{\partial x}\Delta x\right)\left(\frac{\partial y}{\partial x} + \frac{\partial^2 y}{\partial x^2}\Delta x\right) + f\,\Delta x - F\frac{\partial y}{\partial x} = \rho'\,\Delta x\frac{\partial^2 y}{\partial t^2} \tag{9.4}$$

Expanding Equation (9.4) gives:

$$\frac{\partial F}{\partial x}\frac{\partial y}{\partial x}\Delta x + \frac{\partial F}{\partial x}\frac{\partial^2 y}{\partial x^2}\Delta x^2 + F\frac{\partial^2 y}{\partial x^2}\Delta x + f\,\Delta x = \rho'\,\Delta x\frac{\partial^2 y}{\partial t^2} \tag{9.5}$$

For an infinitesimal Δx, the higher terms are negligible, i.e. $\Delta x^2 \cong 0$. Eliminating the term that contains Δx^2 from Equation (9.5), the following simplified form of the equation of motion is obtained as:

$$\frac{\partial}{\partial x}\left[F\frac{\partial y}{\partial x}\right] + f = \rho'\frac{\partial^2 y}{\partial t^2} \qquad (9.6)$$

In case of a uniform cable and constant tension force (F), Equation (9.6) becomes:

$$F\frac{\partial^2 y}{\partial x^2} + f = \rho'\frac{\partial^2 y}{\partial t^2} \qquad (9.7)$$

For free vibration, $f = 0$, and Equation (9.7) reduces to:

$$F\frac{\partial^2 y}{\partial x^2} = \rho'\frac{\partial^2 y}{\partial t^2} \qquad (9.8)$$

which is known as the wave equation. Equation (9.8) can also be written as:

$$c^2\frac{\partial^2 y}{\partial x^2} = \frac{\partial^2 y}{\partial t^2} \qquad (9.9)$$

where $c = \sqrt{\frac{F}{\rho'}}$, which is known as the wave speed.

9.2.2 Free vibration of a string

Equation (9.9), which represents the free vibration equation for a cable, can be solved using the concept of separation of variables, i.e. the function y is replaced by the product of two functions Y and T as:

$$y(x, t) = Y(x)T(t) \qquad (9.10)$$

where the function $Y(x)$ depends only on x and the function $T(t)$ depends only on t. Substituting Equation (9.10) into Equation (9.9) yields:

$$\frac{c^2}{Y}\frac{\partial^2 Y}{\partial x^2} = \frac{1}{T}\frac{\partial^2 T}{\partial t^2} \qquad (9.11)$$

Since the left-hand side of Equation (9.11) is a function only in x and the right-hand side is a function only in t, both sides in Equation (9.11) should be equal to the same constant:

$$\frac{c^2}{Y}\frac{\partial^2 Y}{\partial x^2} = \frac{1}{T}\frac{\partial^2 T}{\partial t^2} = a \qquad (9.12)$$

where a is a constant. Equation (9.12) can be separated into two independent equations as:

$$\frac{\partial^2 Y}{\partial x^2} - \frac{a}{c^2}Y = 0 \qquad (9.13)$$

$$\frac{\partial^2 T}{\partial t^2} - aT = 0 \qquad (9.14)$$

Choosing the constant a equal to $-\omega_n^2$, which is the only choice that gives a physically acceptable solution for the two partial differential Equations (9.13) and (9.14), leads to:

$$\frac{\partial^2 Y}{\partial x^2} + \frac{\omega_n^2}{c^2} Y = 0 \tag{9.15}$$

$$\frac{\partial^2 T}{\partial t^2} + \omega_n^2 T = 0 \tag{9.16}$$

The solutions of Equations (9.15) and (9.16) are:

$$Y(x) = A\cos \frac{\omega_n x}{c} + B\sin \frac{\omega_n x}{c} \tag{9.17}$$

$$T(t) = C\cos \omega_n t + D\sin \omega_n t \tag{9.18}$$

where A, B, C and D are constants and ω_n is the angular frequency of vibration. Substituting Equations (9.17) and (9.18) into Equation (9.10), the solution of y is obtained as:

$$y(x, t) = Y(x)T(t) = \left(A\cos \frac{\omega_n x}{c} + B\sin \frac{\omega_n x}{c}\right)(C\cos \omega_n t + D\sin \omega_n t) \tag{9.19}$$

The constants A, B, C and D in Equation (9.19) can be determined from initial and boundary conditions.

9.2.3 Initial and boundary conditions

In order to determine the four constants in Equation (9.19), two initial and two boundary conditions are required. If the cable is fixed at both ends, then when $x = 0$, $y = Y = 0$ and when $x = L$, $y = Y = 0$ at all times. Therefore, from Equation (9.17), the following conditions should be satisfied:

$$Y(0) = 0 = A\cos 0 + B\sin 0 \tag{9.20}$$

$$Y(L) = 0 = A\cos \frac{\omega_n L}{c} + B\sin \frac{\omega_n L}{c} \tag{9.21}$$

From Equation (9.20), $A = 0$ and, from Equation (9.21), $B\sin \frac{\omega_n L}{c} = 0$. For a nontrivial solution, the following expression should be satisfied:

$$\sin \frac{\omega_n L}{c} = 0 \tag{9.22}$$

This is known as the frequency equation or characteristic equation. The solution of Equation (9.22) is:

$$\frac{\omega_n L}{c} = n\pi, \qquad n = 1, 2, 3, \ldots$$

Thus:

$$\omega_n = \frac{n c \pi}{L}, \qquad n = 1, 2, 3, \ldots \tag{9.23}$$

where the values of ω_n are the angular frequencies of the cable, also called eigenvalues or characteristic values. For each angular frequency ω_n, a displacement time response solution $y_n(x, t)$ can be determined using Equations (9.19), (9.20) and (9.23) as:

$$y_n(x, t) = Y_n(x)T_n(t) = \sin \frac{n\pi x}{L} \left(C_n \cos \frac{nc\pi t}{L} + D_n \sin \frac{nc\pi t}{L} \right) \tag{9.24}$$

where C_n and D_n are constants that can be determined from the initial and boundary conditions. At any time t, i.e. t is constant in Equation (9.24), the function $Y_n(x)$, known as the mode shape function for the nth mode, can be expressed as:

$$Y_n = A_n \sin \frac{n\pi x}{L} \tag{9.25}$$

where A_n is an arbitrary constant. Thus, the mode shape of vibration can be drawn using Equation (9.25). The angular frequency, ω_1, corresponding to $n = 1$ is of particular importance as it is the lowest frequency value of the system and is known as the fundamental angular frequency. For a system with a single degree of freedom, see Chapters 6 and 7, in which only the fundamental angular frequency, ω_1 (denoted as ω_n in Chapters 6 and 7), could be determined. It can easily be shown that the mode shapes are orthogonal so that they satisfy the orthogonal condition given by:

$$\int_0^L Y_i \times Y_j \, dx = 0$$

Example 9.1 Lateral vibration of a cable I

An electric power transmission cable has a length of 2 km as shown in Figure 9.2. The cable has a mass of 4 kg/m and is stretched by a tension of 3000 MN. Determine the first three natural frequencies and draw the corresponding mode shapes.

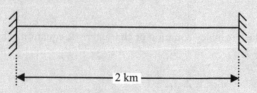

Figure 9.2: Representation of Example 9.1

Solution

The tension force in the cable is $F = 3000$ MN and the mass per unit length is $\rho' = 4$ kg/m. The constant, c, is calculated from Equation (9.9) as: $c = \sqrt{\frac{F}{\rho'}} = \sqrt{\frac{3 \times 10^9}{4}} = 27\,386.128$ m/s.

Using Equation (9.23) for $n = 1, 2$ and 3, the first three angular frequencies (ω_n) can be calculated and then the natural frequencies (f_n) are obtained as:

$$\omega_1 = \frac{c\pi}{L} = 43.018\,\text{rad/s} \Rightarrow f_1 = \frac{\omega_1}{2\pi} = 6.85\,\text{Hz}$$

$$\omega_2 = \frac{2c\pi}{L} = 86.036\,\text{rad/s} \Rightarrow f_2 = \frac{\omega_2}{2\pi} = 13.69\,\text{Hz}$$

$$\omega_3 = \frac{3c\pi}{L} = 129.054\,\text{rad/s} \Rightarrow f_3 = \frac{\omega_3}{2\pi} = 20.54\,\text{Hz}$$

The first three mode shapes are obtained from Equation (9.25) for $n = 1, 2$ and 3 and are shown in Figure 9.3, in which $\frac{Y_n}{A_n}$ is plotted against x/L.

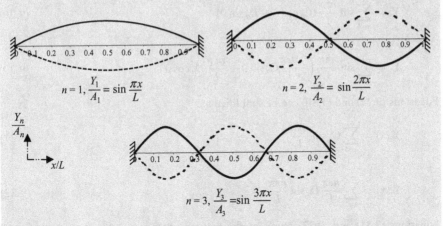

$$n = 1, \frac{Y_1}{A_1} = \sin\frac{\pi x}{L}$$

$$n = 2, \frac{Y_2}{A_2} = \sin\frac{2\pi x}{L}$$

$$n = 3, \frac{Y_3}{A_3} = \sin\frac{3\pi x}{L}$$

Figure 9.3: Mode shapes for Example 9.1

It should be noted that the points at which Y_n is zero are of particular importance. They are called nodes and can be used to suppress vibration. Mode 1 (fundamental mode) has two nodes at $x = 0$ and $x = L$; mode 2 has three nodes, at $x = 0$, $x = L$ and $x = L/2$; and mode 3 has four nodes at $x = 0$, $x = L$, $x = L/3$ and $x = 2L/3$.

9.2.4 Time response of a string

Equation (9.24) represents the global solution for the lateral vibration of a fixed–fixed cable or a string. The total displacement time response can be written in the form of superposition of all mode responses y_n as:

$$y(x, t) = \sum_{n=1}^{\infty} \sin \frac{n\pi x}{L} \left(C_n \cos \frac{nc\pi t}{L} + D_n \sin \frac{nc\pi t}{L} \right) \tag{9.26}$$

Equation (9.26) is called the mode superposition method as the responses of the individual modes (n) are added together to provide the total displacement time response. In order to determine the constants C_n and D_n, initial conditions are required. In a general form, the initial conditions are:

$$y(x, t = 0) = Y_o \tag{9.27}$$

$$\frac{dy}{dt}(x, t = 0) = \dot{Y}_o \tag{9.28}$$

Differentiating Equation (9.26) with respect to time and substituting Equations (9.27) and (9.28) into Equation (9.26) and its derivative gives:

$$Y_o = \sum_{n=1}^{\infty} \sin \frac{n\pi x}{L} (C_n \cos 0 + D_n \sin 0) \tag{9.29}$$

$$\dot{Y}_o = \sum_{n=1}^{\infty} \sin \frac{n\pi x}{L} \left(-\frac{nc\pi}{L} C_n \sin 0 + \frac{nc\pi}{L} D_n \cos 0 \right) \tag{9.30}$$

Equations (9.29) and (9.30) can be simplified as:

$$Y_o(x) = \sum_{n=1}^{\infty} C_n \sin \frac{n\pi x}{L} \tag{9.31}$$

$$\dot{Y}_o(x) = \sum_{n=1}^{\infty} \frac{nc\pi}{L} D_n \sin \frac{n\pi x}{L} \tag{9.32}$$

Equations (9.31) and (9.32) represent the Fourier sine series expansions of $Y_o(x)$ and $\dot{Y}_o(x)$, respectively. In order to determine C_n and D_n, both sides in Equations (9.31) and (9.32) are multiplied by $\sin \frac{n\pi x}{L}$ and integrated from $x = 0$ to $x = L$:

$$\int_0^L Y_o(x) \times \sin \frac{n\pi x}{L} dx = C_n \int_0^L \sum_{n=1}^{\infty} \sin^2 \frac{n\pi x}{L} dx \tag{9.33}$$

$$\int_0^L \dot{Y}_o(x) \times \sin \frac{n\pi x}{L} dx = \frac{nc\pi}{L} D_n \int_0^L \sum_{n=1}^{\infty} \sin^2 \frac{n\pi x}{L} dx \tag{9.34}$$

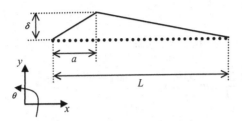

Figure 9.4: A plucked cable

Using $\sin^2 \frac{n\pi x}{L} = \frac{1}{2}\left(1 - \cos\frac{2n\pi x}{L}\right)$ and integrating the right-hand sides in Equations (9.33) and (9.34) yields:

$$C_n = \frac{2}{L}\int_0^L Y_o(x) \times \sin\frac{n\pi x}{L}dx \tag{9.35}$$

$$D_n = \frac{2}{nc\pi}\int_0^L \dot{Y}_o(x) \times \sin\frac{n\pi x}{L}dx \tag{9.36}$$

For a cable plucked at a distance a from its end, as shown in Figure 9.4, the initial conditions are:

$$Y_o(x) = \begin{cases} \dfrac{\delta x}{a} & 0 \le x \le a \\[2mm] \dfrac{\delta(L-x)}{L-a} & a \le x \le L \end{cases} \tag{9.37}$$

$$\dot{Y}_o(x) = 0 \tag{9.38}$$

It follows that $D_n = 0$ and the values of C_n can be calculated from Equations (9.35) as:

$$C_n = \frac{2}{L}\left[\int_0^a \frac{\delta x}{a}\sin\frac{n\pi x}{L}dx + \int_a^L \frac{\delta(L-x)}{L-a}\sin\frac{n\pi x}{L}dx\right] \tag{9.39}$$

Solving Equation (9.39) using integration by parts yields:

$$C_n = \begin{cases} \dfrac{2\delta L}{\pi^2 n^2}\left(\dfrac{1}{a} + \dfrac{1}{L-a}\right)\sin\dfrac{n\pi a}{L}, & n = 1, 3, 5, \ldots \\[2mm] 0, & n = 2, 4, 6, \ldots \end{cases} \tag{9.40}$$

If the cable is plucked at its midpoint, $a = L/2$, the initial conditions become:

$$Y_o(x) = \begin{cases} \dfrac{2\delta x}{L} & 0 \le x \le \frac{L}{2} \\[2mm] \dfrac{2\delta(L-x)}{L} & \frac{L}{2} \le x \le L \end{cases} \tag{9.41}$$

$$\dot{Y}_o(x) = 0 \tag{9.42}$$

and the integration of Equation (9.35) is:

$$C_n = \frac{2}{L}\left[\int_0^{L/2} \frac{2\delta x}{L}\sin\frac{n\pi x}{L}dx + \int_{L/2}^{L} \frac{2\delta(L-x)}{L}\sin\frac{n\pi x}{L}dx\right] \tag{9.43}$$

The solution for C_n is obtained as:

$$C_n = \begin{cases} \dfrac{8\delta}{\pi^2 n^2}\sin\dfrac{n\pi}{2}, & n = 1, 3, 5, \ldots \\ 0, & n = 2, 4, 6, \ldots \end{cases} \tag{9.44}$$

And the total displacement time response for a cable plucked at its midpoint is obtained from Equations (9.26) and (9.44) as:

$$y(x, t) = \sum_{n=1}^{\infty} C_n \sin\frac{n\pi x}{L}\cos\frac{nc\pi t}{L} \tag{9.45}$$

Example 9.2 Lateral vibration of a cable II

Determine the displacement time response of the midpoint ($x = 1000$ m) of the cable in Example 9.1, if an initial deflection of 1 mm is applied as shown in Figure 9.5. Use the mode superposition method with:

a) the first two modes;
b) the first three modes.

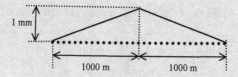

1 mm

1000 m   1000 m

Figure 9.5: Representation of Example 9.2

Solution

Since there is no initial velocity, i.e. $\dot{Y}_o(x) = 0$, from Equation (9.36) D_n becomes:

$$D_n = \frac{2}{nc\pi}\int_0^{L} \dot{Y}_o(x) \times \sin\frac{n\pi x}{L}dx = 0$$

For $x = L/2$, $c = 27\,386.128$ m/s and $L = 2000$ m, using Equation (9.45), the displacement time response is given by:

$$y(L/2, t) = \sum_{n=1}^{\infty} C_n \sin \frac{n\pi}{2} \cos \frac{n \times 27\,386.128 \times \pi t}{2000}$$

$$= \sum_{n=1}^{\infty} C_n \sin \frac{n\pi}{2} \cos 13.69 n\pi t$$

a) Using the mode superposition with two modes ($n = 1$ and $n = 2$), the displacement time response becomes:

$$y(L/2, t) = C_1 \sin \frac{\pi}{2} \cos 13.69 \pi t + C_2 \sin \pi \cos 27.38 \pi t$$

This gives

$$y(L/2, t) = C_1 \cos 13.69 \pi t \qquad \text{(E9.2a)}$$

To calculate C_1, using Equation (9.44) for $n = 1$ and $\delta = 1$ mm, yields:

$$C_1 = \frac{8 \times 1}{\pi^2 (1)^2} \sin \frac{1 \times \pi}{2} = 0.81 \text{ mm}$$

Substituting in Equation (E9.2a) gives:

$$y(L/2) = 0.81 \cos 13.69 \pi t \text{ mm}$$

Thus the total displacement time response is equivalent to the response of the first mode, as shown in Figure 9.6(a).

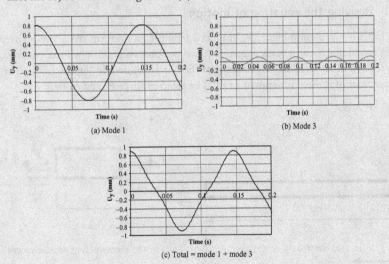

(a) Mode 1

(b) Mode 3

(c) Total = mode 1 + mode 3

Figure 9.6: Displacement time responses for Example 9.2

b) Using the mode superposition with three modes ($n = 1, 2$ and 3), the displacement time response becomes:

$$y(L/2, t) = C_1 \sin \frac{\pi}{2} \cos 13.69 \, \pi t + C_2 \sin \pi \cos 27.38 \, \pi t$$
$$+ C_3 \sin \frac{3\pi}{2} \cos 41.07 \, \pi t$$

This gives:

$$y(L/2, t) = C_1 \cos 13.69 \, \pi t - C_3 \cos 41.07 \, \pi t \qquad \text{(E9.2b)}$$

Calculating C_3 from Equation (9.44), gives:

$$C_3 = \frac{8 \times 1}{\pi^2 (3)^2} \sin \frac{3 \times \pi}{2} = -0.09 \, \text{mm}$$

Using $C_1 = 0.81$ mm and $C_3 = -0.09$ mm, Equation (E9.2b), becomes:

$$y(L/2, t) = 0.81 \cos 13.69 \, \pi t + 0.09 \cos 41.07 \, \pi t \, \text{mm}$$

Thus the total displacement time response in this case is the summation of mode 1 and mode 3 responses, as shown in Figure 9.6(c).

9.3 LONGITUDINAL VIBRATION OF A BAR

9.3.1 Deriving the equation of motion

The elastic bar shown in Figure 9.7(a) has a length L, a uniform cross-section area A, a Young's modulus E and a mass density ρ (mass per unit volume). The forces acting on the cross section of an infinitesimal element of the bar are shown in Figure 9.7(b). The axial displacement of the bar is denoted as u_x. The internal force F is given by:

$$F = \sigma A \qquad (9.46)$$

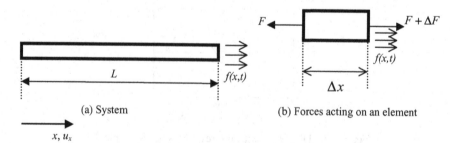

(a) System

(b) Forces acting on an element

Figure 9.7: A vibrating bar

where σ is the axial stress. Assume a linear elastic material $\sigma = E\varepsilon_x$, where ε_x is the axial strain and is equal to $\frac{\partial u_x}{\partial x}$. Substituting $\sigma = E\varepsilon_x$ and $\varepsilon_x = \frac{\partial u_x}{\partial x}$ into Equation (9.46) gives:

$$F = E\varepsilon_x A = EA\frac{\partial u_x}{\partial x} \tag{9.47}$$

Using Newton's second law, the equation of motion is obtained by summing all the forces in the x direction:

$$\sum F_x = m\ddot{u}_x \tag{9.48}$$

where m is the mass of the bar. Substituting m by $\rho A\Delta x$, substituting $\ddot{u}_x$, the acceleration in the x direction, by $\frac{\partial^2 u_x}{\partial t^2}$, and considering Figure 9.7(b), Equation (9.48) becomes:

$$(F + \Delta F) + f\Delta x - F = \rho A\Delta x\frac{\partial^2 u_x}{\partial t^2} \tag{9.49}$$

where f is the external force per unit length and can be a function in location and time, $f(x, t)$. Using $\Delta F = \frac{\partial F}{\partial x}\Delta x$ for an infinitesimal length Δx, Equation (9.49) becomes:

$$\left(F + \frac{\partial F}{\partial x}\Delta x\right) + f\Delta x - F = \rho A\Delta x\frac{\partial^2 u_x}{\partial t^2} \tag{9.50}$$

Substituting Equation (9.47) into Equation (9.50) gives:

$$\frac{\partial}{\partial x}\left[EA\frac{\partial u_x}{\partial x}\right] + f = \rho A\frac{\partial^2 u_x}{\partial t^2} \tag{9.51}$$

In the case of a uniform homogenous bar, Equation (9.51) reduces to:

$$EA\frac{\partial^2 u_x}{\partial x^2} + f = \rho A\frac{\partial^2 u_x}{\partial t^2} \tag{9.52}$$

For free vibration, the force f is equal to zero and Equation (9.52) becomes:

$$E\frac{\partial^2 u_x}{\partial x^2} = \rho\frac{\partial^2 u_x}{\partial t^2} \tag{9.53}$$

which can also be written as:

$$c^2\frac{\partial^2 u_x}{\partial x^2} = \frac{\partial^2 u_x}{\partial t^2} \tag{9.54}$$

where $c = \sqrt{\frac{E}{\rho}}$.

9.3.2 Free vibration of a bar

Equation (9.54) is similar to the wave equation, Equation (9.9), which is the free vibration equation for a cable and it can be solved in a similar way using the method of separation of variables. The solution of $u_x(x, t)$ is in the form:

$$u_x(x, t) = U(x)T(t) = \left(A\cos\frac{\omega_n x}{c} + B\sin\frac{\omega_n x}{c}\right)(C\cos\omega_n t + D\sin\omega_n t) \tag{9.55}$$

Table 9.1: Angular frequencies and mode shapes for a bar in longitudinal vibration

End conditions	Boundary conditions	Angular frequencies	Mode shapes
Fixed–fixed	$u_x(0, t) = 0$ $u_x(L, t) = 0$	$\omega_n = \dfrac{nc\pi}{L},$ $n = 1, 2, 3, \ldots$	$U_n = A_n \sin \dfrac{n\pi x}{L}$
Free–free	$\dfrac{\partial u_x(0, t)}{\partial x} = 0$ $\dfrac{\partial u_x(L, t)}{\partial x} = 0$	$\omega_n = \dfrac{nc\pi}{L},$ $n = 0, 1, 2, 3, \ldots$	$U_n = A_n \cos \dfrac{n\pi x}{L}$
Fixed–free	$u_x(0, t) = 0$ $\dfrac{\partial u_x(L, t)}{\partial x} = 0$	$\omega_{n+1} = \dfrac{(2n + 1)c\pi}{2L},$ $n = 0, 1, 2, 3, \ldots$	$U_{n+1} = A_{n+1} \sin \dfrac{(2n + 1)\pi x}{2L}$

The angular frequencies of the bar and the mode shape function, U_n, can be obtained using the boundary conditions at the ends of the bar. The solutions are summarized in Table 9.1 for various end conditions.

A free–free vibration can represent a flying object, e.g. an aircraft, or an object suspended by light springs.

Example 9.3 Longitudinal Vibration of a bar I

A uniform bar of length 2 m is shown in Figure 9.8. The Young's modulus is 200 GPa and the density is 7800 kg/m³. For free–free vibration, determine the first three natural frequencies and draw the corresponding mode shapes.

Solution

The Young's modulus is $E = 20 \times 10^{10}$ N/m² and the density is $\rho = 7.8 \times 10^3$ kg/m³. The constant c is calculated from Equation (9.54) as:

$$c = \left(\frac{E}{\rho}\right)^{1/2} = \left(\frac{20 \times 10^{10}}{7.8 \times 10^3}\right)^{1/2} = 5063.696 \text{ m/s}$$

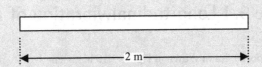

Figure 9.8: Representation of Example 9.3

Using Table 9.1, the first three natural frequencies (for $n = 1$, 2 and 3) are:

$$\omega_1 = \frac{c\pi}{L} = 7954.03 \, \text{rad/s} \Rightarrow f_1 = \frac{\omega_1}{2\pi} = 1265.9 \, \text{Hz}$$

$$\omega_2 = \frac{2c\pi}{L} = 15\,908.07 \, \text{rad/s} \Rightarrow f_2 = \frac{\omega_2}{2\pi} = 2531.9 \, \text{Hz}$$

$$\omega_3 = \frac{3c\pi}{L} = 23\,862.10 \, \text{rad/s} \Rightarrow f_3 = \frac{\omega_3}{2\pi} = 3797.8 \, \text{Hz}$$

Note that for $n = 0$, the rigid body mode is obtained ($\omega_0 = 0$).

The first three mode shapes are obtained from the mode-shape function, $U_n = A_n \cos \frac{n\pi x}{L}$ (see Table 9.1) for $n = 1$, 2 and 3. Figure 9.9 shows the three mode shapes, where $\frac{U_n}{A_n}$ is plotted against x/L. Note that the longitudinal displacement, $\frac{U_n}{A_n}$, is plotted in the transverse direction for clarity. Mode 1 has one node at $x = L/2$; mode 2 has two nodes, at $x = L/4$ and $x = 3L/4$; and mode 3 has three nodes at $x = L/6$, $x = L/2$ and $x = 5L/6$.

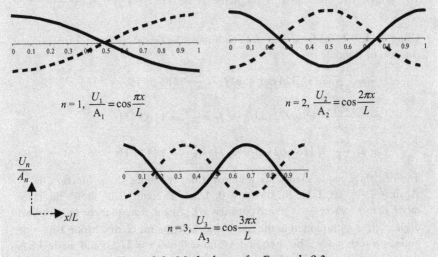

$$n = 1, \quad \frac{U_1}{A_1} = \cos\frac{\pi x}{L}$$

$$n = 2, \quad \frac{U_2}{A_2} = \cos\frac{2\pi x}{L}$$

$$n = 3, \quad \frac{U_3}{A_3} = \cos\frac{3\pi x}{L}$$

Figure 9.9: Mode shapes for Example 9.3

Example 9.4 Longitudinal vibration of a bar II

The bridge support column shown in Figure 9.10 has a length of 20 m. The column is made of concrete with Young's modulus of 21 GPa and density of 2400 kg/m³. Determine the first three natural frequencies for axial vibration of the column and draw the corresponding mode shapes, assuming a constant cross section.

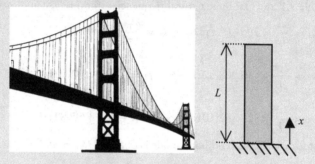

Figure 9.10: Representation of Example 9.4

Solution

The Young's modulus is $E = 2.1 \times 10^{10}$ N/m² and the density is $\rho = 2.4 \times 10^3$ kg/m³. The constant c is the same as before: $c = (\frac{E}{\rho})^{1/2} = (\frac{2.1 \times 10^{10}}{2.4 \times 10^3})^{1/2} = 2958.04$ m/s.

Using Table 9.1, for fixed-free end conditions, the first three natural frequencies (for $n = 0$, 1 and 2) are:

$$\omega_1 = \frac{c\pi}{2L} = 232.324 \text{ rad/s} \Rightarrow f_1 = \frac{\omega_1}{2\pi} = 36.98 \text{ Hz}$$

$$\omega_2 = \frac{3c\pi}{2L} = 696.972 \text{ rad/s} \Rightarrow f_2 = \frac{\omega_2}{2\pi} = 110.93 \text{ Hz}$$

$$\omega_3 = \frac{5c\pi}{2L} = 1161.62 \text{ rad/s} \Rightarrow f_3 = \frac{\omega_3}{2\pi} = 184.88 \text{ Hz}$$

The first three mode shapes are obtained from the mode-shape function, $U_n = A_n \sin \frac{(2n+1)\pi x}{2L}$ (see Table 9.1) for $n = 0$, 1 and 2. Figure 9.11 shows the three mode shapes, where $\frac{U_n}{A_n}$ is plotted against x/L. Again, note that the longitudinal displacement $\frac{U_n}{A_n}$ is plotted in the transverse direction for clarity. Mode 1 has one node at $x = 0$; mode 2 has two nodes, at $x = 0$ and $x = 2L/3$; and mode 3 has three nodes at $x = 0$, $x = 2L/5$ and $x = 4L/5$.

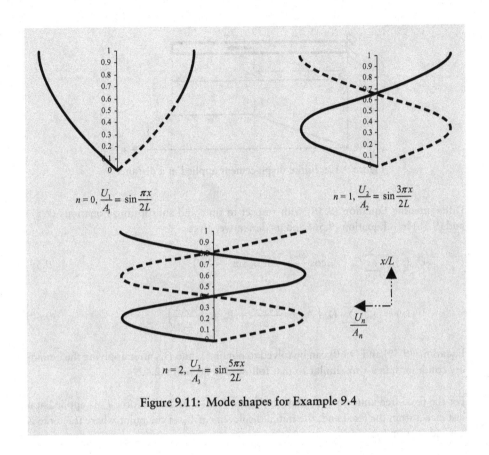

$$n = 0, \frac{U_1}{A_1} = \sin\frac{\pi x}{2L} \qquad\qquad n = 1, \frac{U_2}{A_2} = \sin\frac{3\pi x}{2L}$$

$$n = 2, \frac{U_3}{A_3} = \sin\frac{5\pi x}{2L}$$

Figure 9.11: Mode shapes for Example 9.4

9.3.3 Time response of a bar

The general solution of the axial vibration of a bar was given in Equation (9.55) and can again be written in the superposition form as:

$$u_x(x, t) = \sum_{n=0}^{\infty} \left(A_n \cos\frac{\omega_{n+1}x}{c} + B_n \sin\frac{\omega_{n+1}x}{c} \right) (C_n \cos\omega_{n+1}t + D_n \sin\omega_{n+1}t) \quad (9.56)$$

The general form of initial conditions is:

$$u_x(x, t = 0) = U_{xo}(x) \tag{9.57}$$

$$\frac{du_x}{dt}(x, t = 0) = \dot{U}_{xo} \tag{9.58}$$

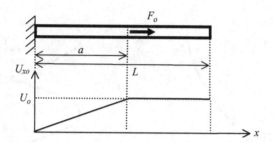

Figure 9.12: Initial displacement applied at a distance a

Differentiating Equation (9.56) with respect to time and substituting Equations (9.57) and (9.58) into Equation (9.56) and its derivative, gives:

$$U_{xo}(x) = \sum_{n=0}^{\infty} C_n \left(A_n \cos \frac{\omega_{n+1} x}{c} + B_n \sin \frac{\omega_{n+1} x}{c} \right) \tag{9.59}$$

$$\dot{U}_{xo}(x) = \omega_{n+1} \sum_{n=0}^{\infty} D_n \left(A_n \cos \frac{\omega_{n+1} x}{c} + B_n \sin \frac{\omega_{n+1} x}{c} \right) \tag{9.60}$$

Equations (9.59) and (9.60) can be solved to obtain C_n and D_n, after applying the boundary conditions, in a way similar to that followed in Section 9.2.4.

For the fixed-free uniform bar shown in Figure 9.12, if an initial force F_o is applied at a distance a from the fixed end, the initial displacement U_o at the point where the force is applied is given by:

$$U_o = \varepsilon_x a = \frac{F_o a}{E A} \tag{9.61}$$

and the initial displacement $U_{xo}(x, t = 0)$ along the bar length is:

$$U_{xo}(x, t = 0) = \begin{cases} \dfrac{U_o x}{a}, & 0 \leq x \leq a \\ U_o, & a \leq x \leq L \end{cases} \tag{9.62}$$

The general solution for a free-fixed bar, using Equation (9.56) and Table 9.1, is:

$$u_x(x, t) = \sum_{n=0}^{\infty} \sin \frac{(2n+1)\pi x}{2L} \left(C_n \cos \frac{(2n+1)\pi ct}{2L} + D_n \sin \frac{(2n+1)\pi ct}{2L} \right) \tag{9.63}$$

As the initial velocity is zero, $\dot{U}_{xo} = 0$ and therefore $D_n = 0$. The constant C_n can be obtained in a way similar to that applied to Equation (9.31), by multiplying both sides in Equation (9.63) (for $t = 0$, $u_x = U_{xo}$) by $\sin \frac{(2n+1)\pi x}{2L}$ and integrating from $x = 0$

to $x = L$:

$$C_n = \frac{2}{L} \int_0^L U_{xo}(x) \times \sin \frac{(2n+1)\pi x}{2L} dx$$

$$= \frac{2}{L} \left[\int_0^a \frac{U_o x}{a} \times \sin \frac{(2n+1)\pi x}{2L} dx + \int_a^L U_o \times \sin \frac{(2n+1)\pi x}{2L} dx \right] \quad (9.64)$$

Integrating Equation (9.64), using integration by parts, yields:

$$C_n = \frac{8 U_o L}{(2n+1)^2 \pi^2 a} \sin \frac{(2n+1)\pi a}{2L} \quad (9.65)$$

If the initial displacement U_o is applied at the free end of the bar, so that $U_o = \frac{F_o L}{EA}$, the initial displacement function $U_{xo}(x, t = 0)$ becomes:

$$U_{xo}(x, t = 0) = \frac{U_o x}{L}, 0 \le x \le L \quad (9.66)$$

And the integration in Equation (9.64) reduces to:

$$C_n = \frac{2}{L} \int_0^L U_{xo}(x) \times \sin \frac{(2n+1)\pi x}{2L} dx = \frac{2}{L} \int_0^L \frac{U_o x}{L} \times \sin \frac{(2n+1)\pi x}{2L} dx \quad (9.67)$$

Solving for C_n gives:

$$C_n = \frac{8 U_o}{\pi^2} \frac{(-1)^n}{(2n+1)^2} \quad (9.68)$$

The total displacement time response of a fixed-free bar with an initial displacement at its free end is obtained as:

$$u_x(x, t) = \sum_{n=0}^\infty C_n \sin \frac{(2n+1)\pi x}{2L} \cos \frac{(2n+1)c\pi t}{2L} \quad (9.69)$$

Example 9.5 Longitudinal vibration of a bar II

For the bridge column in Example 9.4, the column has an approximate uniform cross-sectional area of 2 m^2. If an initial force of 50 MN is applied as shown in Figure 9.13, determine the displacement time response at the top end of the column ($x = L$) using:

a) the first two modes;
b) the first three modes.

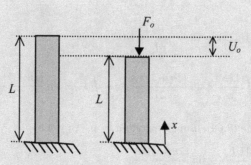

Figure 9.13: **Representation of Example 9.5**

Solution

Since there is no initial velocity, i.e. $\dot{U}_{xo} = 0$, from Equation (9.60) the constant D_n is zero, i.e. $D_n = 0$. For $x = L$, $c = 2958.04$ m/s and $L = 20$ m, Equation (9.69) becomes:

$$u_x(L, t) = \sum_{n=0}^{\infty} C_n \sin \frac{(2n+1)\pi}{2} \cos \frac{(2n+1) \times 2958.04 \times \pi t}{2 \times 20}$$

$$= \sum_{n=0}^{\infty} C_n \sin \frac{(2n+1)\pi}{2} \cos 73.951 \times (2n+1)\pi t$$

a) Using the mode superposition with two modes ($n = 0$ and 1), the displacement time response becomes:

$$u_x(L, t) = C_0 \sin \frac{\pi}{2} \cos 73.951\pi t + C_1 \sin \frac{3\pi}{2} \cos 221.853\pi t \qquad (E9.5a)$$

This gives:

$$u_x(L, t) = C_0 \cos 73.951\pi t - C_1 \cos 221.853\pi t$$

From Equation (9.61), the initial displacement (using $F_o = -5 \times 10^7$ N, $E = 2.1 \times 10^{10}$ N/m^2 and $A = 2$ m^2) is calculated as:

$$U_o = \frac{F_o L}{EA} = \frac{-5 \times 10^7 \times 20}{2.1 \times 10^{10} \times 2} = -0.02381 \, \text{m} = -23.81 \, \text{mm}$$

To calculate the constant C_0, using Equation (9.68) for $n = 0$, gives:

$$C_0 = \frac{8U_o}{\pi^2} \frac{(-1)^n}{(2n+1)^2} = \frac{8 \times (-23.81)}{\pi^2} \frac{(-1)^0}{(2 \times 0 + 1)^2} = -19.3 \, \text{mm}$$

Similarly, to calculate the constant C_1, using Equation (9.68) for $n = 1$, gives:

$$C_1 = \frac{8 \times (-23.81)}{\pi^2} \frac{(-1)^1}{(2 \times 1 + 1)^2} = +2.144 \text{ m} = +2.144 \text{ mm}$$

And the total displacement time response in Equation (E9.5a), becomes:

$$u_x(L, t) = -19.3 \cos 73.951\pi t - 2.144 \cos 221.853\pi t$$

Thus, the total displacement response is the summation of the responses for mode 1 and mode 2, as shown in Figure 9.14.

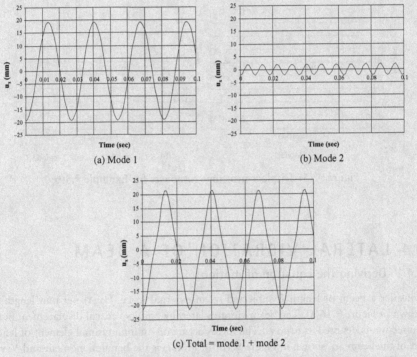

(a) Mode 1

(b) Mode 2

(c) Total = mode 1 + mode 2

Figure 9.14: Displacement time responses for Example 9.5(a)

b) Using the mode superposition with three modes ($n = 0$, 1 and 2) in Equation (9.69), gives:

$$u_x(x, t) = C_0 \cos 73.951\pi t - C_1 \cos 221.853\pi t + C_2 \cos 369.755\pi t \text{ mm} \quad \text{(E9.5b)}$$

To calculate the constant C_2, using Equation (9.68), for $n = 2$, gives:

$$C_2 = \frac{8 \times (-23.81)}{\pi^2} \frac{(-1)^2}{(2 \times 2 + 1)^2} = -0.772 \text{ mm}$$

Substituting $C_0 = -19.3$ mm, $C_1 = 2.144$ mm, and $C_2 = -0.772$ mm in Equation (E9.5b), the displacement time response becomes:

$$u_x(L, t) = -19.3 \cos 73.951\pi t - 2.144 \cos 221.853\pi t - 0.772 \cos 369.755\pi t \text{ mm}$$

Thus the total response is the summation of the responses for modes 1, 2 and 3, as shown in Figure 9.15.

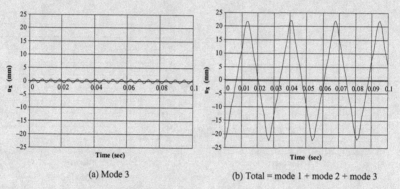

(a) Mode 3   (b) Total = mode 1 + mode 2 + mode 3

Figure 9.15: Displacement time responses for Example 9.5(b)

9.4 LATERAL VIBRATION OF A BEAM

9.4.1 Deriving the equation of motion

Consider a beam of length L subjected to an external force $f(x, t)$ per unit length, as shown in Figure 9.16(a). The beam vibrates laterally and its vertical displacement in the y direction is denoted as $u_y(x, t)$. The forces acting on an infinitesimal element of length Δx of the beam are shown in Figure 9.16(b). $M(x, t)$ is the bending moment and $V(x, t)$ is the shear force. The beam has a Young's modulus E, a mass density ρ, a cross-sectional area A and an area moment of inertia I about the axis perpendicular to the plane $x-y$.

Applying Newton's second law in the y direction to the forces in Figure 9.16(b), and the moment equation equivalent to Newton's second law at point C, gives:

$$\sum F_y = m\ddot{u}_y \tag{9.70}$$

$$\sum M_C \approx 0 \tag{9.71}$$

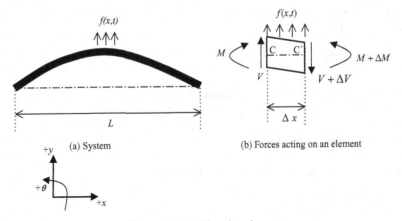

Figure 9.16: A bending beam

where m is the mass, equal to $\rho A \Delta x$ and $\ddot{u}_y$ is the acceleration in the y direction, $\ddot{u}_y = \frac{\partial^2 u_y}{\partial t^2}$. In Equation (9.71), the right-hand side is approximated to zero since the line CC' doesn't rotate ($u_{yC} = u_{yC'}$) for an infinitesimal element of length Δx. Summing the forces in Figure 9.16(b), Equation (9.70) becomes:

$$-(V + \Delta V) + f \Delta x + V = \rho A \Delta x \frac{\partial^2 u_y}{\partial t^2} \qquad (9.72)$$

Applying the moment equation, Equation (9.71), about the axis perpendicular to the plane $x-y$ taken at point C, gives:

$$(M + \Delta M) - (V + \Delta V)\Delta x + f \Delta x \frac{\Delta x}{2} - M = 0 \qquad (9.73)$$

For an infinitesimal length Δx, the terms ΔV and ΔM are approximated as:

$$\Delta V = \frac{\partial V}{\partial x} \Delta x, \qquad \Delta M = \frac{\partial M}{\partial x} \Delta x \qquad (9.74)$$

Substituting Equation (9.74) into Equations (9.72) and (9.73) and ignoring terms involving the second power of Δx, i.e. Δx^2, gives:

$$-\frac{\partial V}{\partial x} + F = \rho A \frac{\partial^2 u_y}{\partial t^2} \qquad (9.75)$$

$$\frac{\partial M}{\partial x} - V = 0 \qquad (9.76)$$

Substituting Equation (9.76) into Equation (9.75) yields:

$$-\frac{\partial^2 M}{\partial x^2} + F = \rho A \frac{\partial^2 u_y}{\partial t^2} \qquad (9.77)$$

From Euler–Bernoulli thin-beam theory, the bending moment M is related to the curvature of the beam $\frac{\partial^2 u_y}{\partial x^2}$ by:

$$M = EI \frac{\partial^2 u_y}{\partial x^2} \tag{9.78}$$

Substituting Equation (9.78) into Equation (9.77), the following equation of motion is obtained:

$$\frac{\partial^2}{\partial x^2} \left[EI \frac{\partial^2 u_y}{\partial x^2} \right] + \rho A \frac{\partial^2 u_y}{\partial t^2} = F \tag{9.79}$$

In case of a uniform beam, Equation (9.79) becomes:

$$EI \frac{\partial^4 u_y}{\partial x^4} + \rho A \frac{\partial^2 u_y}{\partial t^2} = F \tag{9.80}$$

For free vibration, i.e. $F = 0$, Equation (9.80) reduces to:

$$EI \frac{\partial^4 u_y}{\partial x^4} + \rho A \frac{\partial^2 u_y}{\partial t^2} = 0$$

or

$$c^2 \frac{\partial^4 u_y}{\partial x^4} + \frac{\partial^2 u_y}{\partial t^2} = 0 \tag{9.81}$$

where $c = (\frac{EI}{\rho A})^{1/2}$.

9.4.2 Free vibration of a beam

Equation (9.81), the free vibration equation for a beam, can be solved by the method of separation of variables:

$$u_y(x, t) = U(x)T(t) \tag{9.82}$$

where, as before, the function $U(x)$ depends only on x and the function $T(t)$ depends only on t. Substituting Equation (9.82) into Equation (9.81) yields:

$$\frac{c^2}{U} \frac{\delta^4 U}{\partial x^4} = -\frac{1}{T} \frac{\partial^2 T}{\partial t^2} = a = \omega_n^2 \tag{9.83}$$

where the constants a and ω_n^2 are positive numbers. Again, a is chosen to be equal to ω_n^2 because it is the only value that gives a physically acceptable solution for the two partial differential equations in Equation (9.83). Separating Equation (9.83) into two equations, gives:

$$\frac{\delta^4 U}{\partial x^4} - \beta^4 U = 0 \tag{9.84}$$

$$\frac{\partial^2 T}{\partial t^2} + \omega_n^2 T = 0 \tag{9.85}$$

where the constant β^4 is given by:

$$\beta^4 = \frac{\omega_n^2}{c^2} = \frac{\rho A \omega_n^2}{EI} \tag{9.86}$$

where ω_n^2 is the angular frequency of vibration in rad/s. Equation (9.85) is a second-order differential equation and its solution is similar to that of the cable, Equation (9.18). Equation (9.84) is a fourth-order differential equation and its solution is in the form:

$$U(x) = C_1 e^{\beta x} + C_2 e^{-\beta x} + C_3 e^{i\beta x} + C_4 e^{-i\beta x} \tag{9.87}$$

where C_1, C_2, C_3 and C_4 are constants. Equation (9.87) can also be written as:

$$U(x) = C_1 \cos \beta x + C_2 \sin \beta x + C_3 \cosh \beta x + C_4 \sinh \beta x \tag{9.88}$$

The function $U(x)$ is called the normal mode or characteristic function. The angular frequencies of the beam can be determined using Equation (9.86) as:

$$\omega_n = \beta^2 \sqrt{\frac{EI}{\rho A}} = (\beta L)^2 \sqrt{\frac{EI}{\rho A L^4}} \tag{9.89}$$

Using c from Equation (9.81) gives:

$$\omega_n = (\beta L)^2 \frac{c}{L^2}$$

where $c = \sqrt{\frac{EI}{\rho A}}$. For any beam, there is an infinite number of normal modes associated with each natural frequency. The constants C_1, C_2, C_3 and C_4 and the value of β can be determined from the boundary conditions at the ends of the beam.

9.4.3 Boundary conditions

The following boundary conditions at the beam ends are considered:

- Free end, where the bending moment and the shear force are equal to zero:

$$M = EI \frac{\partial^2 u_y}{\partial x^2} = 0 \rightarrow \frac{\partial^2 U}{\partial x^2} = 0$$

$$V = \frac{\partial M}{\partial x} = \frac{\partial}{\partial x} \left(EI \frac{\partial^2 u_y}{\partial x^2} \right) = 0 \rightarrow \frac{\partial^3 U}{\partial x^3} = 0$$

- Simply supported end, where the deflection and bending moment are equal to zero:

$$u_y = 0 \rightarrow U = 0$$

$$M = EI \frac{\partial^2 u_y}{\partial x^2} = 0 \rightarrow \frac{\partial^2 U}{\partial x^2} = 0$$

- Fixed end, where the deflection and slope are equal to zero:

$$u_y = 0 \rightarrow U = 0$$

$$\frac{\partial u_y}{\partial x} = 0 \rightarrow \frac{\partial U}{\partial x} = 0$$

Table 9.2: Values of $\beta_n L$ and mode shapes for a beam in transverse vibration

End conditions	Values of $\beta_n L$	Mode shapes
Free–free	$\beta_0 L = 0$ for rigid body mode	$U_n = C_n \Bigg(\sin \beta_n x + \sinh \beta_n x$
	$\beta_1 L = 4.73$	$+ \left[\dfrac{\sin \beta_n x - \sinh \beta_n x}{\cosh \beta_n x - \cos \beta_n x} \right] (\cos \beta_n x + \cosh \beta_n x) \Bigg)$
	$\beta_2 L = 7.853$	
	$\beta_3 L = 10.995$	
	$\beta_4 L = 14.137$	
Simply supported	$\beta_1 L = \pi$	$U_n = C_n \sin \beta_n x$
	$\beta_2 L = 2\pi$	
	$\beta_3 L = 3\pi$	
	$\beta_4 L = 4\pi$	
Fixed–fixed	$\beta_1 L = 4.73$	$U_n = C_n \Bigg(\sinh \beta_n x - \sin \beta_n x$
	$\beta_2 L = 7.853$	$+ \left[\dfrac{\sinh \beta_n x - \sin \beta_n x}{\cos \beta_n x - \cosh \beta_n x} \right] (\cosh \beta_n x - \cos \beta_n x) \Bigg)$
	$\beta_3 L = 10.995$	
	$\beta_4 L = 14.137$	
Fixed–free	$\beta_1 L = 1.875$	$U_n = C_n \Bigg(\sin \beta_n x - \sinh \beta_n x$
	$\beta_2 L = 4.694$	$- \left[\dfrac{\sin \beta_n x + \sinh \beta_n x}{\cos \beta_n x + \cosh \beta_n x} \right] (\cos \beta_n x - \cosh \beta_n x) \Bigg)$
	$\beta_3 L = 7.854$	
	$\beta_4 L = 10.995$	
Fixed–simply supported	$\beta_1 L = 3.926$	$U_n = C_n \Bigg(\sin \beta_n x - \sinh \beta_n x$
	$\beta_2 L = 7.068$	$+ \left[\dfrac{\sin \beta_n x - \sinh \beta_n x}{\cos \beta_n x - \cosh \beta_n x} \right] (\cosh \beta_n x - \cos \beta_n x) \Bigg)$
	$\beta_3 L = 10.21$	
	$\beta_4 L = 13.351$	

The values of $\beta_n L$ and the mode-shape functions of a beam for different boundary conditions are summarized in Table 9.2.

Example 9.6 Lateral vibration of a beam I

The beam shown in Figure 9.17 has a length of 4 m, Young's modulus of 70 GPa and density of 2800 kg/m³. For free–free vibration, determine the first three natural frequencies and draw the corresponding mode shapes.

Solution

The Young's modulus is $E = 7 \times 10^{10}$ N/m^2, the density is $\rho = 2.8 \times 10^3$ kg/m^3, the cross-sectional area is $A = 0.7 \times 0.4 = 0.28$ m^2 and the area moment of inertia about the z axis (see Appendix B) is $I = \frac{0.7 \times 0.4^3}{12} = 3.733 \times 10^{-3}$ m^4.

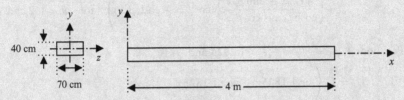

Figure 9.17: Representation of Example 9.6

The constant c is calculated from Equation (9.81) as:

$$c = \left(\frac{EI}{\rho A}\right)^{1/2} = \left(\frac{7 \times 10^{10} \times 3.733 \times 10^{-3}}{2.8 \times 10^3 \times 0.28}\right)^{1/2} = 577.324 \, \text{m}^2/\text{s}$$

There are two rigid-body modes, at zero frequency: translation rigid-body mode and rotation rigid-body mode (see Figure 9.18).

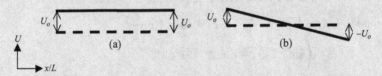

Figure 9.18: Rigid-body modes: a) translation and b) rotation

Using Table 9.2 and for $n = 1$, 2 and 3, the first three natural frequencies are calculated from Equation (9.89) as:

$$\omega_1 = (\beta_1 L)^2 \frac{c}{L^2} = (4.73)^2 \times \frac{577.324}{4^2}$$

$$= 807.27 \, \text{rad/s} \Rightarrow f_1 = \frac{\omega_1}{2\pi} = 128.48 \, \text{Hz}$$

$$\omega_2 = (\beta_2 L)^2 \frac{c}{L^2} = (7.853)^2 \times \frac{577.324}{4^2}$$

$$= 2225.21 \, \text{rad/s} \Rightarrow f_2 = \frac{\omega_2}{2\pi} = 354.15 \, \text{Hz}$$

$$\omega_3 = (\beta_3 L)^2 \frac{c}{L^2} = (10.995)^2 \times \frac{577.324}{4^2}$$

$$= 4362.04 \, \text{rad/s} \Rightarrow f_3 = \frac{\omega_3}{2\pi} = 694.24 \, \text{Hz}$$

The first three mode shapes are obtained from the mode-shape functions given in Table 9.2:

$$U_n = C_n \left(\sin \beta_n x + \sinh \beta_n x \right.$$

$$\left. + \left[\frac{\sin \beta_n x - \sinh \beta_n x}{\cosh \beta_n x - \cos \beta_n x} \right] (\cos \beta_n x + \cosh \beta_n x) \right) \quad \text{for} \quad n = 1, 2 \text{ and } 3.$$

For mode 1, $\beta_1 = \frac{4.73}{L} = \frac{4.73}{4} = 1.1825$ and U_1 is:

$$U_1 = C_1 \left(\sin 1.1825x + \sinh 1.1825x \right.$$

$$\left. + \left[\frac{\sin 1.1825x - \sinh 1.1825x}{\cosh 1.1825x - \cos 1.1825x} \right] (\cos 1.1825x + \cosh 1.1825x) \right)$$

For mode 2, $\beta_2 = \frac{7.853}{L} = \frac{7.853}{4} = 1.963$ and U_2 is:

$$U_2 = C_2 \left(\sin 1.963x + \sinh 1.963x \right.$$

$$\left. + \left[\frac{\sin 1.963x - \sinh 1.963x}{\cosh 1.963x - \cos 1.963x} \right] (\cos 1.963x + \cosh 1.963x) \right)$$

For mode 3, $\beta_3 = \frac{10.995}{L} = \frac{10.995}{4} = 2.749$ and U_3 is:

$$U_3 = C_3 \left(\sin 2.749x + \sinh 2.749x \right.$$

$$\left. + \left[\frac{\sin 2.749x - \sinh 2.749x}{\cosh 2.749x - \cos 2.749x} \right] (\cos 2.749x + \cosh 2.749x) \right)$$

Figure 9.19 shows the three mode shapes, where $\frac{U_n}{C_n}$ are plotted against x/L.

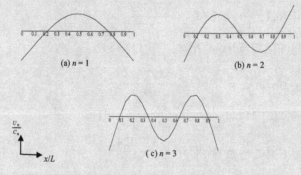

Figure 9.19: Mode shapes for Example 9.6

Example 9.7 Lateral vibration of a beam II

A rotor blade of a helicopter, shown in Figure 9.20, has a constant cross-section of 1 cm × 20 cm and a length of 0.7 m. The material of the blade is aluminium, for which the Young's modulus is 70 GPa and the density is 2800 kg/m^3. Determine the first three natural frequencies of the blade for transverse vibration and draw the corresponding mode shapes.

Rotor blade

Figure 9.20: Representation of Example 9.7

Solution

The rotor blade is idealized to a fixed-free beam as shown in Figure 9.21. The Young's modulus is $E = 7 \times 10^{10}$ N/m^2, the density is $\rho = 2.8 \times 10^3$ kg/m^3, the cross-sectional area is $A = 0.2 \times 0.01 = 0.002$ m^2 and the area moment of inertia about the z axis (see Appendix B) is $I = \frac{0.2 \times 0.01^3}{12} = 1.667 \times 10^{-8}$ m^4.

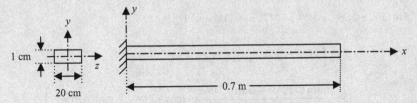

Figure 9.21: Idealized representation of Example 9.7

The constant c is calculated from Equation (9.81) as:

$$c = \left(\frac{EI}{\rho A}\right)^{1/2} = \left(\frac{7 \times 10^{10} \times 1.667 \times 10^{-8}}{2.8 \times 10^3 \times 0.002}\right)^{1/2} = 14.4337 \, \text{m}^2/\text{s}$$

Using Table 9.2 and for $n = 1$, 2 and 3, the first three natural frequencies are calculated from Equation (9.89) as:

$$\omega_1 = (\beta_1 L)^2 \frac{c}{L^2} = (1.875)^2 \times \frac{14.4337}{0.7^2}$$

$$= 103.558 \, \text{rad/s} \Rightarrow f_1 = \frac{\omega_1}{2\pi} = 16.48 \, \text{Hz}$$

$$\omega_2 = (\beta_2 L)^2 \frac{c}{L^2} = (4.694)^2 \times \frac{14.4337}{0.7^2}$$

$$= 649.037 \, \text{rad/s} \Rightarrow f_2 = \frac{\omega_2}{2\pi} = 103.3 \, \text{Hz}$$

$$\omega_3 = (\beta_3 L)^2 \frac{c}{L^2} = (7.854)^2 \times \frac{14.4337}{0.7^2}$$

$$= 1817.04 \, \text{rad/s} \Rightarrow f_3 = \frac{\omega_3}{2\pi} = 289.19 \, \text{Hz}$$

The first three mode shapes are obtained from the mode-shape functions given in Table 9.2:

$$U_n = C_n \Bigg(\sin \beta_n x - \sinh \beta_n x$$

$$- \left[\frac{\sin \beta_n x + \sinh \beta_n x}{\cos \beta_n x + \cosh \beta_n x} \right] (\cos \beta_n x - \cosh \beta_n x) \Bigg) \quad \text{for} \quad n = 1, 2 \text{ and } 3.$$

For mode 1, $\beta_1 = \frac{1.875}{L} = \frac{1.875}{0.7} = 2.6786$ and U_1 is:

$$U_1 = C_1 \Bigg(\sin 2.6786x - \sinh 2.6786x$$

$$- \left[\frac{\sin 2.6786x + \sinh 2.6786x}{\cos 2.6786x + \cosh 2.6786x} \right] (\cos 2.6786x - \cosh 2.6786x) \Bigg)$$

For mode 2, $\beta_2 = \frac{4.694}{L} = \frac{4.694}{0.7} = 6.706$ and U_2 is:

$$U_2 = C_2 \Bigg(\sin 6.706x - \sinh 6.706x$$

$$- \left[\frac{\sin 6.706x + \sinh 6.706x}{\cos 6.706x + \cosh 6.706x} \right] (\cos 6.706x - \cosh 6.706x) \Bigg)$$

For mode 3, $\beta_3 = \frac{7.854}{L} = \frac{7.854}{0.7} = 11.22$ and U_3 is:

$$U_3 = C_3 \Bigg(\sin 11.22x - \sinh 11.22x$$

$$- \left[\frac{\sin 11.22x + \sinh 11.22x}{\cos 11.22x + \cosh 11.22x} \right] (\cos 11.22x - \cosh 11.22x) \Bigg)$$

The mode shapes are shown in Figure 9.22, in which $\frac{U_n}{C_n}$ are plotted against x/L.

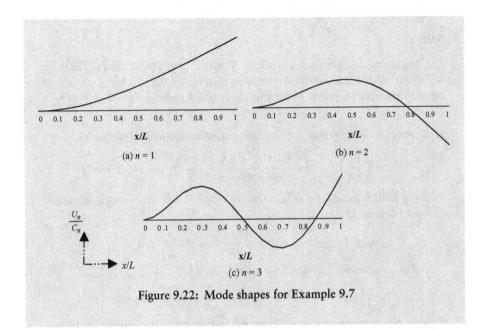

(a) $n = 1$

(b) $n = 2$

(c) $n = 3$

Figure 9.22: Mode shapes for Example 9.7

Example 9.8 Lateral vibration of a beam III

A steel girder (with a Young's modulus of 200 GPa N/m² and density of 7800 kg/m³) is fixed at both ends as shown in Figure 9.23. If a motor of 200 kg is placed at the midpoint of the beam, determine:

a) the fundamental frequency of the girder before placing the motor;
b) the fundamental frequency of the girder after placing the motor.

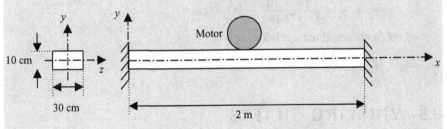

Figure 9.23: Representation of Example 9.8

Solution

The Young's modulus of the beam is $E = 20 \times 10^{10}$ N/m^2, the density of the beam is $\rho = 7.8 \times 10^3$ kg/m^3, the cross-sectional area is $A = 0.3 \times 0.1 = 0.03$ m^2 and the area moment of inertia about the z axis (see Appendix B) is $I = \frac{0.3 \times 0.1^3}{12} = 2.5 \times 10^{-5}$ m^4.

a) Before placing the motor, the constant c is calculated from Equation (9.81) as:

$$c = \left(\frac{EI}{\rho A}\right)^{1/2} = \left(\frac{20 \times 10^{10} \times 2.5 \times 10^{-5}}{7.8 \times 10^3 \times 0.03}\right)^{1/2} = 146.176 \text{ m}^2/\text{s}$$

Using Table 9.1 and for $n = 1$, the fundamental angular frequency is calculated from Equation (9.81) as:

$$\omega_1 = (\beta_1 L)^2 \frac{c}{L^2} = (4.73)^2 \times \frac{146.176}{2^2} = 817.59 \text{ rad/s}$$

and the fundamental natural frequency is given by:

$$f_1 = \frac{\omega_1}{2\pi} = 130.12 \text{ Hz}$$

b) After placing the motor, from Table 6.1 in Chapter 6, the equivalent stiffness (k) is:

$$k = \frac{192EI}{L^3} = \frac{192 \times 20 \times 10^{10} \times 2.5 \times 10^{-5}}{2^3} = 1.2 \times 10^8 \text{ N/m}$$

and the beam's equivalent mass (m_{eq}) is:

$$m_{eq} = 0.37 \times 7.8 \times 10^3 \times 0.03 \times 2 = 173.16 \text{ kg}$$

The total mass (m) is equal to the motor's mass (M) and the equivalent mass (m_{eq}):

$$m = M + m_{eq} = 200 + 173.16 = 373.16 \text{ kg}$$

From Equation (6.7) for an SDOF, the fundamental angular frequency is given by:

$$\omega_1 = \sqrt{\frac{k}{m}} = \sqrt{\frac{1.2 \times 10^8}{373.16}} = 567.08 \text{ rad/s}$$

and the fundamental natural frequency is calculated as:

$$f_1 = \frac{\omega_1}{2\pi} = 90.25 \text{ Hz}$$

9.5 WHIRLING SHAFTS

When a shaft rotates with a speed close to its natural frequency, it can whirl. When a shaft rotates with an increasing angular speed, its lateral deflection starts to build up at a speed

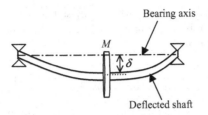

Figure 9.24: A deflected shaft

known as the whirling speed. The main reason for the whirling of shafts phenomenon is that there is always eccentricity between the shaft and the bearings axis. This eccentricity can be due to initial imperfection, static deflection under the shaft's weight or external masses.

To demonstrate why this phenomenon takes place, consider the static equilibrium of the shaft in Figure 9.24, in which the shaft deflects from the bearings axis by an amount δ under the action of a mass M.

When the shaft rotates with a constant angular speed ω, it deflects by the whirl amplitude u_y and an inertia force equivalent to Ma_r, where a_r is the radial acceleration and is equal to $-r\omega^2 = -(u_y + \delta)\,\omega^2$ (see Equations (5.1) and (2.9)) acts towards the bearing axis. Approximating the system to a single degree of freedom and applying Newton's second law (Equation (6.5)) gives

$$-ku_y = -M(u_y + \delta)\omega^2 \tag{9.90}$$

where k is the equivalent stiffness of the shaft (see Table 6.1). Rearranging Equation (9.90) and solving for the whirl amplitude u_y yields:

$$u_y = \frac{\omega^2\delta}{\left(\frac{k}{M} - \omega^2\right)} \tag{9.91}$$

For a SDOF system, the angular frequency of the system is $\omega_n = \sqrt{\frac{k}{m}}$, thus Equation (9.91) becomes:

$$u_y = \frac{\omega^2\delta}{(\omega_n^2 - \omega^2)} = \frac{\delta\left(\dfrac{\omega}{\omega_n}\right)^2}{1 - \left(\dfrac{\omega}{\omega_n}\right)^2} \tag{9.92}$$

It can be seen from Equation (9.92) that when the shaft's rotational speed, ω, approaches the angular frequency of the system, ω_n, the whirl amplitude (deflection) approaches infinity and resonance occurs.

Equation (9.92) can also be derived from the steady-state solution for an undamped SDOF, given in Equation (7.10), by substituting the force amplitude $F_o = m\delta\omega^2$ (see Equation (5.1) for an unbalanced mass):

$$u_y = \frac{F_o/k}{1 - \left(\dfrac{\omega}{\omega_n}\right)^2} = \frac{m\delta\omega^2/k}{1 - \left(\dfrac{\omega}{\omega_n}\right)^2} = \frac{m\delta\omega^2/m\omega_n^2}{1 - \left(\dfrac{\omega}{\omega_n}\right)^2} = \frac{\delta\left(\dfrac{\omega}{\omega_n}\right)^2}{1 - \left(\dfrac{\omega}{\omega_n}\right)^2}$$

Following the same procedures for a damped SDOF system and substituting the force amplitude $F_o = m\delta\omega^2$ in Equation (7.17), the whirl amplitude for a damped shaft is given by:

$$u_y = \frac{\delta\left(\dfrac{\omega}{\omega_n}\right)^2}{\sqrt{\left(1 - \left(\dfrac{\omega}{\omega_n}\right)^2\right)^2 + \left(\dfrac{2\zeta\omega}{\omega_n}\right)^2}} \tag{9.93}$$

It is worth mentioning that ω_n in an SDOF system is equivalent to ω_1 in continuous systems.

9.5.1 Whirling speed of a shaft

For a simply supported shaft, i.e. with short bearings at both ends, the fundamental frequency can be obtained from Table 9.2 and Equation (9.89) by substituting $c = \sqrt{\frac{EI}{\rho A}}$:

$$\omega_1 = (\beta_1 L)^2 \frac{c}{L^2} = \pi^2 \sqrt{\frac{EI}{\rho A L^4}}$$

Substituting $\rho = \dfrac{m}{AL}$, where m is the beam's mass, gives the fundamental angular frequency as:

$$\omega_1 = \pi^2 \sqrt{\frac{EI}{mL^3}} \text{ rad/s} \tag{9.94}$$

The whirling speed of the shaft is obtained by converting Equation (9.94) from rad/s to rev/min:

$$\omega_1 = \frac{60}{2\pi} \times \pi^2 \sqrt{\frac{EI}{mL^3}} \tag{9.95}$$

This gives:

$$\omega_1 = 94.25 \sqrt{\frac{EI}{mL^3}} \text{ rev/min} \tag{9.96}$$

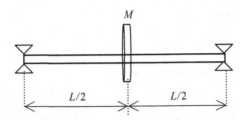

Figure 9.25: A shaft with a mass at its midpoint

9.5.2 Whirling speed of a mass in the middle of a shaft

For a mass M in the middle of a shaft, as shown in Figure 9.25, the equivalent stiffness of the shaft, using a SDOF model, can be obtained from Table 6.1 as:

$$k = \frac{48EI}{L^3} \tag{9.97}$$

The angular frequency of the system is given (from Equation (6.7)) by:

$$\omega_1 = \sqrt{\frac{k}{M}} = \sqrt{\frac{48EI}{ML^3}} = 6.928\sqrt{\frac{EI}{ML^3}} \text{ rad/s} \tag{9.98}$$

Again, the whirling speed of the shaft can be written in revolutions per minute:

$$\omega_1 = \frac{60}{2\pi} \times 6.928\sqrt{\frac{EI}{ML^3}} = 66.15\sqrt{\frac{EI}{ML^3}} \text{ rev/min} \tag{9.99}$$

9.5.3 Whirling speed of a mass shifted from the middle of the shaft

For a mass M at a distance a from the left bearing, as shown in Figure 9.26, an equation similar to Equation (9.99) can be derived. The maximum deflection in the shaft due to a point load F acting at a distance a can be calculated using structural mechanics principles as:

$$\delta_{\max} = \frac{F}{3EI} \times \frac{a^2(L-a)^2}{L} \tag{9.100}$$

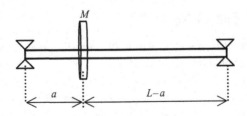

Figure 9.26: A shaft with a mass at a distance a

The equivalent stiffness is:

$$k = \frac{F}{\delta_{max}} = \frac{3EIL}{a^2(L-a)^2} \tag{9.101}$$

The angular frequency of the system becomes:

$$\omega_1 = \sqrt{\frac{k}{M}} = \sqrt{\frac{3EIL}{Ma^2(L-a)^2}} = 1.732\sqrt{\frac{EIL}{Ma^2(L-a)^2}} \text{ rad/s} \tag{9.102}$$

Re-writing Equation (9.102) in revolutions per minute gives:

$$\omega_1 = \frac{60}{2\pi} \times 1.732\sqrt{\frac{EIL}{Ma^2(L-a)^2}} = 16.54\sqrt{\frac{EIL}{Ma^2(L-a)^2}} \text{ rev/min} \tag{9.103}$$

9.5.4 Dunkerley's formula

The whirling speed derived in the previous sections applies only to a shaft or a shaft with a single mass. In order to calculate the whirling speed of a shaft carrying several masses, a superposition method known as Dunkerley's formula can be used. The whirling speeds of the shaft and the masses are calculated separately and then combined to give the whirling speed of the system. Dunkerley's formula can be derived by considering a discrete system of masses and springs and finding the relationship between its natural frequencies. To prove Dunkerley's formula, recalling the solution of the system with two degrees of freedom (Figures 8.3 and 8.4), the characteristic or eigenvalue equation, Equation (8.42), can be written as:

$$a\omega_n^4 - b\omega_n^2 + c = 0 \tag{9.104}$$

where the constants a, b and c are as given in Equation (8.43). Since the roots of Equation (8.43) or (9.104) are ω_1 and ω_2, Equation (9.104) can be written as:

$$\left(\omega_n^2 - \omega_1^2\right)\left(\omega_n^2 - \omega_2^2\right) = 0 \tag{9.105}$$

Expanding Equation (9.105) gives:

$$\omega_n^4 - \omega_1^2\omega_n^2 - \omega_2^2\omega_n^2 + \omega_1^2\omega_2^2 = 0$$

or

$$\omega_n^4 - \left(\omega_1^2 + \omega_2^2\right)\omega_n^2 + \omega_1^2\omega_2^2 = 0 \tag{9.106}$$

Comparing Equation (9.104) with Equation (9.106), since they should be identical, the constants a, b and c should be:

$$a = 1 \tag{9.107}$$
$$b = (\omega_1^2 + \omega_2^2) \tag{9.108}$$
$$c = \omega_1^2\omega_2^2 \tag{9.109}$$

Dividing Equation (9.108) by Equation (9.109) leads to:

$$\frac{b}{c} = \frac{\left(\omega_1^2 + \omega_2^2\right)}{\omega_1^2 \omega_2^2} = \frac{1}{\omega_1^2} + \frac{1}{\omega_2^2} \tag{9.110}$$

For a system of n degrees of freedom, Equation (9.110) can be extended as follows:

$$\frac{b}{c} = \frac{1}{\omega_1^2} + \frac{1}{\omega_2^2} + \frac{1}{\omega_3^2} + \cdots + \frac{1}{\omega_n^2} \tag{9.111}$$

The approximation made by Dunkerley is that, since the fundamental angular frequency of the system, ω_1 is dominant and is the lowest angular frequency, i.e. $\omega_1 < \omega_2 < \omega_3 < \cdots < \omega_n$, then $\frac{1}{\omega_1^2} >> \frac{1}{\omega_2^2} >> \frac{1}{\omega_3^2} >> \cdots >> \frac{1}{\omega_n^2}$. In such a case, the first term in the right-hand side of Equation (9.111), $\frac{1}{\omega_1^2}$, gives a good approximation of the whole term:

$$\frac{1}{\omega_1^2} \approx \frac{1}{\omega_1^2} + \frac{1}{\omega_2^2} + \frac{1}{\omega_3^2} + \cdots + \frac{1}{\omega_n^2} \tag{9.112}$$

If each angular frequency in the right-hand side of Equation (9.112) is regarded to be for an independent system, Equation (9.112) can be used to approximate the fundamental angular frequency of the system in Figure 9.27 as follows:

$$\frac{1}{\omega_A^2} = \frac{1}{\omega_B^2} + \frac{1}{\omega_C^2} + \frac{1}{\omega_D^2} \tag{9.113}$$

where ω_A is the angular frequency or the whirling speed of the system in Figure 9.27(a), which takes into account the shaft's weight, mass M_1 and mass M_2; ω_B is the whirling

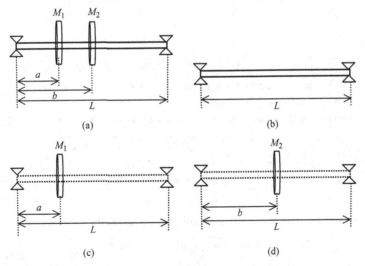

Figure 9.27: Application of Dunkerley's formula to a shaft carrying two masses (case (a) = case (b) + case (c) + case (d))

speed of the system in Figure 9.27(b), which takes into account only the shaft's weight; ω_C is the whirling speed of the system in Figure 9.27(c), which takes into account only mass M_1; and ω_D is the whirling speed of the system in Figure 9.27(d), which takes into account only mass M_2.

Example 9.9 Whirling of shafts I

The shaft shown in Figure 9.28 is 1 m long and carries a mass of 2 kg at its midpoint. The shaft is made of steel that has a Young's modulus of 200 GPa and density of 7800 kg/m³. If the shaft's diameter is 12 mm, determine:

a) the whirling speed due to the weight of the shaft alone;
b) the whirling speed due to the 2 kg mass alone;
c) the whirling speed of the system;
d) the whirl amplitude if the shaft operates at 600 rev/min and the eccentricity of the mass is 0.1 mm (ignoring damping).

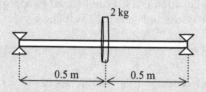

Figure 9.28: Representation of Example 9.9

Solution

a) The shaft's mass is $m = \rho A L = 7800 \times \frac{\pi \times 0.012^2}{4} \times 1 = 0.88216\,\text{kg}$ and the area moment of inertia I about the x axis or the y axis (see Appendix B) is $I = \frac{\pi d^4}{64} = \frac{\pi \times 0.012^4}{64} = 1.017876 \times 10^{-9}\,\text{m}^4$.

From Equation (9.96), the whirling speed due to the shaft's weight alone, ω_B is:

$$\omega_B = 94.25\sqrt{\frac{EI}{mL^3}} = 94.25\sqrt{\frac{200 \times 10^9 \times 1.017876 \times 10^{-9}}{0.88216 \times 1^3}}$$

$$= 1431.7596 = 1432\,\text{rev/min}$$

b) For the 2 kg mass alone, using Equation (9.99) gives the whirling speed, ω_C:

$$\omega_C = 66.15\sqrt{\frac{200 \times 10^9 \times 1.017876 \times 10^{-9}}{2 \times 1^3}} = 667.3863 = 667\,\text{rev/min}$$

c) Using Dunkerley's formula, Equation (9.113), the system's whirling speed, ω_A is obtained as:

$$\frac{1}{\omega_A^2} = \frac{1}{\omega_B^2} + \frac{1}{\omega_C^2} \Rightarrow \frac{1}{\omega_A^2} = \frac{1}{1431.7596^2}$$

$$+ \frac{1}{667.3863^2} \Rightarrow \omega_A = 604.898 = 605 \text{ rev/min}$$

d) The whirl amplitude is obtained from Equation (9.92) for an undamped system using $\delta = 0.1$ mm and $\omega = 600$ rev/min:

$$u_y = \frac{\delta \left(\frac{\omega}{\omega_n}\right)^2}{1 - \left(\frac{\omega}{\omega_n}\right)^2} = \frac{0.1 \times \left(\frac{600}{604.898}\right)^2}{1 - \left(\frac{600}{604.898}\right)^2} = 6.099 = 6.1 \text{ mm}$$

Example 9.10 Whirling of shafts II

A flywheel of mass 25 kg and eccentricity of 2 mm is mounted at the midpoint of a shaft of mass 4 kg, length 1 m and flexural rigidity $EI = 3500$ N/m. If the shaft has a damping ratio of 0.01 and operates at 1100 rev/min, determine:

a) the whirl amplitude at operating speed;
b) the whirl amplitude at whirling speed (critical speed).

Solution

From Equation (9.96), the whirling speed due to the shaft's weight alone, ω_B, is:

$$\omega_B = 94.25 \sqrt{\frac{EI}{mL^3}} = 94.25 \sqrt{\frac{3500}{4 \times 1^3}} = 2787.9526 \text{ rev/min}$$

For the flywheel, Equation (9.99) gives the whirling speed, ω_C:

$$\omega_C = 66.15 \sqrt{\frac{3500}{25 \times 1^3}} = 782.68736 \text{ rev/min}$$

Using Dunkerley's formula, Equation (9.113), gives the system's whirling speed, ω_A, as:

$$\frac{1}{\omega_A^2} = \frac{1}{\omega_B^2} + \frac{1}{\omega_C^2} \Rightarrow \frac{1}{\omega_A^2}$$

$$= \frac{1}{2787.9526^2} + \frac{1}{782.68736^2} \Rightarrow \omega_A = 753.5639 \text{ rev/min}$$

a) To calculate the whirl amplitude at operating speed, Equation (9.93) with $\omega = 1100$ rev/min, $\omega_n = 753.5639$ rev/min, $\delta = 2$ mm and $\zeta = 0.01$, gives:

$$u_y = \frac{\delta \left(\frac{\omega}{\omega_n}\right)^2}{\sqrt{\left(1 - \left(\frac{\omega}{\omega_n}\right)^2\right)^2 + \left(\frac{2\zeta\omega}{\omega_n}\right)^2}}$$

$$= \frac{2 \times \left(\frac{1100}{753.5639}\right)^2}{\sqrt{\left(1 - \left(\frac{1100}{753.5639}\right)^2\right)^2 + \left(\frac{2 \times 0.01 \times 1100}{753.5639}\right)^2}}$$

$$= 3.767 = 3.8 \, \text{mm}$$

b) At whirling or critical speed, $\frac{\omega}{\omega_n} = 1$ and Equation (9.93) becomes $u_y = \frac{\delta}{2\zeta}$, from which the whirl amplitude is calculated as:

$$u_y = \frac{2}{2 \times 0.01} = 100 \, \text{mm}$$

9.6 Tutorial Sheet

9.6.1 Lateral vibration of cables or strings

Q9.1 A string of length 3 m and mass 1 kg/m is stretched with a tension of 500 N and is fixed at both ends. Find the first three natural frequencies and draw the corresponding mode shapes.

[3.73 Hz, 7.45 Hz, 11.18 Hz]

Q9.2 A steel wire has a length L and a mass of 2 kg/m and is fixed at both ends. If the wire is stretched with a tension of 200 N, determine:

a) the fundamental frequency when $L = 2$ m;

[2.5 Hz]

b) the minimum wire length required to give a fundamental frequency above 5 Hz.

[1 m]

c) Sketch the sixth mode shape. How many nodes are there in this mode?

[7 nodes]

Q9.3   If an initial displacement of 20 mm was applied at the midpoint of the string in Question 9.1, use the mode superposition method to determine the response at 1/3 of the string length ($x = L/3$) at time $t = 0.07$ s. Consider the following two cases:

a) the first two modes;

[−0.96 mm]

b) the first three modes.

[−0.96 mm]

c) Compare these results and comment on the contribution of each mode in the displacement time response.

Q9.4   A string ABC, shown in Figure 9.29(a), of length 2 m and mass 5 kg/m is stretched with a tension of 100 N and is fixed at both ends.

a) Determine the first three natural frequencies and draw the corresponding mode shapes

[1.12 Hz, 2.24 Hz, 3.35 Hz]

b) Determine the time response at the midpoint B of the string if an initial deflection of 3 mm is applied at B as shown in Figure 9.29(b).

$$\left[u_y(L/2, t) = \sum_{n=1}^{\infty} C_n \sin \frac{n\pi}{2} \cos 2.236 n\pi t \right]$$

c) Sketch the displacement amplitude versus time curve, considering the first three modes and using the mode superposition method.

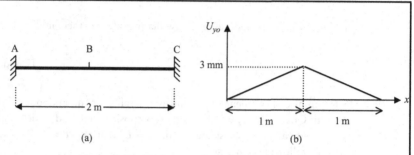

(a)   (b)

Figure 9.29: Representation of Question 9.4

Q9.5 A steel wire of 2 mm diameter and density $\rho = 7800\,\text{kg/m}^3$ is fixed between two points 2 m apart.

a) If the tensile force in the wire is 250 N, determine the fundamental frequency of vibration.

[25.25 Hz]

b) If an initial deflection of 25 mm is applied in the middle of the wire, use the first three modes to determine:
   i) the displacement in the middle of the wire at time $t = 0.01$ s;

[−0.21 mm]

   ii) the velocity at $x = 0.3$ m and time $t = 0.015$ s.

[−207.3 mm/s]

9.6.2 Longitudinal vibration of bars

Q9.6 A uniform bar has a length of 1 m. The beam is made of cast iron, which has a Young's modulus of 180 GPa and density of 7000 kg/m^3. Determine the first three natural frequencies and draw the corresponding mode shapes assuming free-free axial vibration.

[2535.5 Hz, 5070.9 Hz, 7606.4 Hz]

Q9.7 A fixed-free cast-iron bar of 10 m length has $E = 180$ GPa and $\rho = 7000$ kg/m^3. The bar is subjected to an initial axial force of 40 MN at the free edge as shown in Figure 9.30 and then it is left free to vibrate. The bar has a cross-sectional area of 0.25 m^2. Determine the displacement at the midpoint of the bar ($x = L/2$) at time $t = 1.97 \times 10^{-4}$ s, using mode superposition with:

a) the first two modes;

[4.53 mm]

b) the first three modes.

[4.38 mm]

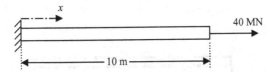

Figure 9.30: Representation of Question 9.7

Q9.8 An aluminium bar has a Young's modulus of 69 GPa and a density of 2770 kg/m³. The bar has a length of 2 m. For free-free axial vibration as shown in Figure 9.31(a), determine:

a) the first three natural frequencies;

[1247.7 Hz, 2495.5 Hz, 3743.2 Hz]

b) the position of the maximum displacement for each of the modes in a).

[For mode 1, $n = 1$ and $x = 0$, L; for mode 2, $n = 2$ and $x = 0$, $L/2$, L; for mode 3, $n = 3$ and $x = 0$, $L/3$, $2L/3$, L]

c) If the rod is constrained to move axially at a distance 0.5 m from the left end, as shown in Figure 9.31(b), what is the change in the fundamental frequency?

[831.8 Hz]

d) Sketch the first mode shape of the rod in Figure 9.31(b).

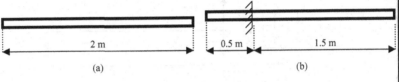

(a)   (b)

Figure 9.31: Representation of Question 9.8

Q9.9 A steel shaft of 5 cm diameter and 1 m length, shown in Figure 9.32(a), has a Young's modulus of 200 GPa and a density of 7800 kg/m³. Determine:

a) the first three natural frequencies;

[1265.9 Hz, 3797.8 Hz, 6329.6 Hz]

b) the displacement at $x = 0.3$ m and time $t = 0.02$ s, using the mode superposition method and considering the first two modes, for the initial conditions shown in Figure 9.32(b) (the initial velocity is zero).

[−0.48 mm]

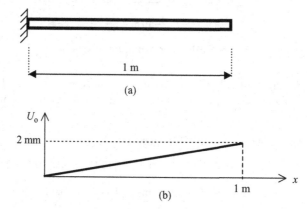

(a)

(b)

Figure 9.32: Representation of Question 9.9

9.6.3 Lateral vibration of beams

Q9.10 A beam of length 1.5 m has a Young's modulus of 69 GPa and density of 2700 kg/m³. For simply supported ends, as shown in Figure 9.33, determine the first three natural frequencies and draw the corresponding mode shapes for transverse vibration.

[91.69 Hz, 366.77 Hz, 825.23 Hz]

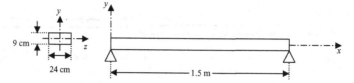

Figure 9.33: Representation of Question 9.10

Q9.11 A shaft of length 2.5 m has an annular cross section as shown in Figure 9.34, a Young's modulus of 70 GPa and a density of 2700 kg/m³. For fixed-simply supported ends and lateral vibration:

a) Determine the first three natural frequencies.

[134.1 Hz, 434.6 Hz, 906.8 Hz]

b) Sketch the first three mode shapes.

c) If it is a requirement of the shaft's design that the fundamental frequency is greater than 130 Hz, which of the materials listed in Table 9.3 will you choose? Justify your choice.

[Steel or aluminium has $f_1 > 130$ Hz]

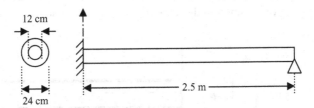

Figure 9.34: Representation of Question 9.11

Table 9.3: Materials for the shaft
in Question 9.11

	E(GPa)	ρ(kg/m³)
Aluminium	70	2700
Steel	200	7800
Concrete	40	2400

Q9.12 A beam should be designed so that its fundamental frequency, for lateral vibration in the x-y plane, as shown in Figure 9.35, should be below 9 Hz. The beam is 4 m long and has a cross-sectional area of 25 cm × 15 cm.

a) For simply supported boundary conditions, determine the minimum ratio of the Young's modulus to the density (E/ρ) required in order to satisfy the fundamental frequency condition. Choose a suitable material for the beam from the list below (where E is in N/m^2 and ρ in kg/m^3):

- Steel $E/\rho = 2.5 \times 10^6$
- Steel/concrete composite $E/\rho = 2.0 \times 10^6$
- High strength concrete $E/\rho = 1.6 \times 10^6$

$$[1.61 \times 10^6 \text{ N/m}^2\text{/kg/m}^3]$$

b) If the beam is made of steel, what boundary conditions at both ends would you propose in order to satisfy the fundamental frequency requirement?

[Fixed–free]

c) Explain how to determine the position of nodes in each mode of vibration of a beam. For the case of simply supported ends, calculate the position of the nodes for the first three modes.

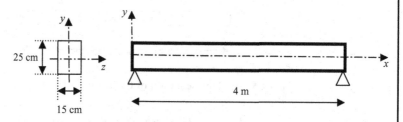

Figure 9.35: Representation of Question 9.12

Q9.13 A beam has a length of 3 m as shown in Figure 9.36. The beam is made of reinforced concrete and has a Young's modulus of 20 GPa and density of 2400 kg/m^3. The width of the beam is one half its depth, b. The fundamental natural frequency for lateral vibration in the x-y plane should be equal to 50 Hz. Find the cross-sectional area A and the area moment of inertia I about the z-axis, if:

a) both ends are simply supported;

$$[591 \text{ cm}^2, 58\ 144 \text{ cm}^4]$$

b) both ends are fixed.

$$[115 \text{ cm}^2, 2201 \text{ cm}^4]$$

c) Compare the results of parts (a) and (b) and comment on them.

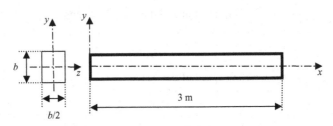

Figure 9.36: Representation of Question 9.13

Q9.14   The average cross section of an aeroplane wing can be idealized to a hollow box 300 mm × 2000 mm with 2 mm thickness, as shown in Figure 9.37. The wing is made of aluminium, which has a Young's modulus $E = 70$ GPa and density $\rho = 2770$ kg/m^3. If the wing is considered to be fixed to the fuselage at one end and free at the other, determine the first three natural frequencies of the wing.

[4 Hz, 25.1 Hz, 70.28 Hz]

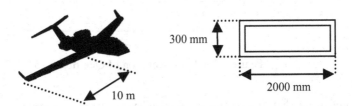

Figure 9.37: Representation of Question 9.14

9.6.4 Whirling of shafts

Q9.15   The shaft shown in Figure 9.38, carries two masses of 1 kg and 2 kg, at 0.4 m and 0.6 m, respectively, from the left bearing. The shaft has a solid circular cross section of diameter 15 mm, $E = 200$ GPa and $\rho = 7800$ kg/m^3. It is supported by two bearings 1.2 m apart. Determine:

a) the whirling speed due to the shaft's weight alone;

[1243 rev/min]

b) the whirling speed due to the 1 kg mass alone;

[1262 rev/min]

c) the whirling speed due to the 2 kg mass alone;

[793 rev/min]

d) the whirling speed of the system;

[591 rev/min]

e) the eccentricity of the mass if the whirl amplitude is 5 mm when the shaft operates at 600 rev/min (ignoring damping).

[0.15 mm]

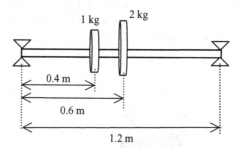

Figure 9.38: Representation of Question 9.15

Q9.16　A shaft of tubular cross section has an external diameter of 3.5 cm and an internal diameter of 2.5 cm. It is made of steel with Young's modulus of 200 GPa and density of 7800 kg/m³. If the shaft has a length of 2 m, determine its whirling speed.

[1283 rev/min]

Q9.17　A steel shaft of length 0.5 m has a solid cross section of diameter 2.5 cm and is supported at both ends, as shown in Figure 9.39. If a wheel of mass 50 kg is attached to the shaft at point C, determine the whirling speed of the system. The Young's modulus of steel is 200 GPa and the density is 7800 kg/m³.

[1690 rev/min]

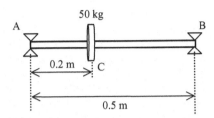

Figure 9.39: Representation of Question 9.17

Q9.18 Determine the whirl amplitude of a shaft that operates at 1200 rev/min and has a mass of 3 kg, length of 0.9 m and flexural rigidity of $EI = 3000$ N/m, if a rotor of mass 20 kg and eccentricity of 3 mm is mounted at its midpoint. The damping ratio of the shaft is 0.02.

[7.2 mm]

Q9.19 Recalculate the whirl amplitude of the shaft in Question 9.18, ignoring the shaft's weight.

[8 mm]

Q9.20 A flywheel of eccentricity 2.5 mm is mounted on a shaft of damping ratio 0.05. Determine the whirl amplitude at the critical speed.

[25 mm]

Q9.21 A rotor of eccentricity 2 mm is mounted on a shaft of negligible damping at its midpoint as shown in Figure 9.40. When the shaft operates at 900 rev/min, a whirl amplitude of 7 mm is observed; when it operates at 950 rev/min, a whirl amplitude of 13 mm is observed. Determine the whirling speed of the system.

[1021 rev/min]

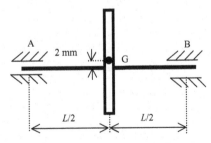

Figure 9.40: Representation of Question 9.21

CHAPTER 10

Finite-Element Method

10.1 INTRODUCTION

In Chapter 9, analytical solutions for continuous systems were developed. If the analytical solution cannot be obtained, numerical techniques provide an alternative way to solve the problem. Numerical techniques can be used to obtain approximate solutions which are, in most cases, acceptable. One of those numerical techniques is the finite-element method, which can be used for static and dynamic analyses. The finite-element method is very useful in modelling complex structures which cannot be modelled otherwise. It is very easy to program and implement in computer software and has, therefore, gained lots of popularity. Many powerful finite element packages, such as ANSYS and ABAQUS, are widely used in industry and academia.

In the finite-element technique, the physical structure is divided into small (finite) elements. Each finite element is assumed to behave as a continuous structural member. The elements are connected to each other at points, known as nodes. A finite element mesh or grid consists of a number of finite elements and nodes. An approximate solution for the degrees of freedom, such as displacements, is assumed within each finite element. The total number of equations that are required to describe the entire structure is obtained through the assembly of all elements. During the assembly process, force equilibrium equations and displacement compatibility equations are enforced so that the whole structure responds as one body. The finite element solution converges to the exact solution as the element size becomes smaller. In vibration analysis, the vibration of the structure is approximated through the displacement of the nodes as a function of time.

In this chapter, we consider the theoretical development of the finite-element method and its application to simple vibration problems with respect to one-dimensional bar and beam elements. The element stiffness and mass matrices are derived and solutions for the natural frequencies and vibration mode shapes are obtained. Section 10.4 gives some guidelines on using ANSYS to model and analyze vibration problems.

10.2 BAR ELEMENT

10.2.1 Mass and stiffness matrices

For the bar element shown in Figure 10.1, the nodes i and j are located at the two ends. The axial forces F_i and F_j, acting at nodes i and j, respectively, are called nodal forces and are, in general, functions of time. The element has a Young's modulus E, a uniform cross-sectional area A and a density ρ.

If it is assumed that the displacement variation in the x direction is linear, the displacement function, u_x, is given by:

$$u_x = C_1 + C_2 x \tag{10.1}$$

where the constants C_1 and C_2 are, in general, functions of time and can be determined from the boundary conditions. Assuming that the boundary conditions are at $x = 0$,

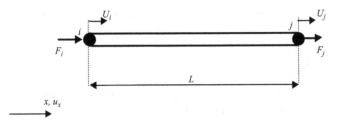

Figure 10.1: A bar element

$u_x = U_i$ and $x = L$, $u_x = U_j$, where U_i and U_j, are the nodal displacements at nodes i and j, respectively, the constants C_1 and C_2 can be found as:

$$C_1 = U_i, \tag{10.2}$$

$$C_2 = \frac{U_j - U_i}{L} \tag{10.3}$$

Substituting Equations (10.2) and (10.3) into Equation (10.1) yields:

$$u_x = U_i + \frac{U_j - U_i}{L} x \tag{10.4}$$

Rearranging Equation (10.4) so that the terms for U_i and U_j are separated gives:

$$u_x = \left[1 - \frac{x}{L}\right] U_i + \frac{x}{L} U_j \tag{10.5}$$

or in another form:

$$u_x = N_i U_i + N_j U_j \tag{10.6}$$

where N_i and N_j are known as the shape functions for nodes i and j, respectively, and are equal to:

$$N_i = \left[1 - \frac{x}{L}\right], $$
$$N_j = \frac{x}{L} \tag{10.7}$$

Note that the shape function N_i is equal to 1 at node i and 0 at node j as shown in Figure 10.2(a). Similarly, N_j is equal to 1 at node j and 0 at node i, as shown in Figure 10.2(b). N_i and N_j are also called interpolation functions, as they interpolate the displacement between nodes i and j.

(a) N_i (b) N_j

Figure 10.2: Shape functions of a bar element

In order to derive the element's mass matrix, the kinetic energy is obtained as:

$$KE = \frac{1}{2} \int_0^L m_l v_x^2 \, dx \tag{10.8}$$

where m_l is the mass per unit length and v_x is the velocity in the x direction. Substituting $m_l = \rho A$ and $v_x = \frac{\partial u_x}{\partial t}$ into Equation (10.8) gives:

$$KE = \frac{1}{2} \int_0^L \rho A \left[\frac{\partial u_x}{\partial t} \right]^2 dx \tag{10.9}$$

Differentiating Equation (10.5) with respect to time and substituting into Equation (10.9) yields:

$$KE = \frac{1}{2} \int_0^L \rho A \left[\left(1 - \frac{x}{L}\right) \frac{\partial U_i}{\partial t} + \frac{x}{L} \frac{\partial U_j}{\partial t} \right]^2 dx \tag{10.10}$$

Integrating Equation (10.10) gives:

$$KE = \frac{\rho A L}{6} \left(\dot{U}_i^2 + \dot{U}_i \dot{U}_j + \dot{U}_j^2 \right) \tag{10.11}$$

where $\dot{U}_i$ and $\dot{U}_j$ are the nodal velocities:

$$\dot{U}_i = \frac{\partial U_i}{\partial t}, \qquad \dot{U}_j = \frac{\partial U_j}{\partial t}$$

Re-writing Equation (10.11) in a matrix form yields:

$$KE = \frac{1}{2} \{\dot{U}\}^T [m] \{\dot{U}\} \tag{10.12}$$

where the velocity vector $\{\dot{U}\}$ is given by:

$$\{\dot{U}\} = \begin{Bmatrix} \dot{U}_i \\ \dot{U}_j \end{Bmatrix}, \qquad \{\dot{U}\}^T = [\dot{U}_i \quad \dot{U}_j] \tag{10.13}$$

and the element mass matrix $[m]$ is defined as:

$$[m] = \frac{\rho A L}{6} \begin{bmatrix} 2 & 1 \\ 1 & 2 \end{bmatrix} \tag{10.14}$$

In order to derive the element stiffness matrix, similar procedures can be followed by writing the strain energy of the bar element as:

$$SE = \frac{1}{2} \int_V \sigma_x \varepsilon_x \, dV \tag{10.15}$$

where σ_x and ε_x are the stress and strain, respectively, in the x direction and dV is the element volume. For a constant cross section A, $dV = A\,dx$ and Equation (10.15) becomes:

$$SE = \frac{1}{2} \int_0^L A\sigma_x \varepsilon_x\, dx \tag{10.16}$$

Using Hook's law for linear elastic material, $\sigma_x = E\varepsilon_x$, and using the relationship between strain and displacement for small deformation, $\varepsilon_x = \frac{\partial u_x}{\partial x}$, thus Equation (10.16) becomes:

$$SE = \frac{1}{2} \int_0^L EA\left(\frac{\partial u_x}{\partial x}\right)^2 dx \tag{10.17}$$

Differentiating the displacement function u_x, Equation (10.5), with respect to the coordinate x and substituting into Equation (10.17) yields:

$$SE = \frac{1}{2} \int_0^L EA\left[-\frac{1}{L}U_i + \frac{1}{L}U_j\right]^2 dx \tag{10.18}$$

Integrating Equation (10.18) gives:

$$SE = \frac{EA}{2L}\left(U_i^2 - 2U_iU_j + U_j^2\right) \tag{10.19}$$

Re-writing Equation (10.19) in matrix form leads to:

$$SE = \frac{1}{2}\{U\}^T[k]\{U\} \tag{10.20}$$

where the displacement vector $\{U\}$ is given by:

$$\{U\} = \begin{Bmatrix} U_i \\ U_j \end{Bmatrix}, \qquad \{U\}^T = [\,U_i \quad U_j\,] \tag{10.21}$$

and the bar element stiffness matrix $[k]$ is defined as:

$$[k] = \frac{EA}{L}\begin{bmatrix} 1 & -1 \\ -1 & 1 \end{bmatrix} \tag{10.22}$$

10.2.2 Equations of motion

The equations of motion for the bar element can be obtained in a way similar to that applied to systems with two degrees of freedom in Chapter 8, using Newton's second law as:

$$[m]_{2\times2}\{\ddot{U}\}_{2\times1} + [k]_{2\times2}\{U\}_{2\times1} = \{F\}_{2\times1} \tag{10.23}$$

where $[m]$ is the mass matrix, $[k]$ is the stiffness matrix, $\{U\}$ is the displacement vector, $\{\ddot{U}\}$ is the acceleration vector and $\{F\}$ is the force vector. Equation (10.23) can also be derived using Lagrange's equation, which is given by:

$$\frac{d}{dt}\left(\frac{\partial(KE)}{\partial \dot{U}}\right) - \frac{\partial(KE)}{\partial U} + \frac{\partial(SE)}{\partial U} = F \qquad (10.24)$$

Substituting the kinetic energy (KE) from Equation (10.12) and the strain energy (SE) from Equation (10.20) into Equation (10.24) and performing the differentiation leads to:

$$[m]\{\ddot{U}\} + [k]\{U\} = \{F\}$$

which is identical to Equation (10.23) in its general form. For a structure containing several bar elements, the equations of motion are written as:

$$[m]_{n\times n}\{\ddot{U}\}_{n\times 1} + [k]_{n\times n}\{U\}_{n\times 1} = \{F\}_{n\times 1}$$

where n is the number of degrees of freedom in the structure. In a bar element, this is the same as the number of nodes since each node has only one degree of freedom. For free vibration, the equations of motion become:

$$[m]_{n\times n}\{\ddot{U}\}_{n\times 1} + [k]_{n\times n}\{U\}_{n\times 1} = 0 \qquad (10.25)$$

The displacement time response (or transient) solutions of these differential equations are in the following form (see Equation (6.20) for SDOF in Chapter 6):

$$\{U\}_{n\times 1} = \{U_m\}_{n\times 1}\sin(\omega_n t + \psi) \qquad (10.26)$$

where the vector $\{U_m\}$ contains the maximum displacement (amplitude) for each degree of freedom and ω_n is the angular frequency. Differentiating Equation (10.26) twice with respect to time and substituting into Equation (10.25) yields:

$$\left([k]_{n\times n} - \omega_n^2[m]_{n\times n}\right)\{U_m\}_{n\times 1} = 0 \qquad (10.27)$$

Equation (10.27) is a problem with n eigenvalues (angular frequencies) and eigenvectors (mode shapes).

10.2.3 Boundary conditions

The boundary conditions can be simply considered in Equation (10.27) by modifying the matrices $[m]$ and $[k]$. For example, zero displacement is incorporated by eliminating the corresponding rows and columns from $[m]$ and $[k]$. Consider the following equations of motion:

$$\begin{bmatrix} k_{11} & k_{12} & k_{13} & .. & .. & k_{1n} \\ k_{21} & k_{22} & k_{23} & .. & .. & k_{2n} \\ k_{31} & k_{32} & k_{33} & .. & .. & k_{3n} \\ .. & & & & & .. \\ .. & & & & & .. \\ k_{n1} & k_{n2} & k_{n3} & .. & .. & k_{nn} \end{bmatrix} - \omega^2 \begin{bmatrix} m_{11} & m_{12} & m_{13} & .. & .. & m_{1n} \\ m_{21} & m_{22} & m_{23} & .. & .. & m_{2n} \\ m_{31} & m_{32} & m_{33} & .. & .. & m_{3n} \\ .. & & & & & .. \\ .. & & & & & .. \\ m_{n1} & m_{n2} & m_{n3} & .. & .. & m_{nn} \end{bmatrix} \begin{Bmatrix} U_{m_1} \\ U_{m_2} \\ U_{m_3} \\ .. \\ .. \\ U_{m_n} \end{Bmatrix} = 0 \quad (10.28)$$

Assuming that the structure is constrained at node 2, i.e. $U_{m_2} = 0$. In order to implement this boundary condition in Equation (10.28), the second row and the second column in $[m]$ and $[k]$ should be eliminated:

$$\begin{bmatrix} k_{11} & k_{12} & k_{13} & .. & .. & k_{1n} \\ k_{21} & k_{22} & k_{23} & .. & .. & k_{2n} \\ k_{31} & k_{32} & k_{33} & .. & .. & k_{3n} \\ .. & & & & & .. \\ .. & & & & & .. \\ k_{n1} & k_{n2} & k_{n3} & .. & .. & k_{nn} \end{bmatrix} - \omega^2 \begin{bmatrix} m_{11} & m_{12} & m_{13} & .. & .. & m_{1n} \\ m_{21} & m_{22} & m_{23} & .. & .. & m_{2n} \\ m_{31} & m_{32} & m_{33} & .. & .. & m_{3n} \\ .. & & & & & .. \\ .. & & & & & .. \\ m_{n1} & m_{n2} & m_{n3} & .. & .. & m_{nn} \end{bmatrix} \begin{Bmatrix} U_{m_1} \\ U_{m_2} \\ U_{m_3} \\ .. \\ .. \\ U_{m_n} \end{Bmatrix} = 0$$

And the system of equations is reduced to:

$$\begin{bmatrix} k_{11} & k_{13} & k_{14} & .. & .. & k_{1n} \\ k_{31} & k_{33} & k_{34} & .. & .. & k_{3n} \\ k_{41} & k_{43} & k_{44} & .. & .. & k_{4n} \\ .. & & & & & .. \\ .. & & & & & .. \\ k_{n1} & k_{n3} & k_{n4} & .. & .. & k_{nn} \end{bmatrix} - \omega^2 \begin{bmatrix} m_{11} & m_{13} & m_{14} & .. & .. & m_{1n} \\ m_{31} & m_{33} & m_{34} & .. & .. & m_{3n} \\ m_{41} & m_{43} & m_{44} & .. & .. & m_{4n} \\ .. & & & & & .. \\ .. & & & & & .. \\ m_{n1} & m_{n3} & m_{n4} & .. & .. & m_{nn} \end{bmatrix} \begin{Bmatrix} U_{m_1} \\ U_{m_3} \\ U_{m_4} \\ .. \\ .. \\ U_{m_n} \end{Bmatrix} = 0 \quad (10.29)$$

Example 10.1 Bar element I

A uniform bar of length 2 m has a Young's modulus of 200 GPa and a density of 7800 kg/m^3. For free–free vibration, use a bar finite element as shown in Figure 10.3 to determine the first natural frequency.

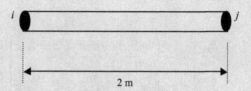

Figure 10.3: Representation of Example 10.1

Solution

Using Equation (10.14), the mass matrix is obtained as $[m] = \frac{\rho A L}{6} \begin{bmatrix} 2 & 1 \\ 1 & 2 \end{bmatrix}$. Using Equation (10.22), the stiffness matrix is $[k] = \frac{EA}{L} \begin{bmatrix} 1 & -1 \\ -1 & 1 \end{bmatrix}$.

The eigenvalue equation of free vibration, Equation (10.27), becomes:

$$\left(\frac{EA}{L} \begin{bmatrix} 1 & -1 \\ -1 & 1 \end{bmatrix} - \omega^2 \frac{\rho A L}{6} \begin{bmatrix} 2 & 1 \\ 1 & 2 \end{bmatrix} \right) \begin{Bmatrix} U_{mi} \\ U_{mj} \end{Bmatrix} = 0$$

or

$$\left(\begin{bmatrix} 1 & -1 \\ -1 & 1 \end{bmatrix} - \lambda \begin{bmatrix} 2 & 1 \\ 1 & 2 \end{bmatrix} \right) \begin{Bmatrix} U_{mi} \\ U_{mj} \end{Bmatrix} = 0$$

where $\lambda = \frac{\rho L^2 \omega^2}{6E}$.

This gives

$$\begin{bmatrix} 1 - 2\lambda & -1 - \lambda \\ -1 - \lambda & 1 - 2\lambda \end{bmatrix} \begin{Bmatrix} U_{mi} \\ U_{mj} \end{Bmatrix} = 0$$

The solution for this eigenvalue problem is obtained by equating the determinant of the matrix to zero:

$$\begin{vmatrix} 1 - 2\lambda & -1 - \lambda \\ -1 - \lambda & 1 - 2\lambda \end{vmatrix} = 0$$

Solving for λ, yields $(1 - 2\lambda)(1 - 2\lambda) - (1 + \lambda)(1 + \lambda) = 0 \Rightarrow 3\lambda^2 - 6\lambda = 0 \Rightarrow \lambda_o = 0, \lambda_1 = 2$. The rigid body mode corresponds to $\lambda_o = 0$, and the first mode of vibration (the fundamental mode) to $\lambda_1 = 2$.

The first angular frequency is:

$$\omega_1 = \sqrt{\frac{6E\lambda_1}{\rho L^2}} = \sqrt{\frac{12E}{\rho L^2}}$$

For $E = 20 \times 10^{10}$ N/m^2, $\rho = 7.8 \times 10^3$ kg/m^3 and $L = 2$ m, the first angular frequency becomes:

$$\omega_1 = \sqrt{\frac{12 \times 20 \times 10^{10}}{7.8 \times 10^3 \times 2^2}} = 8770.58 \text{ rad/s}$$

and the first natural frequency is:

$$f_1 = \frac{\omega_1}{2\pi} = 1395.88 \text{ Hz}$$

To calculate the eigenvector for the rigid-body mode, using $\lambda = 0$ gives

$$\begin{bmatrix} 1-2\lambda & -1-\lambda \\ -1-\lambda & 1-2\lambda \end{bmatrix} \begin{Bmatrix} U_{mi} \\ U_{mj} \end{Bmatrix} = 0 \Rightarrow \begin{bmatrix} 1-0 & -1-0 \\ -1-0 & 1-0 \end{bmatrix} \begin{Bmatrix} U_{mi} \\ U_{mj} \end{Bmatrix} = 0$$

$$\begin{bmatrix} 1 & -1 \\ -1 & 1 \end{bmatrix} \begin{Bmatrix} U_{mi} \\ U_{mj} \end{Bmatrix} = 0$$

Thus, $U_{mj} = U_{mi}$ and the eigenvector is $\begin{Bmatrix} U_{mi} \\ U_{mj} \end{Bmatrix} = \begin{Bmatrix} 1 \\ 1 \end{Bmatrix}$, which is plotted as rigid-body mode in Figure 10.4(a).

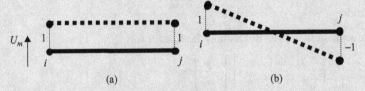

Figure 10.4: Mode shapes for Example 10.1: a) rigid-body mode and b) mode shape 1

To calculate the eigenvector for mode 1, using $\lambda = 2$ gives:

$$\begin{bmatrix} 1-2\lambda & -1-\lambda \\ -1-\lambda & 1-2\lambda \end{bmatrix} \begin{Bmatrix} U_{mi} \\ U_{mj} \end{Bmatrix} = 0 \Rightarrow \begin{bmatrix} 1-4 & -1-2 \\ -1-2 & 1-4 \end{bmatrix} \begin{Bmatrix} U_{mi} \\ U_{mj} \end{Bmatrix} = 0$$

$$\begin{bmatrix} -3 & -3 \\ -3 & -3 \end{bmatrix} \begin{Bmatrix} U_{mi} \\ U_{mj} \end{Bmatrix} = 0$$

Thus, $U_{mj} = -U_{mi}$ and the eigenvector is $\left\{ \begin{array}{c} U_{mi} \\ U_{mj} \end{array} \right\} = \left\{ \begin{array}{c} 1 \\ -1 \end{array} \right\}$, which is plotted in Figure 10.4(b). It should be noted that the actual mode shape 1 was obtained using the analytical solution in Example 9.3 (see Figure E9.9); Figure 10.4(b) shows the approximate mode shape using one element.

Example 10.2 Bar element II

For the bridge support column in Figure 10.5, determine the first natural frequency and calculate the eigenvector. For concrete, use $E = 21$ GPa and $\rho = 2400$ kg/m^3.

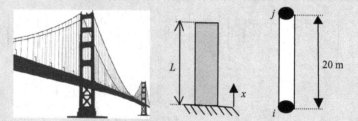

Figure 10.5: Representation of Example 10.2

Solution

Following same steps as in Example 10.1, leads to:

$$\begin{bmatrix} 1 - 2\lambda & -1 - \lambda \\ -1 - \lambda & 1 - 2\lambda \end{bmatrix} \left\{ \begin{array}{c} U_{mi} \\ U_{mj} \end{array} \right\} = 0$$

Applying the boundary conditions, $U_{mi} = 0$, by eliminating the row and column related to U_{mi} yields:

$$\begin{bmatrix} 1 - 2\lambda & -1 - \lambda \\ -1 - \lambda & 1 - 2\lambda \end{bmatrix} \left\{ \begin{array}{c} U_{mi} \\ U_{mj} \end{array} \right\} = 0$$

This reduces to:

$$(1 - 2\lambda)(U_{mj}) = 0$$

The solution for the eigenvalue problem is simple: $|1 - 2\lambda| = 0 \Rightarrow \lambda_1 = 0.5$.

The first angular frequency is:

$$\omega_1 = \sqrt{\frac{6E\lambda_1}{\rho L^2}} = \sqrt{\frac{3E}{\rho L^2}}$$

For $E = 2.1 \times 10^{10}$ N/m^2, $\rho = 2.4 \times 10^3$ kg/m^3 and $L = 20$ m, the fundamental angular frequency becomes:

$$\omega_1 = \sqrt{\frac{3 \times 2.1 \times 10^{10}}{2.4 \times 10^3 \times 20^2}} = 256.174 \, \text{rad/s}$$

and the fundamental natural frequency is calculated as:

$$f_1 = \frac{\omega_1}{2\pi} = 40.77 \, \text{Hz}$$

The eigenvector for mode 1 is simply (from boundary condition $U_{mi} = 0$), $\begin{Bmatrix} U_{mi} \\ U_{mj} \end{Bmatrix} = \begin{Bmatrix} 0 \\ c \end{Bmatrix}$ where c is a scalar. Figure 10.6 shows a plot for mode shape 1. It should be noted that the actual mode shape 1, was obtained using the analytical solution in Example 9.4 (see Figure 9.11); Figure 10.6 shows the approximate mode shape using one element.

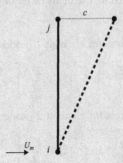

Figure 10.6: Mode shape 1 for Example 10.2

Example 10.3 Bar element III

For the fixed-fixed bar shown in Figure 10.7, use two bar finite elements to determine the first natural frequency. Assume $E = 200$ GPa and $\rho = 7800$ kg/m^3.

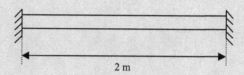

Figure 10.7: Representation of Example 10.3

Solution

The bar is divided into two elements, three nodes, as shown in Figure 10.8. The structure has three degrees of freedom, i.e. one degree of freedom (axial displacement) at each node.

Figure 10.8: Elements of Example 10.3

The mass and stiffness matrices for element 1 (nodes i and j) are:

$$[m] = \frac{\rho AL}{6} \begin{bmatrix} 2 & 1 \\ 1 & 2 \end{bmatrix} \begin{matrix} i \\ j \end{matrix} \quad \text{and} \quad [k] = \frac{EA}{L} \begin{bmatrix} 1 & -1 \\ -1 & 1 \end{bmatrix} \begin{matrix} i \\ j \end{matrix}$$

The mass and stiffness matrices for element 2 (nodes j and k) are:

$$[m] = \frac{\rho AL}{6} \begin{bmatrix} 2 & 1 \\ 1 & 2 \end{bmatrix} \begin{matrix} j \\ k \end{matrix} \quad \text{and} \quad [k] = \frac{EA}{L} \begin{bmatrix} 1 & -1 \\ -1 & 1 \end{bmatrix} \begin{matrix} j \\ k \end{matrix}$$

The global mass matrix is obtained by assembling the mass matrices for elements 1 and 2 as:

$$[m] = \frac{\rho AL}{6} \begin{bmatrix} 2 & 1 & 0 \\ 1 & 2+2 & 1 \\ 0 & 1 & 2 \end{bmatrix} \begin{matrix} i \\ j \\ k \end{matrix} = \frac{\rho AL}{6} \begin{bmatrix} 2 & 1 & 0 \\ 1 & 4 & 1 \\ 0 & 1 & 2 \end{bmatrix} \begin{matrix} i \\ j \\ k \end{matrix}$$

Similarly, the global stiffness matrix is:

$$[k]_i = \frac{EA}{L} \begin{bmatrix} 1 & -1 & 0 \\ -1 & 1+1 & -1 \\ 0 & -1 & 1 \end{bmatrix} \begin{matrix} i \\ j \\ k \end{matrix} = \frac{EA}{L} \begin{bmatrix} 1 & -1 & 0 \\ -1 & 2 & -1 \\ 0 & -1 & 1 \end{bmatrix} \begin{matrix} i \\ j \\ k \end{matrix}$$

The global equations of motion are then obtained as:

$$\left(\frac{EA}{L} \begin{bmatrix} 1 & -1 & 0 \\ -1 & 2 & -1 \\ 0 & -1 & 1 \end{bmatrix} - \omega^2 \frac{\rho AL}{6} \begin{bmatrix} 2 & 1 & 0 \\ 1 & 4 & 1 \\ 0 & 1 & 2 \end{bmatrix} \right) \begin{Bmatrix} U_{mi} \\ U_{mj} \\ U_{mk} \end{Bmatrix} = 0$$

or

$$\left(\begin{bmatrix} 1 & -1 & 0 \\ -1 & 2 & -1 \\ 0 & -1 & 1 \end{bmatrix} - \lambda \begin{bmatrix} 2 & 1 & 0 \\ 1 & 4 & 1 \\ 0 & 1 & 2 \end{bmatrix} \right) \begin{Bmatrix} U_{mi} \\ U_{mj} \\ U_{mk} \end{Bmatrix} = 0$$

where $\lambda = \frac{\rho L^2 \omega^2}{6E}$.

This gives:

$$\begin{bmatrix} 1 - 2\lambda & -1 - \lambda & 0 \\ -1 - \lambda & 2 - 4\lambda & -1 - \lambda \\ 0 & -1 - \lambda & 1 - 2\lambda \end{bmatrix} \begin{Bmatrix} U_{mi} \\ U_{mj} \\ U_{mk} \end{Bmatrix} = 0$$

Applying the boundary conditions, $U_{mi} = U_{mk} = 0$ by eliminating the rows and columns related to U_{mi} and U_{mk} leads to:

$$\begin{bmatrix} 1 - 2\lambda & -1 - \lambda & 0 \\ -1 - \lambda & 2 - 4\lambda & -1 - \lambda \\ 0 & -1 - \lambda & 1 - 2\lambda \end{bmatrix} \begin{Bmatrix} U_{mi} \\ U_{mj} \\ U_{mk} \end{Bmatrix} = 0$$

Thus $(2 - 4\lambda)(U_{mj}) = 0$

The solution for the eigenvalue problem is $|2 - 4\lambda| = 0 \Rightarrow \lambda_1 = 0.5$.

The first angular frequency is:

$$\omega_1 = \sqrt{\frac{6E\lambda_1}{\rho L^2}} = \sqrt{\frac{3E}{\rho L^2}}$$

For $E = 20 \times 10^{10}$ N/m^2, $\rho = 7.8 \times 10^3$ kg/m^3 and $L = 1$ m, the fundamental angular frequency becomes:

$$\omega_1 = \sqrt{\frac{3 \times 20 \times 10^{10}}{7.8 \times 10^3 \times 1^2}} = 8770.58 \, \text{rad/s}$$

and the fundamental natural frequency is calculated as:

$$f_1 = \frac{\omega_1}{2\pi} = 1395.88 \, \text{Hz}$$

The eigenvector for mode 1 (see the plot in Figure 10.9) is simply $\begin{Bmatrix} U_{mi} \\ U_{mj} \\ U_{mk} \end{Bmatrix} = \begin{Bmatrix} 0 \\ c \\ 0 \end{Bmatrix}$, where c is a scalar and $U_{mi} = U_{mk} = 0$ (from the boundary condition).

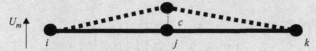

Figure 10.9: Mode shape 1 for Example 10.3

Again, it should be noted that the actual mode shape 1 is a sine function, $\sin \frac{n\pi\pi}{L}$ from Table 9.1, using the analytical solution and it is approximated using two elements as shown in Figure 10.9.

Example 10.4 Bar element IV

If a spring of stiffness K_s is attached to a fixed-free bar as shown in Figure 10.10, write the equation of motion in matrix form using two finite elements to model the bar.

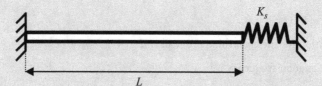

Figure 10.10: Representation of Example 10.4

Solution

The global stiffness matrix is obtained by assembling the two bars (as in Example 10.3) and the spring stiffness contribution to node k:

$$[k] = \frac{EA}{L} \begin{bmatrix} 1 & -1 & 0 \\ -1 & 1+1 & -1 \\ 0 & -1 & 1+\frac{K_s L}{EA} \end{bmatrix} \begin{matrix} i \\ j \\ k \end{matrix} = \frac{EA}{L} \begin{bmatrix} 1 & -1 & 0 \\ -1 & 2 & -1 \\ 0 & -1 & 1+\frac{K_s L}{EA} \end{bmatrix} \begin{matrix} i \\ j \\ k \end{matrix}$$

The global mass matrix is obtained, as before, by assembling the mass matrices for elements 1 and 2 as:

$$[m] = \frac{\rho A L}{6} \begin{bmatrix} 2 & 1 & 0 \\ 1 & 2+2 & 1 \\ 0 & 1 & 2 \end{bmatrix} \begin{matrix} i \\ j \\ k \end{matrix} = \frac{\rho A L}{6} \begin{bmatrix} 2 & 1 & 0 \\ 1 & 4 & 1 \\ 0 & 1 & 2 \end{bmatrix} \begin{matrix} i \\ j \\ k \end{matrix}$$

The global equations of motion are then obtained as:

$$\left(\frac{E A}{L} \begin{bmatrix} 1 & -1 & 0 \\ -1 & 2 & -1 \\ 0 & -1 & 1+\frac{K_s L}{E A} \end{bmatrix} - \omega^2 \frac{\rho A L}{6} \begin{bmatrix} 2 & 1 & 0 \\ 1 & 4 & 1 \\ 0 & 1 & 2 \end{bmatrix} \right) \begin{Bmatrix} U_{mi} \\ U_{mj} \\ U_{mk} \end{Bmatrix} = 0$$

Example 10.5 Bar element V

a) For the two bar finite elements in Figure 10.11, write the equations of motion in matrix form.
b) For free–free vibration, write expressions for the first and second angular frequencies in terms of E, ρ and L.
c) Compare the expressions obtained in b) with the analytical solution (use Table 9.1). Comment on the results.
d) Calculate the eigenvectors.

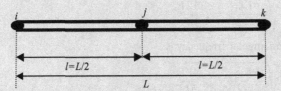

Figure 10.11: Representation of Example 10.5

Solution

a) The global equations of motions are obtained as:

$$\left(\frac{E A}{L} \begin{bmatrix} 1 & -1 & 0 \\ -1 & 2 & -1 \\ 0 & -1 & 1 \end{bmatrix} - \omega^2 \frac{\rho A L}{6} \begin{bmatrix} 2 & 1 & 0 \\ 1 & 4 & 1 \\ 0 & 1 & 2 \end{bmatrix} \right) \begin{Bmatrix} U_{mi} \\ U_{mj} \\ U_{mk} \end{Bmatrix} = 0$$

or

$$\left(\begin{bmatrix} 1 & -1 & 0 \\ -1 & 2 & -1 \\ 0 & -1 & 1 \end{bmatrix} - \lambda \begin{bmatrix} 2 & 1 & 0 \\ 1 & 4 & 1 \\ 0 & 1 & 2 \end{bmatrix} \right) \begin{Bmatrix} U_{mi} \\ U_{mj} \\ U_{mk} \end{Bmatrix} = 0, \text{ where } \lambda = \frac{\rho l^2 \omega^2}{6E}.$$

b) Adding the matrices gives:

$$\begin{bmatrix} 1-2\lambda & -1-\lambda & 0 \\ -1-\lambda & 2-4\lambda & -1-\lambda \\ 0 & -1-\lambda & 1-2\lambda \end{bmatrix} \begin{Bmatrix} U_{mi} \\ U_{mj} \\ U_{mk} \end{Bmatrix} = 0$$

Solving for the determinant yields:

$$\begin{vmatrix} 1-2\lambda & -1-\lambda & 0 \\ -1-\lambda & 2-4\lambda & -1-\lambda \\ 0 & -1-\lambda & 1-2\lambda \end{vmatrix} = 0$$

from which

$$(1-2\lambda)\left[(2-4\lambda)(1-2\lambda)-(1+\lambda)^2\right]-(1+\lambda)(1-2\lambda)(1+\lambda)$$
$$= 0 \Rightarrow -\lambda(1-2\lambda)(12-6\lambda)$$
$$= 0$$

Solving for λ gives:

$$\lambda_o = 0, \quad \lambda_1 = 0.5, \quad \lambda_2 = 2$$

The first angular frequency is:

$$\omega_1 = \sqrt{\frac{6E\lambda_1}{\rho(L/2)^2}} = \sqrt{\frac{12E}{\rho L^2}}$$

The second angular frequency is:

$$\omega_2 = \sqrt{\frac{6E\lambda_2}{\rho(L/2)^2}} = \sqrt{\frac{48E}{\rho L^2}}$$

c) Using Table 9.1, the analytical solution for mode 1 is given by:

$$\omega_1 = \frac{\pi c}{L} = \frac{\pi}{L}\sqrt{\frac{E}{\rho}} = \sqrt{\frac{\pi^2 E}{\rho L^2}}$$

Dividing the finite-element (FE) solution by the analytical solution of mode 1 gives:

$$\frac{\omega_1(FE)}{\omega_1(\text{analytical})} = \sqrt{\frac{12E}{\rho L^2}} \bigg/ \sqrt{\frac{\pi^2 E}{\rho L^2}} = 1.10 \,(\text{a } 10\% \text{ difference})$$

The analytical solution for mode 2 is given by:

$$\omega_2 = \frac{2\pi c}{L} = \frac{2\pi}{L}\sqrt{\frac{E}{\rho}} = \sqrt{\frac{4\pi^2 E}{\rho L^2}}$$

Again, dividing the FE solution by the analytical solution for mode 2 gives:

$$\frac{\omega_2(\text{FE})}{\omega_2(\text{analytical})} = \sqrt{\frac{48 E}{\rho L^2}} \bigg/ \sqrt{\frac{4\pi^2 E}{\rho L^2}} = 1.10 \,(\text{a 10\% difference})$$

The difference is the same for both modes 1 and 2 (see approximated mode shapes in Figures 10.12(b) and 10.12(c)), i.e. both modes have a similar level of approximation because of the free–free boundary conditions (see Figure 9.9 in Example 9.3 for true mode shapes). Calculating the first angular frequency using only one bar element results in the same percentage difference (10%) as can be deduced from Example 10.1.

d) The eigenvector for $\lambda = 0$ is:

$$\begin{bmatrix} 1-2\lambda & -1-\lambda & 0 \\ -1-\lambda & 2-4\lambda & -1-\lambda \\ 0 & -1-\lambda & 1-2\lambda \end{bmatrix} \begin{Bmatrix} U_{mi} \\ U_{mj} \\ U_{mk} \end{Bmatrix} = 0 \Rightarrow \begin{Bmatrix} U_{mi} \\ U_{mj} \\ U_{mk} \end{Bmatrix} = \begin{Bmatrix} 1 \\ 1 \\ 1 \end{Bmatrix},$$

see Figure 10.12(a).

The eigenvector for $\lambda = 0.5$ is:

$$\begin{bmatrix} 1-2\lambda & -1-\lambda & 0 \\ -1-\lambda & 2-4\lambda & -1-\lambda \\ 0 & -1-\lambda & 1-2\lambda \end{bmatrix} \begin{Bmatrix} U_{mi} \\ U_{mj} \\ U_{mk} \end{Bmatrix} = 0 \Rightarrow \begin{Bmatrix} U_{mi} \\ U_{mj} \\ U_{mk} \end{Bmatrix} = \begin{Bmatrix} 1 \\ 0 \\ -1 \end{Bmatrix},$$

see Figure 10.12(b).

The eigenvector for $\lambda = 2$ is:

$$\begin{bmatrix} 1-2\lambda & -1-\lambda & 0 \\ -1-\lambda & 2-4\lambda & -1-\lambda \\ 0 & -1-\lambda & 1-2\lambda \end{bmatrix} \begin{Bmatrix} U_{mi} \\ U_{mj} \\ U_{mk} \end{Bmatrix} = 0 \Rightarrow \begin{Bmatrix} U_{mi} \\ U_{mj} \\ U_{mk} \end{Bmatrix} = \begin{Bmatrix} -1 \\ 1 \\ -1 \end{Bmatrix},$$

see Figure 10.12(c).

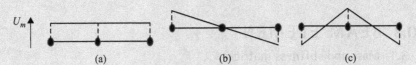

Figure 10.12: Mode shapes for Example 10.5

Example 10.6 Bar element VI

For the two finite elements in Figure 10.13, write the global stiffness and mass matrices for the three degrees of freedom at nodes i, j and k.

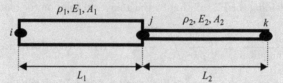

Figure 10.13: Representation of Example 10.6

Solution

a) The mass and stiffness matrices for element 1 (nodes i and j) are:

$$[m] = \frac{\rho_1 A_1 L_1}{6} \begin{bmatrix} 2 & 1 \\ 1 & 2 \end{bmatrix} \begin{matrix} i \\ j \end{matrix} \quad \text{and} \quad [k] = \frac{E_1 A_1}{L_1} \begin{bmatrix} 1 & -1 \\ -1 & 1 \end{bmatrix} \begin{matrix} i \\ j \end{matrix}$$

The mass and stiffness matrices for element 2 (nodes j and k) are:

$$[m] = \frac{\rho_2 A_2 L_2}{6} \begin{bmatrix} 2 & 1 \\ 1 & 2 \end{bmatrix} \begin{matrix} j \\ k \end{matrix} \quad \text{and} \quad [k] = \frac{E_2 A_2}{L_2} \begin{bmatrix} 1 & -1 \\ -1 & 1 \end{bmatrix} \begin{matrix} j \\ k \end{matrix}$$

The global mass matrix is obtained by assembling the mass matrices for elements 1 and 2:

$$[m] = \begin{bmatrix} \frac{\rho_1 A_1 L_1}{3} & \frac{\rho_1 A_1 L_1}{6} & 0 \\ \frac{\rho_1 A_1 L_1}{6} & \frac{\rho_1 A_1 L_1}{3} + \frac{\rho_2 A_2 L_2}{3} & \frac{\rho_2 A_2 L_2}{6} \\ 0 & \frac{\rho_2 A_2 L_2}{6} & \frac{\rho_2 A_2 L_2}{3} \end{bmatrix} \begin{matrix} i \\ j \\ k \end{matrix}$$

Similarly, the global stiffness matrix is:

$$[k] = \begin{bmatrix} \frac{E_1 A_1}{L_1} & -\frac{E_1 A_1}{L_1} & 0 \\ -\frac{E_1 A_1}{L_1} & \frac{E_1 A_1}{L_1} + \frac{E_2 A_2}{L_2} & -\frac{E_2 A_2}{L_2} \\ 0 & -\frac{E_2 A_2}{L_2} & \frac{E_2 A_2}{L_2} \end{bmatrix} \begin{matrix} i \\ j \\ k \end{matrix}$$

10.3 BEAM ELEMENT

10.3.1 Mass and stiffness matrices

Consider the uniform beam element shown in Figure 10.14 with nodes i and j located at the beam ends. Each node has two degrees of freedom: translation in the y direction,

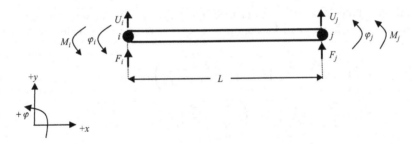

Figure 10.14: A beam element

u_y, and rotation φ. The nodal degrees of freedom at node i are denoted as U_i and φ_i and those at node j as U_j and φ_j. The external force and the bending moment acting at node i are F_i and M_i, respectively, and those acting at node j are F_j and M_j. The element has a Young's modulus E, a uniform cross-sectional area A, an area moment of inertia about the axis perpendicular to the x-y plane I and density ρ.

It is assumed that the transverse displacement variation in the x direction along the beam length is a cubic function in x:

$$u_y = C_1 + C_2 x + C_3 x^2 + C_4 x^3 \tag{10.30}$$

where the constants C_1, C_2, C_3 and C_4 are, in general, functions of time and can be determined from the boundary conditions. The boundary conditions, in general form, are:

$$x = 0 \Rightarrow u_y = U_i \tag{10.31}$$

$$x = 0 \Rightarrow \varphi = \frac{\partial u_y}{\partial x} = \varphi_i \tag{10.32}$$

$$x = L \Rightarrow u_y = U_j \tag{10.33}$$

$$x = L \Rightarrow \varphi = \frac{\partial u_y}{\partial x} = \varphi_j \tag{10.34}$$

Differentiating Equation (10.30) with respect to x and substituting the boundary conditions, Equations (10.31) to (10.34), into Equation (10.30) and its derivative the four constants C_1 to C_4 can be found as:

$$C_1 = U_i \tag{10.35}$$

$$C_2 = \varphi_i \tag{10.36}$$

$$C_3 = \frac{1}{L^2}(-3U_i - 2\varphi_i L + 3U_j - \varphi_j L) \tag{10.37}$$

$$C_4 = \frac{1}{L^3}(2U_i + \varphi_i L - 2U_j + \varphi_j L) \tag{10.38}$$

Substituting Equations (10.35) to (10.38) into Equation (10.30), and rearranging the terms so that each degree of freedom is in a separate term, the following expression is obtained:

$$u_y = \left(1 - \frac{3x^2}{L^2} + \frac{2x^3}{L^3}\right) U_i + \left(x - \frac{2x^2}{L} + \frac{x^3}{L^2}\right) \varphi_i$$

$$+ \left(\frac{3x^2}{L^2} - \frac{2x^3}{L^3}\right) U_j + \left(-\frac{x^2}{L} + \frac{x^3}{L^2}\right) \varphi_j \tag{10.39}$$

or

$$u_y = N_i U_i + N_i' \varphi_i + N_j U_j + N_j' \varphi_j \tag{10.40}$$

where N_i, N_i', N_j, N_j' are the shape functions (see Figure 10.15):

$$N_i = 1 - \frac{3x^2}{L^2} + \frac{2x^3}{L^3},$$

$$N_i' = x - \frac{2x^2}{L} + \frac{x^3}{L^2},$$

$$N_j = \frac{3x^2}{L^2} - \frac{2x^3}{L^3}, \tag{10.41}$$

$$N_j' = -\frac{x^2}{L} + \frac{x^3}{L^2}$$

It should be noted that for the shape function N_i, $N_i = 1$ at node i, $N_i = 0$ at node j and $\frac{dN_i}{dx} = 0$ at nodes i and j. Similarly, for the shape function N_j, $N_j = 1$ at node j, $N_j = 0$ at node i and $\frac{dN_j}{dx} = 0$ at nodes i and j; for the shape function N_i', $\frac{dN_i'}{dx} = 1$ at node i, $\frac{dN_i'}{dx} = 0$ at node j, $N_i' = 0$ at nodes i and j; finally, for the shape function N_j', $\frac{dN_j'}{dx} = 1$ at node j, $\frac{dN_j'}{dx} = 0$ at node i and $N_j' = 0$ at nodes i and j. In Table 10.1, the values of the shape functions and their derivatives with respect to x at nodes i and j are summarized. It can be seen from Table 10.1 and Figure 10.15, that the shape functions can be used to interpolate the

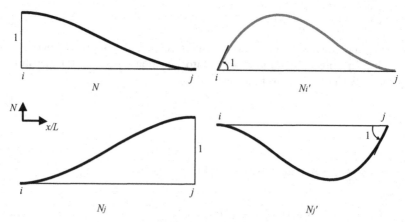

Figure 10.15: Shape functions of a beam element

Table 10.1: Shape functions and their derivatives at nodes i and j

Node	N_i	N_j	N_i'	N_j'	$\frac{dN_i}{dx}$	$\frac{dN_j}{dx}$	$\frac{dN_i'}{dx}$	$\frac{dN_j'}{dx}$
i	1	0	0	0	0	0	1	0
j	0	1	0	0	0	0	0	1

displacement at any point along the beam using only the four degrees of freedom U_i, φ_i, U_j and φ_j.

The kinetic energy, KE, of the beam element can be obtained as:

$$KE = \frac{1}{2} \int_0^L m_l v_y^2 \, dx \tag{10.42}$$

where m_l is the mass per unit length and v_y is the velocity in the y direction. Substituting $m_l = \rho A$ and $v_y = \frac{\partial u_y}{\partial t}$ into Equation (10.42) gives

$$KE = \frac{1}{2} \int_0^L \rho A \left[\frac{\partial u_y}{\partial t} \right]^2 dx \tag{10.43}$$

Differentiating Equation (10.39) with respect to time and substituting into Equation (10.43) yields:

$$KE = \frac{1}{2} \int_0^L \rho A \left[\left(1 - \frac{3x^2}{L^2} + \frac{2x^3}{L^3} \right) \dot{U}_i + \left(x - \frac{2x^2}{L} + \frac{x^3}{L^2} \right) \dot{\varphi}_i \right.$$
$$\left. + \left(\frac{3x^2}{L^2} - \frac{2x^3}{L^3} \right) \dot{U}_j + \left(-\frac{x^2}{L} + \frac{x^3}{L^2} \right) \dot{\varphi}_j \right]^2 dx \tag{10.44}$$

where

$$\dot{U}_i = \frac{\partial U_i}{\partial t}, \quad \dot{\varphi}_i = \frac{\partial \phi_i}{\partial t}, \quad \dot{U}_j = \frac{\partial U_j}{\partial t}, \quad \dot{\varphi}_j = \frac{\partial \phi_j}{\partial t}$$

Integrating Equation (10.44) and writing it in matrix form, gives:

$$KE = \frac{1}{2} \{\dot{U}\}^T [m] \{\dot{U}\} \tag{10.45}$$

where the velocity vector $\{\dot{U}\}$ is given by:

$$\{\dot{U}\} = \begin{Bmatrix} \dot{U}_i \\ \dot{\varphi}_i \\ \dot{U}_j \\ \dot{\varphi}_j \end{Bmatrix}, \quad \{\dot{U}\}^T = \begin{Bmatrix} \dot{U}_i & \dot{\varphi}_i & \dot{U}_j & \dot{\varphi}_j \end{Bmatrix} \tag{10.46}$$

And the mass matrix $[m]$ is given by:

$$[m] = \frac{\rho AL}{420} \begin{bmatrix} 156 & 22L & 54 & -13L \\ 22L & 4L^2 & 13L & -3L^2 \\ 54 & 13L & 156 & -22L \\ -13L & -3L^2 & -22L & 4L^2 \end{bmatrix} \tag{10.47}$$

The strain energy, SE, of the beam element due to bending moment can be expressed as:

$$SE = \frac{1}{2} \int_V \sigma_x \varepsilon_x \, dV \tag{10.48}$$

where σ_x and ε_x are the stress and strain, respectively, in the x direction and dV is the element volume. Using small deflection, Euler–Bernoulli beam theory, the stress σ_x due to bending moment is:

$$\sigma_x = \frac{My}{I} \tag{10.49}$$

where M is the bending moment, y is the co-ordinate of the point at which stress is calculated and I is the area moment of inertia about the axis perpendicular to the x–y plane. For linear elastic material, Hook's law is applicable, i.e. $\sigma_x = E\varepsilon_x$ so that the strain ε_x is given by:

$$\varepsilon_x = \frac{My}{EI} \tag{10.50}$$

Substituting Equations (10.49) and (10.50) into Equation (10.48), the strain energy becomes:

$$SE = \frac{1}{2} \int_V \frac{M^2 y^2}{EI^2} \, dV \tag{10.51}$$

Again, from Euler–Bernoulli beam theory, the relationship between the bending moment M and the curvature of the beam $\frac{\partial^2 u_y}{\partial x^2}$ is:

$$M = EI \frac{\partial^2 u_y}{\partial x^2} \tag{10.52}$$

Substituting Equation (10.52) into Equation (10.51), gives:

$$SE = \frac{1}{2} \int_V E \left(\frac{\partial^2 u_y}{\partial x^2} \right)^2 y^2 \, dV \tag{10.53}$$

Replacing the element volume $\int_V dV$ by $\int_0^L \int_A dA dx$ and using $I = \int_A y^2 \, dA$, Equation (10.53) becomes:

$$SE = \frac{1}{2} \int_0^L EI \left(\frac{\partial^2 u_y}{\partial x^2} \right)^2 dx \tag{10.54}$$

Differentiating Equation (10.39) twice with respect to x, substituting into Equation (10.54), carrying out the integration and writing in matrix form yields:

$$SE = \frac{1}{2} \{U\}^T [k] \{U\} \tag{10.55}$$

where the nodal degrees of freedom vector $\{U\}$ is given by:

$$\{U\} = \begin{Bmatrix} U_i \\ \varphi_i \\ U_j \\ \varphi_j \end{Bmatrix}, \qquad \{U\}^T = \begin{Bmatrix} U_i & \varphi_i & U_j & \varphi_j \end{Bmatrix} \tag{10.56}$$

The stiffness matrix $[k]$ is calculated as:

$$[k] = \frac{EI}{L^3} \begin{bmatrix} 12 & 6L & -12 & 6L \\ 6L & 4L^2 & -6L & 2L^2 \\ -12 & -6L & 12 & -6L \\ 6L & 2L^2 & -6L & 4L^2 \end{bmatrix} \tag{10.57}$$

10.3.2 Equations of motion

The equations of motion for the beam element are obtained in a way similar to that for the bar element in Section 10.2.2, using Newton's second law or Lagrange's equation. Since the beam has four degrees of freedom, the equations of motion become:

$$[m]_{4\times4} \{\ddot{U}\}_{4\times1} + [k]_{4\times4} \{U\}_{4\times1} = \{F\}_{4\times1} \tag{10.58}$$

For a structure containing several beam elements, the equations of motion are:

$$[m]_{n\times n} \{\ddot{U}\}_{n\times1} + [k]_{n\times n} \{U\}_{n\times1} = \{F\}_{n\times1} \tag{10.59}$$

where n is the number of degree of freedom (or twice the number of nodes, as each node has two degrees of freedom) in the structure. The solution of Equation (10.59) for free vibration is the same as that for the bar, Equation (10.27):

$$\left([k]_{n\times n} - \omega^2 [m]_{n\times n}\right) \{U_m\}_{n\times1} = 0 \tag{10.60}$$

This is a problem that has n eigenvalues and eigenvectors.

10.3.3 Boundary conditions

The boundary conditions are applied in a way similar to that explained in Section 10.2.3 for the bar element. Again, they can be considered by modifying the matrices $[m]$ and $[k]$. For example, zero displacement is incorporated by eliminating the corresponding rows and columns from the $[m]$ and $[k]$ matrices.

Example 10.7 Beam element I

The beam shown in Figure 10.16 is simply supported at its ends. The beam has a length of 4 m, Young's modulus of 70 GPa and density of 2800 kg/m^3. Use one beam finite element to determine the first two natural frequencies.

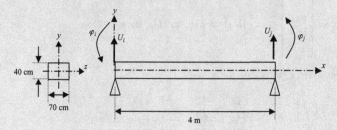

Figure 10.16: Representation of Example 10.7

Solution

From Equation (10.47), the mass matrix $[m]$ is:

$$[m] = \frac{\rho\,AL}{420} \begin{bmatrix} 156 & 22L & 54 & -13L \\ 22L & 4L^2 & 13L & -3L^2 \\ 54 & 13L & 156 & -22L \\ -13L & -3L^2 & -22L & 4L^2 \end{bmatrix}$$

From Equation (10.57), the stiffness matrix $[k]$ is:

$$[k] = \frac{EI}{L^3} \begin{bmatrix} 12 & 6L & -12 & 6L \\ 6L & 4L^2 & -6L & 2L^2 \\ -12 & -6L & 12 & -6L \\ 6L & 2L^2 & -6L & 4L^2 \end{bmatrix}$$

The eigenvalue equation of free vibration $\left([k]_{n\times n} - \omega^2 [m]_{n\times n}\right)\{U_m\}_{n\times 1} = 0$ becomes:

$$\left(\frac{EI}{L^3} \begin{bmatrix} 12 & 6L & -12 & 6L \\ 6L & 4L^2 & -6L & 2L^2 \\ -12 & -6L & 12 & -6L \\ 6L & 2L^2 & -6L & 4L^2 \end{bmatrix} - \omega^2 \frac{\rho\,AL}{420} \begin{bmatrix} 156 & 22L & 54 & -13L \\ 22L & 4L^2 & 13L & -3L^2 \\ 54 & 13L & 156 & -22L \\ -13L & -3L^2 & -22L & 4L^2 \end{bmatrix} \right) \begin{Bmatrix} U_{mi} \\ \varphi_{mi} \\ U_{mj} \\ \varphi_{mj} \end{Bmatrix} = 0$$

or

$$\left(\begin{bmatrix} 12 & 6L & -12 & 6L \\ 6L & 4L^2 & -6L & 2L^2 \\ -12 & -6L & 12 & -6L \\ 6L & 2L^2 & -6L & 4L^2 \end{bmatrix} - \lambda \begin{bmatrix} 156 & 22L & 54 & -13L \\ 22L & 4L^2 & 13L & -3L^2 \\ 54 & 13L & 156 & -22L \\ -13L & -3L^2 & -22L & 4L^2 \end{bmatrix} \right) \begin{Bmatrix} U_{mi} \\ \varphi_{mi} \\ U_{mj} \\ \varphi_{mj} \end{Bmatrix} = 0$$

where $\lambda = \dfrac{\rho A L^4 \omega^2}{420EI}$.

This gives:

$$
\begin{bmatrix}
12 - 156\lambda & 6L - 22L\lambda & -12 - 54\lambda & 6L + 13L\lambda \\
6L - 22L\lambda & 4L^2 - 4L^2\lambda & -6L - 13L\lambda & 2L^2 + 3L^2\lambda \\
-12 - 54\lambda & -6L - 13L\lambda & 12 - 156\lambda & -6L + 22L\lambda \\
6L + 13L\lambda & 2L^2 + 3L^2\lambda & -6L + 22L\lambda & 4L^2 - 4L^2\lambda
\end{bmatrix}
\begin{Bmatrix}
U_{mi} \\ \varphi_{mi} \\ U_{mj} \\ \varphi_{mj}
\end{Bmatrix} = 0
$$

Applying boundary conditions, $U_{mi} = U_{mj} = 0$ by eliminating the rows and columns related to U_{mi} and U_{mj}, leads to:

$$
\begin{bmatrix}
12 - 156\lambda & 6L - 22L\lambda & -12 - 54\lambda & 6L + 13L\lambda \\
6L - 22L\lambda & 4L^2 - 4L^2\lambda & -6L - 13L\lambda & 2L^2 + 3L^2\lambda \\
-12 - 54\lambda & -6L - 13L\lambda & 12 - 156\lambda & -6L + 22L\lambda \\
6L + 13L\lambda & 2L^2 + 3L^2\lambda & -6L + 22L\lambda & 4L^2 - 4L^2\lambda
\end{bmatrix}
\begin{Bmatrix}
U_{mi} \\ \varphi_{mi} \\ U_{mj} \\ \varphi_{mj}
\end{Bmatrix} = 0
$$

The problem is reduced to:

$$
L^2 \begin{bmatrix}
4 - 4\lambda & 2 + 3\lambda \\
2 + 3\lambda & 4 - 4\lambda
\end{bmatrix}
\begin{Bmatrix}
\varphi_{mi} \\ \varphi_{mj}
\end{Bmatrix} = 0
\tag{E10.7}
$$

The solution for the eigenvalue problem is obtained by equating the determinant of the matrix to zero:

$$
\begin{vmatrix}
4 - 4\lambda & 2 + 3\lambda \\
2 + 3\lambda & 4 - 4\lambda
\end{vmatrix} = 0
$$

Solving for λ gives:

$$(4 - 4\lambda)(4 - 4\lambda) - (2 + 3\lambda)(2 + 3\lambda) = 0 \Rightarrow 7\lambda^2 - 44\lambda + 12 = 0 \Rightarrow \lambda_1 = \frac{2}{7}, \lambda_2 = 6$$

The first angular frequency is:

$$
\omega_1 = \sqrt{\frac{420EI\lambda_1}{\rho A L^4}}
$$

For $E = 7 \times 10^{10}$ N/m^2, $\rho = 2.8 \times 10^3$ kg/m^3, $A = 0.7 \times 0.4 = 0.28$ m^2, $I = \frac{0.7 \times 0.4^3}{12} = 3.733 \times 10^{-3}$ m^4 (about z-axis, see Appendix B) and $L = 4$ m, the fundamental angular frequency becomes:

$$
\omega_1 = \sqrt{\frac{420 \times 7 \times 10^{10} \times 3.733 \times 10^{-3} \times \frac{2}{7}}{2.8 \times 10^3 \times 0.28 \times 4^4}} = 395.26 \text{ rad/s}
$$

The fundamental natural frequency is:

$$f_1 = \frac{\omega_1}{2\pi} = 62.9\,\text{Hz}$$

The second angular frequency is calculated as:

$$\omega_2 = \sqrt{\frac{420 E I \lambda_2}{\rho A L^4}} = \sqrt{\frac{420 \times 7 \times 10^{10} \times 3.733 \times 10^{-3} \times 6}{2.8 \times 10^3 \times 0.28 \times 4^4}} = 1811.34\,\text{rad/s}$$

The second natural frequency is:

$$f_2 = \frac{\omega_2}{2\pi} = 288.28\,\text{Hz}$$

The eigenvectors are obtained by substituting the values of λ in Equation (E10.7).

For mode 1, $\lambda_1 = {}^2/_7$:

$$\begin{bmatrix} 4 - 4 \times {}^2/_7 & 2 + 3 \times {}^2/_7 \\ 2 + 3 \times {}^2/_7 & 4 - 4 \times {}^2/_7 \end{bmatrix} \begin{bmatrix} \varphi_{mi} \\ \varphi_{mj} \end{bmatrix} = \begin{bmatrix} 2.857 & 2.857 \\ 2.857 & 2.857 \end{bmatrix} \begin{bmatrix} \varphi_{mi} \\ \varphi_{mj} \end{bmatrix} = 0$$

Thus $\varphi_{mi} = -\varphi_{mj}$ and the eigenvector is:

$$\begin{bmatrix} U_{mi} \\ \varphi_{mi} \\ U_{mj} \\ \varphi_{mj} \end{bmatrix} = \begin{bmatrix} 0 \\ -1 \\ 0 \\ 1 \end{bmatrix}$$

The significance of the eigenvector for mode 1 can be understood from Figure 10.17(a). If the rotation of node j is $\varphi_{mj} = 1$ (anti-clockwise), the rotation of node i is $\varphi_{mi} = -1$ (clockwise).

For mode 2, $\lambda_2 = 6$:

$$\begin{bmatrix} 4 - 4 \times 6 & 2 + 3 \times 6 \\ 2 + 3 \times 6 & 4 - 4 \times 6 \end{bmatrix} \begin{bmatrix} \varphi_{mi} \\ \varphi_{mj} \end{bmatrix} = \begin{bmatrix} -20 & 20 \\ 20 & -20 \end{bmatrix} \begin{bmatrix} \varphi_{mi} \\ \varphi_{mj} \end{bmatrix} = 0$$

from which $\varphi_1 = \varphi_2$ and the eigenvector is:

$$\begin{bmatrix} U_{mi} \\ \varphi_{mi} \\ U_{mj} \\ \varphi_{mj} \end{bmatrix} = \begin{bmatrix} 0 \\ 1 \\ 0 \\ 1 \end{bmatrix}$$

Again, the significance of the eigenvector for mode 2 can be understood from the sign of the rotation at nodes i and j as shown in Figure 10.17(b). If the rotation of node j is $\varphi_{mj} = 1$ (anti-clockwise), the rotation of node i is $\varphi_{mi} = 1$ (anti-clockwise).

Figure 10.17: Eigenvectors for Example 10.7: a) Mode 1 b) Mode 2

Example 10.8 Beam element II

For the helicopter rotor blade in Example 9.7, use one beam finite element to determine the first two natural frequencies. The rotor blade can be idealized as a fixed free beam (Figure 10.18). It has a constant cross section of 1 cm × 20 cm, a length of 0.7 m and is made of aluminium that has a Young's modulus of 70 GPa and a density of 2800 kg/m³.

Figure 10.18: Representation of Example 10.8

Solution

Following the same procedure as in Example 10.7, the eigenvalue equation, $([k]_{n \times n} - \omega^2 [m]_{n \times n}) \{U_m\}_{n \times 1} = 0$ is:

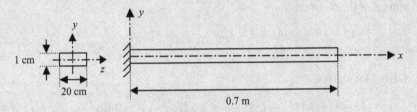

$$\left(\frac{EI}{L^3} \begin{bmatrix} 12 & 6L & -12 & 6L \\ 6L & 4L^2 & -6L & 2L^2 \\ -12 & -6L & 12 & -6L \\ 6L & 2L^2 & -6L & 4L^2 \end{bmatrix} - \omega^2 \frac{\rho AL}{420} \begin{bmatrix} 156 & 22L & 54 & -13L \\ 22L & 4L^2 & 13L & -3L^2 \\ 54 & 13L & 156 & -22L \\ -13L & -3L^2 & -22L & 4L^2 \end{bmatrix} \right) \begin{Bmatrix} U_{mi} \\ \varphi_{mi} \\ U_{mj} \\ \varphi_{mj} \end{Bmatrix} = 0$$

and

$$\begin{bmatrix} 12 - 156\lambda & 6L - 22L\lambda & -12 - 54\lambda & 6L + 13L\lambda \\ 6L - 22L\lambda & 4L^2 - 4L^2\lambda & -6L - 13L\lambda & 2L^2 + 3L^2\lambda \\ -12 - 54\lambda & -6L - 13L\lambda & 12 - 156\lambda & -6L + 22L\lambda \\ 6L + 13L\lambda & 2L^2 + 3L^2\lambda & -6L + 22L\lambda & 4L^2 - 4L^2\lambda \end{bmatrix} \begin{Bmatrix} U_{mi} \\ \varphi_{mi} \\ U_{mj} \\ \varphi_{mj} \end{Bmatrix} = 0$$

where

$$\lambda = \frac{\rho A L^4 \omega^2}{420 E I}$$

Applying boundary conditions, $U_{mi} = \varphi_{mi} = 0$, by eliminating the rows and columns related to U_{mi} and φ_{mi} leads to:

$$\begin{bmatrix} 12 + 156\lambda & 6L - 22L\lambda & -12 - 54\lambda & 6L + 13L\lambda \\ 6L - 22L\lambda & 4L^2 + 4L^2\lambda & -6L - 13L\lambda & 2L^2 + 3L^2\lambda \\ -12 - 54\lambda & -6L + 13L\lambda & 12 - 156\lambda & -6L + 22L\lambda \\ 6L + 13L\lambda & 2L^2 + 3L^2\lambda & -6L + 22L\lambda & 4L^2 - 4L^2\lambda \end{bmatrix} \begin{Bmatrix} U_{mi} \\ \varphi_{mi} \\ U_{mj} \\ \varphi_{mj} \end{Bmatrix} = 0$$

The problem is reduced to:

$$\begin{bmatrix} 12 - 156\lambda & -6L + 22L\lambda \\ -6L + 22L\lambda & 4L^2 - 4L^2\lambda \end{bmatrix} \begin{Bmatrix} U_{mj} \\ \varphi_{mj} \end{Bmatrix} = 0 \tag{E10.8}$$

Again, the solution for the eigenvalue problem is obtained by equating the determinant of the matrix to zero:

$$\begin{vmatrix} 12 - 156\lambda & -6L + 22L\lambda \\ -6L + 22L\lambda & 4L^2 - 4L^2\lambda \end{vmatrix} = 0$$

Solving for λ gives:

$$(12 - 156\lambda)(4L^2 - 4L^2\lambda) - (-6 + 22L\lambda)^2 = 0 \Rightarrow 140\lambda^2 - 408\lambda + 12 = 0$$

from which:

$$\lambda_1 = 0.0297, \lambda_2 = 2.8845$$

For $E = 7 \times 10^{10}$ N/m^2, $\rho = 2.8 \times 10^3$ kg/m^3, the cross-sectional area $A = 0.2 \times 0.01 = 0.002$ m^2, the area moment of inertia about the z axis (see Appendix B) $I = \frac{0.2 \times 0.01^3}{12} = 1.667 \times 10^{-8}$ m^4 and $L = 0.7$ m, the first two angular frequencies are:

$$\omega_1 = \sqrt{\frac{420 E I \lambda_1}{\rho A L^4}} = \sqrt{\frac{420 \times 7 \times 10^{10} \times 1.667 \times 10^{-8} \times 0.0297}{2.8 \times 10^3 \times 0.002 \times 0.7^4}} = 104.047 \,\text{rad/s}$$

$$\omega_2 = \sqrt{\frac{420 E I \lambda_2}{\rho A L^4}} = \sqrt{\frac{420 \times 7 \times 10^{10} \times 1.667 \times 10^{-8} \times 2.8845}{2.8 \times 10^3 \times 0.002 \times 0.7^4}} = 1025.384 \,\text{rad/s}$$

The first two natural frequencies are:

$$f_1 = \frac{\omega_1}{2\pi} = 16.56\,\text{Hz}$$

$$f_2 = \frac{\omega_2}{2\pi} = 163.19\,\text{Hz}$$

The eigenvectors are obtained by substituting the values of λs in Equation (E10.8).

For mode 1, $\lambda_1 = 0.0297$:

$$\begin{bmatrix} 12 - 156 \times 0.0297 & -6L + 22L \times 0.0297 \\ -6L + 22L \times 0.0297 & 4L^2 - 4L^2 \times 0.0297 \end{bmatrix} \begin{Bmatrix} U_{mj} \\ \varphi_{mj} \end{Bmatrix} = 0$$

$$\begin{bmatrix} 7.3668 & -5.3466L \\ -5.3466L & 3.88L^2 \end{bmatrix} \begin{Bmatrix} U_{mj} \\ \varphi_{mj} \end{Bmatrix} = 0$$

from which $U_{mj} = 0.7257L\varphi_{mj}$:

$$\begin{bmatrix} U_{mi} \\ \varphi_{mi} \\ U_{mj} \\ \varphi_{mj} \end{bmatrix} = \begin{bmatrix} 0 \\ 0 \\ 0.726L \\ 1 \end{bmatrix}$$

The significance of the eigenvector for mode 1 can be understood from Figure E10.19(a). If the rotation of node j is $\varphi_{mj} = 1$ (anti-clockwise), the displacement at node j is $U_{mj} = 0.726L$ (upwards).

For mode 2, $\lambda_2 = 2.8845$:

$$\begin{bmatrix} 12 - 156 \times 2.8845 & -6L + 22L \times 2.8845 \\ -6L + 22L \times 2.8845 & 4L^2 - 4L^2 \times 2.8845 \end{bmatrix} \begin{Bmatrix} U_{mj} \\ \varphi_{mj} \end{Bmatrix} = 0$$

$$\begin{bmatrix} -437.982 & 57.459L \\ 57.459L & -7.538L^2 \end{bmatrix} \begin{Bmatrix} U_{mj} \\ \varphi_{mj} \end{Bmatrix} = 0$$

from which $U_{mj} = 0.1312L\varphi_{mj}$ and the eigenvector is:

$$\begin{bmatrix} U_{mi} \\ \varphi_{mi} \\ U_{mj} \\ \varphi_{mj} \end{bmatrix} = \begin{bmatrix} 0 \\ 0 \\ 0.131L \\ 1 \end{bmatrix}$$

Again, the significance of the eigenvector for mode 2 can be understood from the ratio between the displacement and the rotation at node j as shown in Figure 10.19(b). If the rotation of node j $\varphi_{mj} = 1$ (anti-clockwise), the displacement at node j $U_{mj} = 0.131L$ (upwards).

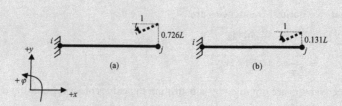

Figure 10.19: Eigenvectors for Example 10.8: a) Mode 1 b) Mode 2

Example 10.9 Beam element III

A beam of a length 4 m has a Young's modulus of 7×10^{10} N/m^2 and density of 2.8×10^3 kg/m^3. For fixed-simply supported ends as shown in Figure 10.20, use one beam finite element to determine the fundamental natural frequency.

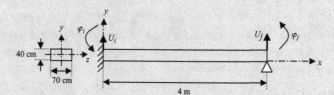

Figure 10.20: Representation of Example 10.9

Solution

Again, following the same procedure as in the Example 10.7, and applying boundary conditions, $U_{mi} = \varphi_{mi} = U_{mi} = 0$ by eliminating the rows and columns related to U_{mi}, U_{mj} and φ_{mi}, leads to:

$$\begin{bmatrix} 12 + 156\lambda & 6L + 22L\lambda & -12 - 54\lambda & 6L + 13L\lambda \\ 6L + 22L\lambda & 4L^2 + 4L^2\lambda & -6L + 13L\lambda & 2L^2 + 3L^2\lambda \\ -12 - 54\lambda & -6L - 13L\lambda & 12 + 156\lambda & -6L + 22L\lambda \\ 6L + 13L\lambda & 2L^2 + 3L^2\lambda & -6L + 22L\lambda & 4L^2 - 4L^2\lambda \end{bmatrix} \begin{Bmatrix} U_{mi} \\ \varphi_{mi} \\ U_{mj} \\ \varphi_{mj} \end{Bmatrix} = 0$$

The problem is reduced to:

$$(4L^2 - 4L^2\lambda)\left[\varphi_{mj}\right] = 0$$

Solving for the eigenvalue, $\lambda_1 = 1$, for $E = 7 \times 10^{10}$ N/m^2, $\rho = 2.8 \times 10^3$ kg/m^3, $A = 0.7 \times 0.4 = 0.28$ m^2, $I = \frac{0.7 \times 0.4^3}{12} = 3.733 \times 10^{-3}$ m^4 (about z-axis, see

Appendix B) and $L = 4$ m, the first natural frequency is:

$$\omega_1 = \sqrt{\frac{420EI\lambda_1}{\rho AL^4}} = \sqrt{\frac{420 \times 7 \times 10^{10} \times 3.733 \times 10^{-3} \times 1}{2.8 \times 10^3 \times 0.28 \times 4^4}} = 739.47 \text{ rad/s}$$

$$\Rightarrow f_1 = \frac{\omega_1}{2\pi} = 117.69 \text{ Hz} \tag{10.63}$$

The eigenvector for mode 1 (from boundary condition $U_{mi} = U_{mj} = \varphi_{mi} = 0$) is:

$$\begin{bmatrix} U_{mi} \\ \varphi_{mi} \\ U_{mj} \\ \varphi_{mj} \end{bmatrix} = \begin{bmatrix} 0 \\ 0 \\ 0 \\ c \end{bmatrix}$$

where c is a scalar. The significance of the mode 1 eigenvector is shown in Figure 10.21.

Figure 10.21: Mode 1 eigenvector for Example 10.9

Example 10.10 Beam element IV

The beam shown in Figure 10.22 is fixed at both ends. If the Young's modulus is 200 GPa and density is 7800 kg/m³, use two beam finite elements to determine the first two natural frequencies.

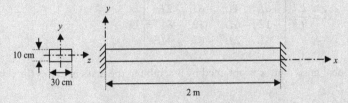

Figure 10.22: Representation of Example 10.10

Solution

The beam is divided into two elements and three nodes as shown in Figure E10.23. The structure has six degrees of freedom, i.e. two degrees of freedom (U and φ) at each node.

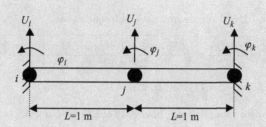

Figure 10.23: Finite elements of Example 10.10

The mass and stiffness matrices for element 1 (nodes i and j) are:

$$[m] = \frac{\rho A L}{420} \begin{bmatrix} 156 & 22L & 54 & -13L \\ 22L & 4L^2 & 13L & -3L^2 \\ 54 & 13L & 156 & -22L \\ -13L & -3L^2 & -22L & 4L^2 \end{bmatrix} \begin{matrix} U_i \\ \varphi_i \\ U_j \\ \varphi_j \end{matrix}$$

and

$$[k] = \frac{E I}{L^3} \begin{bmatrix} 12 & 6L & -12 & 6L \\ 6L & 4L^2 & -6L & 2L^2 \\ -12 & -6L & 12 & -6L \\ 6L & 2L^2 & -6L & 4L^2 \end{bmatrix} \begin{matrix} U_i \\ \varphi_i \\ U_j \\ \varphi_j \end{matrix}$$

And for element 2 (nodes j and k):

$$[m] = \frac{\rho A L}{420} \begin{bmatrix} 156 & 22L & 54 & -13L \\ 22L & 4L^2 & 13L & -3L^2 \\ 54 & 13L & 156 & -22L \\ -13L & -3L^2 & -22L & 4L^2 \end{bmatrix} \begin{matrix} U_j \\ \varphi_j \\ U_k \\ \varphi_k \end{matrix},$$

and

$$[k] = \frac{E I}{L^3} \begin{bmatrix} 12 & 6L & -12 & 6L \\ 6L & 4L^2 & -6L & 2L^2 \\ -12 & -6L & 12 & -6L \\ 6L & 2L^2 & -6L & 4L^2 \end{bmatrix} \begin{matrix} U_j \\ \varphi_j \\ U_k \\ \varphi_k \end{matrix}$$

The global mass matrix is obtained by assembling the mass matrices for elements 1 and 2:

$$[m]_{6\times6} = \frac{\rho A L}{420} \begin{bmatrix} 156 & 22L & 54 & -13L & 0 & 0 \\ 22L & 4L^2 & 13L & -3L^2 & 0 & 0 \\ 54 & 13L & 156+156 & -22L+22L & 54 & -13L \\ -13L & -3L^2 & -22L+22L & 4L^2+4L^2 & 13L & -3L^2 \\ 0 & 0 & 54 & 13L & 156 & -22L \\ 0 & 0 & -13L & -3L^2 & -22L & 4L^2 \end{bmatrix} \begin{matrix} U_{mi} \\ \varphi_{mi} \\ U_{mj} \\ \varphi_{mj} \\ U_{mk} \\ \varphi_{mk} \end{matrix}$$

$$
[m]_{6\times6} = \frac{\rho A L}{420}
\begin{bmatrix}
156 & 22L & 54 & -13L & 0 & 0 \\
22L & 4L^2 & 13L & -3L^2 & 0 & 0 \\
54 & 13L & 312 & 0 & 54 & -13L \\
-13L & -3L^2 & 0 & 8L^2 & 13L & -3L^2 \\
0 & 0 & 54 & 13L & 156 & -22L \\
0 & 0 & -13L & -3L^2 & -22L & 4L^2
\end{bmatrix}
\begin{matrix}
U_{mi} \\ \varphi_{mi} \\ U_{mj} \\ \varphi_{mj} \\ U_{mk} \\ \varphi_{mk}
\end{matrix}
$$

Similarly, the global stiffness matrix is:

$$
[k]_{6\times6} = \frac{EI}{L^3}
\begin{bmatrix}
12 & 6L & -12 & 6L & 0 & 0 \\
6L & 4L^2 & -6L & 2L^2 & 0 & 0 \\
-12 & -6L & 12+12 & -6L+6L & -12 & 6L \\
6L & 2L^2 & -6L+6L & 4L^2+4L^2 & -6L & 2L^2 \\
0 & 0 & -12 & -6L & 12 & -6L \\
0 & 0 & 6L & 2L^2 & -6L & 4L^2
\end{bmatrix}
\begin{matrix}
U_{mi} \\ \varphi_{mi} \\ U_{mj} \\ \varphi_{mj} \\ U_{mk} \\ \varphi_{mk}
\end{matrix}
$$

$$
[k]_{6\times6} = \frac{EI}{L^3}
\begin{bmatrix}
12 & 6L & -12 & 6L & 0 & 0 \\
6L & 4L^2 & -6L & 2L^2 & 0 & 0 \\
-12 & -6L & 24 & 0 & -12 & 6L \\
6L & 2L^2 & 0 & 8L^2 & -6L & 2L^2 \\
0 & 0 & -12 & -6L & 12 & -6L \\
0 & 0 & 6L & 2L^2 & -6L & 4L^2
\end{bmatrix}
\begin{matrix}
U_{mi} \\ \varphi_{mi} \\ U_{mj} \\ \varphi_{mj} \\ U_{mk} \\ \varphi_{mk}
\end{matrix}
$$

The global equations of motion are then obtained as:

$$
\left(
\frac{EI}{L^3}
\begin{bmatrix}
12 & 6L & -12 & 6L & 0 & 0 \\
6L & 4L^2 & -6L & 2L^2 & 0 & 0 \\
-12 & -6L & 24 & 0 & -12 & 6L \\
6L & 2L^2 & 0 & 8L^2 & -6L & 2L^2 \\
0 & 0 & -12 & -6L & 12 & -6L \\
0 & 0 & 6L & 2L^2 & -6L & 4L^2
\end{bmatrix}
\right.
$$

$$
\left.
-\omega^2 \frac{\rho A L}{420}
\begin{bmatrix}
156 & 22L & 54 & -13L & 0 & 0 \\
22L & 4L^2 & 13L & -3L^2 & 0 & 0 \\
54 & 13L & 312 & 0 & 54 & -13L \\
-13L & -3L^2 & 0 & 8L^2 & 13L & -3L^2 \\
0 & 0 & 54 & 13L & 156 & -22L \\
0 & 0 & -13L & -3L^2 & -22L & 4L^2
\end{bmatrix}
\right)
\begin{Bmatrix}
U_{mi} \\ \varphi_{mi} \\ U_{mj} \\ \varphi_{mj} \\ U_{mk} \\ \varphi_{mk}
\end{Bmatrix} = 0
$$

and

$$
\begin{bmatrix}
12-156\lambda & 6L-22L\lambda & -12-54\lambda & 6L+13L\lambda & 0 & 0 \\
6L-22L\lambda & 4L^2-4L^2\lambda & -6L-13L\lambda & 2L^2+3L^2\lambda & 0 & 0 \\
-12-54\lambda & -6L-13L\lambda & 24-312\lambda & 0 & -12-54\lambda & 6L+13L\lambda \\
6L+13L\lambda & 2L^2+3L^2\lambda & 0 & 8L^2-8L^2\lambda & -6L-13L\lambda & 2L^2+3L^2\lambda \\
0 & 0 & -12-54\lambda & -6L-13L\lambda & 12-156\lambda & -6L-22L\lambda \\
0 & 0 & 6L+13L\lambda & 2L^2+3L^2\lambda & -6L-22L\lambda & 4L^2-4L^2\lambda
\end{bmatrix}
\begin{Bmatrix}
U_{mi} \\ \varphi_{mi} \\ U_{mj} \\ \varphi_{mj} \\ U_{mk} \\ \varphi_{mk}
\end{Bmatrix} = 0
$$

where $\lambda = \dfrac{\rho A L^4 \omega^2}{420 E I}$.

Applying boundary conditions, $U_{mi} = \varphi_{mi} = U_{mk} = \varphi_{mk} = 0$, by eliminating the rows and columns related to U_{mi}, $U_{mk}\varphi_{mi}$ and φ_{mk} leads to:

$$\begin{bmatrix} 12 + 156\lambda & 6L + 22L\lambda & -12 - 54\lambda & 6L + 13L\lambda & 0 & 0 \\ 6L + 22L\lambda & 4L^2 + 4L^2\lambda & -6L - 13L\lambda & 2L^2 + 3L^2\lambda & 0 & 0 \\ -12 - 54\lambda & -6L - 13L\lambda & 24 - 312\lambda & 0 & -12 - 54\lambda & 6L + 13L\lambda \\ 6L + 13L\lambda & 2L^2 + 3L^2\lambda & 0 & 8L^2 - 8L^2\lambda & -6L - 13L\lambda & 2L^2 + 3L^2\lambda \\ 0 & 0 & -12 - 54\lambda & -6L - 13L\lambda & 12 + 156\lambda & -6L - 22L\lambda \\ 0 & 0 & 6L + 13L\lambda & 2L^2 + 3L^2\lambda & -6L - 22L\lambda & 4L^2 + 4L^2\lambda \end{bmatrix} \begin{Bmatrix} U_{mi} \\ \varphi_{mi} \\ U_{mj} \\ \varphi_{mj} \\ U_{mk} \\ \varphi_{mk} \end{Bmatrix} = 0$$

The problem reduces to:

$$\begin{bmatrix} 24 - 312\lambda & 0 \\ 0 & 8L^2 - 8L^2\lambda \end{bmatrix} \begin{Bmatrix} U_{mj} \\ \varphi_{mj} \end{Bmatrix} = 0 \tag{E10.10}$$

The eigenvalues are obtained by equating the determinant of the matrix to zero:

$$\begin{vmatrix} 24 - 312\lambda & 0 \\ 0 & 8L^2 - 8L^2\lambda \end{vmatrix} = 0$$

Solving for λ:

$$(24 - 312\lambda)(8L^2 - 8L^2\lambda) = 0 \Rightarrow \lambda_1 = \frac{1}{13}, \lambda_2 = 1$$

For $E = 20 \times 10^{10}$ N/m^2, $\rho = 7.8 \times 10^3$ kg/m^3, $A = 0.3 \times 0.1 = 0.03$ m^2, $I = \dfrac{0.3 \times 0.1^3}{12} = 2.5 \times 10^{-5}$ m^4 (about z-axis, see Appendix B) and $L = 1$ m, the first two angular frequencies are:

$$\omega_1 = \sqrt{\frac{420 E I \lambda_1}{\rho A L^4}} = \sqrt{\frac{420 \times 20 \times 10^{10} \times 2.5 \times 10^{-5} \times \frac{1}{13}}{7.8 \times 10^3 \times 0.03 \times 1^4}} = 830.86 \text{ rad/s}$$

$$\omega_2 = \sqrt{\frac{420 E I \lambda_2}{\rho A L^4}} = \sqrt{\frac{420 \times 20 \times 10^{10} \times 2.5 \times 10^{-5} \times 1}{7.8 \times 10^3 \times 0.03 \times 1^4}} = 2995.72 \text{ rad/s}$$

The first two natural frequencies are:

$$f_1 = \frac{\omega_1}{2\pi} = 132.23 \text{ Hz}$$

$$f_2 = \frac{\omega_2}{2\pi} = 476.78 \text{ Hz}$$

The eigenvectors are obtained from Equation (E10.10) as:

$$\begin{bmatrix} 24 - 312\lambda & 0 \\ 0 & 8L^2 - 8L^2\lambda \end{bmatrix} \begin{Bmatrix} U_{mj} \\ \varphi_{mj} \end{Bmatrix} = 0$$

The first equation can be written as:

$$(24 - 312\lambda)U_{mj} + 0 \times \varphi_{mj} = 0$$

from which $U_{mj} = 0$. For a non-trivial solution $\varphi_{mj} = c$, the eigenvector is:

$$\begin{bmatrix} U_{mi} \\ \varphi_{mi} \\ U_{mj} \\ \varphi_{mj} \\ U_{mk} \\ \varphi_{mk} \end{bmatrix} = \begin{bmatrix} 0 \\ 0 \\ 0 \\ c \\ 0 \\ 0 \end{bmatrix}$$

The second equation can be written as:

$$0 \times U_{mj} + (8L^2 - 8L^2\lambda)\varphi_{mj} = 0$$

from which $\varphi_{mj} = 0$ and, again for a non-trivial solution, $U_{mj} = c$ and the eigenvector is:

$$\begin{bmatrix} U_{mi} \\ \varphi_{mi} \\ U_{mj} \\ \varphi_{mj} \\ U_{mk} \\ \varphi_{mk} \end{bmatrix} = \begin{bmatrix} 0 \\ 0 \\ c \\ 0 \\ 0 \\ 0 \end{bmatrix}$$

The significance of the second eigenvector is shown in Figure E10.24(a). The rotation of node j is $\varphi_{mj} = 0$ and the displacement at node j is $U_{mj} = c$, which corresponds to the first mode of vibration (mode 1).

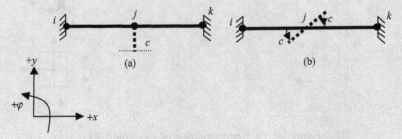

Figure 10.24: Eigenvectors for Example 10.10: a) Mode 1 b) Mode 2

The significance of the first eigenvector is shown in Figure E10.24(b), where the rotation of node j is $\varphi_{mj} = c$ (anti-clockwise) and the displacement at node j $U_{mj} = 0$, which corresponds to the second mode of vibration (mode 2).

Example 10.11 Beam element V

A steel beam is supported by a steel column as shown in Figure 10.25. The steel has a Young's modulus of 200 GPa and a density of 7800 kg/m³. If the beam and the column are modelled using one beam finite element each:

a) What is the overall global number of degrees of freedom (before applying boundary conditions)?
b) Determine the fundamental natural frequency of the system for the constraints shown in Figure 10.25 (the vertical and horizontal displacements are constrained at the corner).

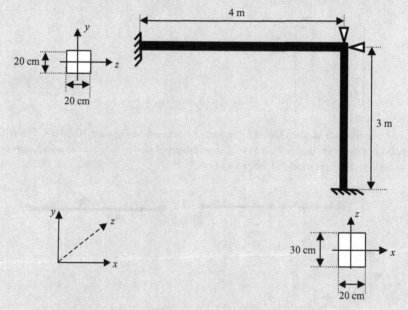

Figure 10.25: Representation of Example 10.11

Solution

a) There are seven overall global degrees of freedom, as shown in Figure 10.26.

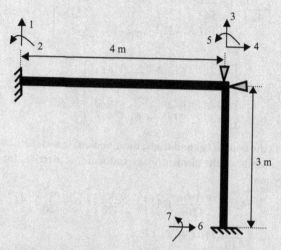

Figure 10.26: Degrees of freedom in Example 10.11

b) The data for the beams are (denote the horizontal beam as beam 1 and the vertical beam as beam 2):

$$E = 20 \times 10^{10} \, \text{N/m}^2;$$
$$\rho = 7.8 \times 10^3 \, \text{kg/m}^3;$$

$A_1 = 0.2 \times 0.2 = 0.04 \ \text{m}^2, \ I_1 = \frac{0.2 \times 0.2^3}{12} = 1.33 \times 10^{-4} \, \text{m}^4$ (about z-axis, see Appendix B) and $L_1 = 4$ m;

$A_2 = 0.3 \times 0.2 = 0.06 \ \text{m}^2, \ I_2 = \frac{0.3 \times 0.2^3}{12} = 2 \times 10^{-4} \, \text{m}^4$ (about z-axis, see Appendix B) and $L_2 = 3$ m.

The mass and stiffness matrices for element 1 are:

$$[m] = \frac{\rho A_1 L_1}{420} \begin{bmatrix} 156 & 22L_1 & 54 & -13L_1 \\ 22L_1 & 4L^2 & 13L_1 & -3L^2 \\ 54 & 13L_1 & 156 & -22L_1 \\ -13L_1 & -3L^2 & -22L_1 & 4L^2 \end{bmatrix} \begin{matrix} 1 \\ 2 \\ 3 \\ 5 \end{matrix}$$

and

$$[k] = \frac{E I_1}{L_1^3} \begin{bmatrix} 12 & 6L_1 & -12 & 6L_1 \\ 6L_1 & 4L^2 & -6L_1 & 2L^2 \\ -12 & -6L_1 & 12 & -6L_1 \\ 6L_1 & 2L^2 & -6L_1 & 4L^2 \end{bmatrix} \begin{matrix} 1 \\ 2 \\ 3 \\ 5 \end{matrix}$$

For element 2, the mass and stiffness matrices are:

$$[m] = \frac{\rho A_2 L_2}{420}
\begin{bmatrix}
156 & 22L_2 & 54 & -13L_2 \\
22L_2 & 4L^2 & 13L_2 & -3L^2 \\
54 & 13L_2 & 156 & -22L_2 \\
-13L_2 & -3L^2 & -22L_2 & 4L^2
\end{bmatrix}
\begin{matrix} 4 \\ 5 \\ 6 \\ 7 \end{matrix}$$

and

$$[k] = \frac{E I_2}{L_2^3}
\begin{bmatrix}
12 & 6L_2 & -12 & 6L_2 \\
6L_2 & 4L^2 & -6L_2 & 2L^2 \\
-12 & -6L_2 & 12 & -6L_2 \\
6L_2 & 2L^2 & -6L_2 & 4L^2
\end{bmatrix}
\begin{matrix} 4 \\ 5 \\ 6 \\ 7 \end{matrix}$$

After applying the boundary conditions, the global mass and stiffness matrices are obtained by assembling the element mass and stiffness matrices for only the 5th degree of freedom:

$$[m]_{1\times1} = \left[4L_1^2 \times \frac{\rho A_1 L_1}{420} + 4L_2^2 \times \frac{\rho A_2 L_2}{420} \right] = \frac{\rho A_1 L_1}{420} \left[4L_1^2 + 4L_2^2 \times \frac{A_2 L_2}{A_1 L_1} \right]$$

$$= \frac{\rho A_1 L_1}{420} [104.5]$$

$$[k]_{1\times1} = \left[4L_1^2 \times \frac{E I_1}{L_1^3} + 4L_2^2 \times \frac{E I_2}{L_2^3} \right] = \frac{E I_1}{L_1^3} \left[4L_1^2 + 4L_2^2 \times \frac{I_2 L_1^3}{I_1 L_2^3} \right]$$

$$= \frac{E I_1}{L_1^3} [192.32]$$

Writing the equation of motion and solving the eigenvalue problem gives:

$$\left[\frac{E I_1}{L_1^3} [192.32] - \omega^2 \frac{\rho A_1 L_1}{420} [104.5] \right] [\varphi_5] = 0$$

Equating the determinant of the matrix to zero yields:

$$|192.32 - 104.5\lambda| = 0$$

Solving for λ gives $\lambda_1 = 1.84$.

The fundamental angular frequency is:

$$\omega_1 = \sqrt{\frac{420 E I_1 \lambda_1}{\rho A_1 L_1^4}}$$

$$\omega_1 = \sqrt{\frac{420 \times 20 \times 10^{10} \times 1.33 \times 10^{-4} \times 1.84}{7.8 \times 10^3 \times 0.04 \times 4^4}} = 507.314 \, \text{rad/s}$$

The fundamental natural frequency is:

$$f_1 = \frac{\omega_1}{2\pi} = 80.74 \, \text{Hz}$$

Example 10.12 Beam element VI

An aluminium beam has a Young's modulus of 70 GPa and a density of 2800 kg/m³. Use one finite element to determine the fundamental natural frequency:

a) for fixed-simply supported ends as shown in Figure 10.27;

b) if the support at the right-hand edge is replaced by a concrete column with Young's modulus of 21 GPa and a density of 2400 kg/m³, as shown in Figure 10.28 (use one beam finite element for the aluminium beam and one bar finite element for the concrete column).

c) Compare the results in cases (a) and (b). Comment on the difference in fundamental frequency.

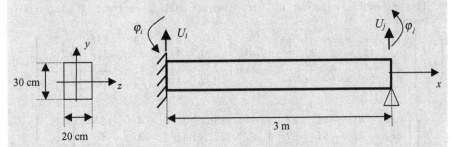

Figure 10.27: Representation of Question 10.12(a)

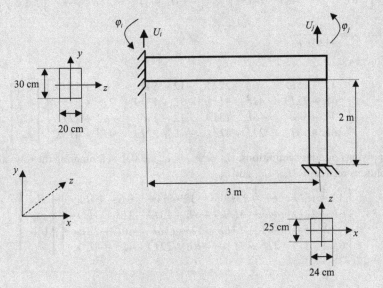

Figure 10.28: Representation of Question 10.12(b)

Solution

a) The mass and stiffness matrices for one element are:

$$[m] = \frac{\rho A L}{420} \begin{bmatrix} 156 & 22L & 54 & -13L \\ 22L & 4L^2 & 13L & -3L^2 \\ 54 & 13L & 156 & -22L \\ -13L & -3L^2 & -22L & 4L^2 \end{bmatrix}$$

and

$$[k] = \frac{EI}{L^3} \begin{bmatrix} 12 & 6L & -12 & 6L \\ 6L & 4L^2 & -6L & 2L^2 \\ -12 & -6L & 12 & -6L \\ 6L & 2L^2 & -6L & 4L^2 \end{bmatrix}$$

The eigenvalue equation of free vibration $([k]_{n \times n} - \omega^2 [m]_{n \times n}) \{U_m\}_{n \times 1} = 0$ becomes:

$$\left(\frac{EI}{L^3} \begin{bmatrix} 12 & 6L & -12 & 6L \\ 6L & 4L^2 & -6L & 2L^2 \\ -12 & -6L & 12 & -6L \\ 6L & 2L^2 & -6L & 4L^2 \end{bmatrix} - \omega^2 \frac{\rho A L}{420} \begin{bmatrix} 156 & 22L & 54 & -13L \\ 22L & 4L^2 & 13L & -3L^2 \\ 54 & 13L & 156 & -22L \\ -13L & -3L^2 & -22L & 4L^2 \end{bmatrix} \right) \begin{Bmatrix} U_{mi} \\ \varphi_{mi} \\ U_{mj} \\ \varphi_{mj} \end{Bmatrix} = 0$$

or

$$\left(\begin{bmatrix} 12 & 6L & -12 & 6L \\ 6L & 4L^2 & -6L & 2L^2 \\ -12 & -6L & 12 & -6L \\ 6L & 2L^2 & -6L & 4L^2 \end{bmatrix} - \lambda \begin{bmatrix} 156 & 22L & 54 & -13L \\ 22L & 4L^2 & 13L & -3L^2 \\ 54 & 13L & 156 & -22L \\ -13L & -3L^2 & -22L & 4L^2 \end{bmatrix} \right) \begin{Bmatrix} U_{mi} \\ \varphi_{mi} \\ U_{mj} \\ \varphi_{mj} \end{Bmatrix} = 0$$

where $\lambda = \dfrac{\rho A L^4 \omega^2}{420 EI}$.

This gives:

$$\begin{bmatrix} 12 - 156\lambda & 6L - 22L\lambda & -12 - 54\lambda & 6L + 13L\lambda \\ 6L - 22L\lambda & 4L^2 - 4L^2\lambda & -6L - 13L\lambda & 2L^2 + 3L^2\lambda \\ -12 - 54\lambda & -6L - 13L\lambda & 12 - 156\lambda & -6L + 22L\lambda \\ 6L + 13L\lambda & 2L^2 + 3L^2\lambda & -6L + 22L\lambda & 4L^2 - 4L^2\lambda \end{bmatrix} \begin{Bmatrix} U_{mi} \\ \varphi_{mi} \\ U_{mj} \\ \varphi_{mj} \end{Bmatrix} = 0$$

Applying boundary conditions, $U_{mi} = \varphi_{mi} = U_{mj} = 0$, by eliminating the rows and columns related to U_{mi}, U_{mj} and φ_{mi}:

$$\begin{bmatrix} 12 - 156\lambda & 6L - 22L\lambda & -12 - 54\lambda & 6L + 13L\lambda \\ 6L + 22L\lambda & 4L^2 - 4L^2\lambda & -6L - 13L\lambda & 2L^2 + 3L^2\lambda \\ -12 - 54\lambda & -6L - 13L\lambda & 12 - 156\lambda & -6L + 22L\lambda \\ 6L + 13L\lambda & 2L^2 + 3L^2\lambda & -6L + 22L\lambda & 4L^2 - 4L^2\lambda \end{bmatrix} \begin{Bmatrix} U_{mi} \\ \varphi_{mi} \\ U_{mj} \\ \varphi_{mj} \end{Bmatrix} = 0$$

The problem is reduced to $(4L^2 - 4L^2\lambda)\left[\varphi_{mj}\right] = 0$.

Solving for the eigenvalue, gives $\lambda_1 = 1$.

For $E = 7 \times 10^{10}$ N/m^2, $\rho = 2.8 \times 10^3$ kg/m^3, $A = 0.3 \times 0.2 = 0.06$ m^2, $I = \frac{0.2 \times 0.3^3}{12} = 4.5 \times 10^{-4}$ m^4 (about z-axis, see Appendix B) and $L = 3$ m, the fundamental angular frequency is:

$$\omega_1 = \sqrt{\frac{420EI\lambda_1}{\rho A L^4}} = \sqrt{\frac{420 \times 7 \times 10^{10} \times 4.5 \times 10^{-4} \times 1}{2.8 \times 10^3 \times 0.06 \times 3^4}} = 986.013 \,\text{rad/s}$$

and the fundamental natural frequency is:

$$f_1 = \frac{\omega_1}{2\pi} = 156.93 \,\text{Hz}$$

b) The contribution of the bar's mass and stiffness are included in the degree of freedom U_{mj} as follows (mass' contribution of $\frac{(\rho_b A_b L_b)}{3}$ and stiffness' contribution of $\frac{E_b A_b}{L_b}$, see Equations (10.14) and (10.22)):

$$\left(\frac{EI}{L^3} \begin{bmatrix} 12 & 6L & -12 & 6L \\ 6L & 4L^2 & -6L & 2L^2 \\ -12 & -6L & 12 + \dfrac{E_b A_b L^3}{L_b EI} & -6L \\ 6L & 2L^2 & -6L & 4L^2 \end{bmatrix} \right.$$
$$\left. -\omega^2 \frac{\rho AL}{420} \begin{bmatrix} 156 & 22L & 54 & -13L \\ 22L & 4L^2 & 13L & -3L^2 \\ 54 & 13L & 156 + \dfrac{140\rho_b A_b L_b}{\rho AL} & -22L \\ -13L & -3L^2 & -22L & 4L^2 \end{bmatrix} \right) \begin{Bmatrix} U_{mi} \\ \varphi_{mi} \\ U_{mj} \\ \varphi_{mj} \end{Bmatrix} = 0$$

where E_b, ρ_b and L_b are the bar's E-modulus, density and length, respectively. Applying boundary conditions, the problem is reduced to:

$$\left(\frac{EI}{L^3} \begin{bmatrix} 12 + \frac{E_b A_b L^3}{L_b EI} & -6L \\ -6L & 4L^2 \end{bmatrix} - \omega^2 \frac{\rho AL}{420} \begin{bmatrix} 156 + \frac{140\rho_b A_b L_b}{\rho AL} & -22L \\ -22L & 4L^2 \end{bmatrix} \right) \begin{bmatrix} U_{mj} \\ \varphi_{mj} \end{bmatrix} = 0$$

Equating the determinant of the matrix to zero yields:

$$\begin{vmatrix} \left(12 + \frac{E_b A_b L^3}{L_b EI}\right) - \lambda \left(156 + \frac{140\rho_b A_b L_b}{\rho AL}\right) & -6L + 22L\lambda \\ -6L + 22L\lambda & 4L^2 - 4L^2\lambda \end{vmatrix} = 0$$

Substituting by the properties of the beam and the column ($E_b = 21 \times 10^9$ N/m^2, $\rho_b = 2.4 \times 10^3$ kg/m^3, $A_b = 0.24 \times 0.25 = 0.06$ m^2, $L_b = 2$ m) leads to:

$$\begin{vmatrix} \left(12 + \dfrac{21 \times 10^9 \times 0.06 \times 3^3}{2 \times 70 \times 10^9 \times 4.5 \times 10^{-4}}\right) - \lambda \left(156 + \dfrac{140 \times 2400 \times 0.06 \times 2}{2800 \times 0.06 \times 3}\right) & -6L + 22L\lambda \\ -6L + 22L\lambda & 4L^2 - 4L^2\lambda \end{vmatrix} = 0$$

This gives:

$$\begin{vmatrix} 552 - 236\lambda & -6L + 22L\lambda \\ -6L + 22L\lambda & 4L^2 - 4L^2\lambda \end{vmatrix} = 0 \Rightarrow (552 - 236\lambda)(4L^2 - 4L^2\lambda) - (-6L + 22L\lambda)^2$$

$$= 0 \Rightarrow \lambda_1 = 0.87365, \lambda_2 = 5.4046$$

Thus, the fundamental angular frequency is:

$$\omega_1 = \sqrt{\frac{420EI\lambda_1}{\rho AL^4}} = \sqrt{\frac{420 \times 7 \times 10^{10} \times 4.5 \times 10^{-4} \times 0.87365}{2.8 \times 10^3 \times 0.06 \times 3^4}} = 921.619 \, \text{rad/s}$$

and the fundamental natural frequency is:

$$f_1 = \frac{\omega_1}{2\pi} = 146.68 \, \text{Hz}$$

c) Comparison between the results in cases (a) and (b):

$$\frac{f_1(b)}{f_1(a)} = \frac{146.68}{156.93} = 0.9347$$

The fundamental frequency in case (b) is about 6.5% lower than that in case (a). The concrete column support in case (b) is more flexible, so the frequencies are expected to drop off.

10.4 GUIDELINES FOR USING ANSYS

Start Ansys

1. From the Start, Programs menu, select Ansys 11.0 (or higher version), Configure Ansys Product, File management.
2. Select your working directory (e.g. C:\ temp).
3. Enter a job name.
4. Select Run.

Create the finite element model

1. To define the element type, from the menu, select Preprocessor, Element Type, Add Edit Delete, Add.
2. Select "Beam 2D elastic 3", then click OK and Close.
3. To define the properties of the area, from the menu, select Preprocessor, Real Constants, Add Edit Delete, Add.

4. "Type 1 Beam 3", then click OK.

5. Specify the cross-sectional area (AREA in m^2), area moments of inertia (IZZ in m^4), total beam height (HEIGHT in m).

6. Click OK and Close.

7. To define the material properties, from the menu, select Preprocessor, Material Props, Material Models.

8. Double-click Structural – Linear – Elastic – Isotropic, specify Young's modulus (EX in N/m^2) and click OK.

9. Double-click Density, specify density (DENS in kg/m^3) and click OK.

10. To create the model, from the menu, select Preprocessor, Modeling Create, Keypoints, On Working Plane.

11. Specify the location of the beam ends in the input window, e.g. for keypoint 1 specify 0,0 and for keypoint 2 specify L,0 (where L is the beam length). Click OK.

12. From the menu, select Preprocessor, Modeling Create, Lines, Lines, Straight Line.

13. Pick keypoints 1 and 2 and click OK.

14. From the menu, select Preprocessor, Meshing, Size Cntrls, ManualSize, Global, Size.

15. Use 10 elements, define the element size as $L/10$ and click OK.

16. From the menu, select Preprocessor, Meshing, Mesh, Lines, Pick all.

17. On the toolbar, click SAVE_DB.

Apply constraints and run the analysis

1. To define the analysis type and option, from the menu, select Solution, Analysis Type, New Analysis.

2. Select Modal and then click OK.

3. From the menu, select Solution, Analysis Type, Analysis Option.

4. Specify 3 for both the No. of modes to extract and the No. of modes to expand. Click OK.

5. A new window is displayed. Set the Start frequency (in Hz) to 0 and the End frequency (in Hz) to 1000. (This will cover at least the first two natural frequencies in Questions 10.7 and 10.8.)

6. To define boundary conditions (not applicable for a free-free beam), from the Utility menu, select Plot, Lines.

7. From the main menu, select Solution, Define Loads, Apply, Structural, Displacement, On Keypoints.

8. Select the keypoint at one end of the beam and click OK.

9. Select All DOF (for fixed end) and UY+UX (for simply supported end).

10. Select Apply as constant values.

11. Set value to 0 and Expand disp to nodes? to No. Click OK.

12. Repeat steps 9, 10 and 11 for the other keypoint.

13. On the toolbar, click SAVE_DB.

14. From the menu, select Solution, Solve, Current LS.

15. Check and close the window that opens.
16. Click OK and wait for successful completion.

Review the results

1. From the menu, select General Postproc, Results Summary.
2. Write down the natural frequencies. Close the summary window.
3. From the menu, select General Postproc, Read Results, First Set.
4. From the menu, select General Postproc, Plot Results, Deformed Shape.
5. Select Def+Undeformed and click OK.
6. From the Utility menu, select PlotCtrls, Animate, Mode Shapes and click OK.
7. To view mode shape 2, from the menu, select General Postproc, Read Results, Next Set and repeat steps 4 to 6.
8. To make a hard copy, from the Utility menu, select PlotCtrls, Hard Copy, To File and specify a file name.

Quit ANSYS

1. On the ANSYS toolbar, click QUIT and then OK.

10.5 Tutorial Sheet

10.5.1 Bar elements

Q10.1   A uniform bar of length 1 m is fixed at one end and free at the other. The bar is made of cast iron, which has a Young's modulus of 180 GPa and a density of 7000 kg/m^3. Use one bar finite element to determine the first natural frequency of the bar for axial vibration.

[1397.9 Hz]

Q10.2   Solve Question 10.1 using two finite elements and determine the first two natural frequencies, then calculate the eigenvectors and hence sketch the approximate mode shapes.

$$\left[1300.5\,\text{Hz}, 4543.1\text{Hz}, \begin{bmatrix} 0.707 \\ 1 \end{bmatrix}, \begin{bmatrix} -0.707 \\ 1 \end{bmatrix} \right]$$

Q10.3   Compare the fundamental natural frequencies obtained from Question 10.2 and Question 10.1 to the analytical solution (use Table 9.1). Comment on the results.

Q10.4 If a spring of stiffness $K_s = 10^9$ N/m is attached to a fixed-free bar as shown in Figure 10.29, determine (use for the bar $E = 70$ GPa, $\rho = 2800$ kg/m^3 and $A = 0.03$ m^2):

a) the first two natural frequencies using two finite elements to model the bar;

[600.73 Hz, 1655.4 Hz]

b) the eigenvectors.

$$\left[\begin{bmatrix} 1.06 \\ 1 \end{bmatrix}, \begin{bmatrix} -0.584 \\ 1 \end{bmatrix} \right]$$

c) Sketch the mode shapes.

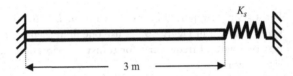

Figure 10.29: Representation of Question 10.4

Q10.5 For the free-free bar shown in Figure 10.30, $E = 200$ GPa and $\rho = 7800$ kg/m^3, $L = 2.5$ m, use two finite elements to calculate the natural frequencies of longitudinal vibration and their corresponding eigenvectors.

$$\left[0, 1117 \text{ Hz}, 2233 \text{ Hz}, \begin{Bmatrix} 1 \\ 1 \\ 1 \end{Bmatrix}, \begin{Bmatrix} 1 \\ 0 \\ -1 \end{Bmatrix}, \begin{Bmatrix} -1 \\ 1 \\ -1 \end{Bmatrix} \right]$$

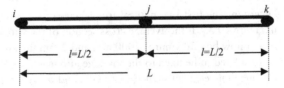

Figure 10.30: Representation of Question 10.5

Q10.6   For the fixed-fixed two bar finite elements model in Figure 10.31, determine the fundamental frequency. Use:

$$\rho_1 = 2400 \, \text{kg/m}^3, \, E_1 = 20 \, \text{GPa}, \, A_1 = 0.25 \, \text{m}^2, \, L_1 = 1 \, \text{m}$$
$$\rho_2 = 2770 \, \text{kg/m}^3, \, E_2 = 70 \, \text{GPa}, \, A_2 = 0.125 \, \text{m}^2, \, L_2 = 2 \, \text{m}$$

[742.4 Hz]

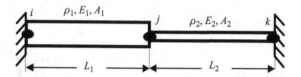

Figure 10.31: Representation of Question 10.6

10.5.2 Beam elements

Q10.7   For a simply supported beam ($L = 1.5$ m, $E = 69$ GPa and $\rho = 2700$ kg/m^3), shown in Figure 10.32, use one finite element to determine the first two natural frequencies for transverse vibration and calculate the eigenvectors.

$$\left[101.77 \, \text{Hz}, \, 466.37 \, \text{Hz}, \begin{bmatrix} -1 \\ 1 \end{bmatrix}, \begin{bmatrix} 1 \\ 1 \end{bmatrix} \right]$$

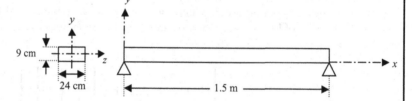

Figure 10.32: Representation of Question 10.7

Q10.8   Determine the first two natural frequencies of the aircraft wing shown in Figure 10.33. If the average cross section of the wing can be idealized to a hollow box 300 mm × 2000 mm and 2 mm thickness, and the wing is considered to be fixed to the fuselage at one end and free at the other, use one finite element with $E = 70$ GPa and $\rho = 2770$ kg/m^3.

[4.02 Hz, 39.66 Hz]

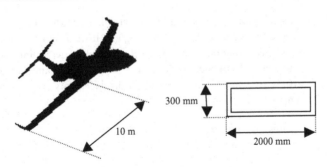

300 mm

10 m

2000 mm

Figure 10.33: Representation of Question 10.8

Q10.9 A steel beam is supported by two springs of stiffness $K_s = 1$ kN/m and $2K_s$, as shown in Figure 10.34, has a Young's modulus of 200 GPa and a density of 7800 kg/m³. If the rotations at both ends are constrained (i.e. $\varphi_i = \varphi_j = 0$), use one beam finite element to determine:

a) the equation of motion in matrix form for the two vertical displacements (U_{mi} and U_{mj});

$$\left[\left(\frac{EI}{L^3} \begin{bmatrix} 12 + \frac{K_s L^3}{EI} & -12 \\ -12 & 12 + \frac{2K_s L^3}{EI} \end{bmatrix} \right) - \omega^2 \frac{\rho A L}{420} \begin{bmatrix} 156 & 54 \\ 54 & 156 \end{bmatrix} \right) \right] \begin{Bmatrix} U_{mi} \\ U_{mj} \end{Bmatrix} = 0$$

b) the first two natural frequencies of the system;

[2.78 Hz, 15 Hz]

c) the first two eigenvectors.

$$\left[\begin{Bmatrix} 1 \\ 0.952 \end{Bmatrix} \begin{Bmatrix} 1 \\ -1.024 \end{Bmatrix} \right]$$

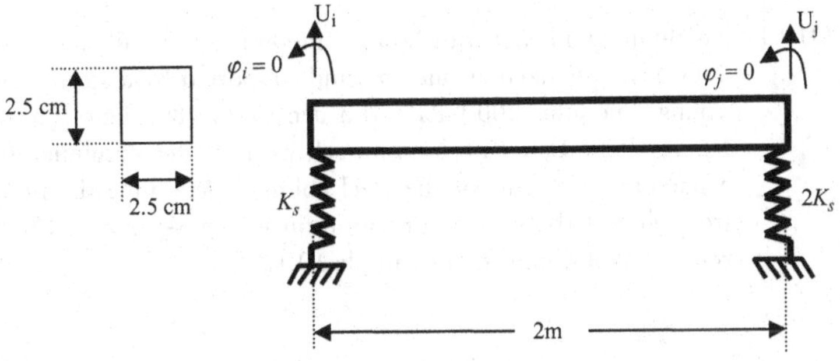

U_i

$\varphi_i = 0$

U_j

$\varphi_j = 0$

2.5 cm

2.5 cm

K_s

$2K_s$

2m

Figure 10.34: Representation of Question 10.9

Q10.10 A steel beam is supported by a steel column with a Young's modulus of 200 GPa and a density of 7800 kg/m³. If the beam and the column are modelled using one beam finite element each, determine the fundamental natural frequency of the system for the constraints shown in Figure 10.35 (the vertical and horizontal displacements are constrained at the corner).

[105.8 Hz]

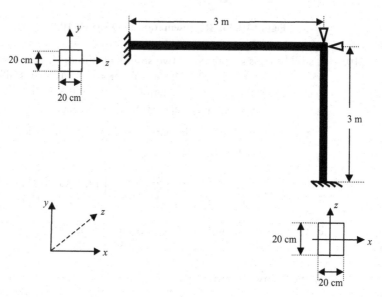

Figure 10.35: Representation of Question 10.10

Q10.11 An aluminium beam with Young's modulus of 70 GPa and a density of 2800 kg/m³ is fixed at one end and supported by a steel column with Young's modulus 200 GPa and a density of 7800 kg/m³, as shown in Figure 10.36. Using a beam finite element for the aluminium beam and a bar finite element for the steel column, determine the fundamental frequency of the system for vibration in the x-y plane. Compare the result to that obtained in Example 10.12.

[155.9 Hz]

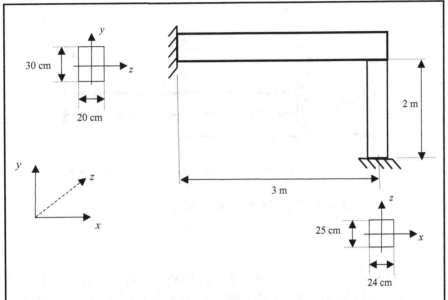

Figure 10.36: Representation of Question 10.11

Q10.12 A fixed-fixed steel beam, shown in Figure 10.37, has a Young's modulus of 200 GPa and a density of 7800 kg/m³. The width of the beam is one half its depth, b. An electric motor of mass $M = 100$ kg and speed 1500 rev/min is fixed in the middle of the beam. Use two beam finite elements to determine:

a) the equations of motion in matrix form for the two degrees of freedom in the middle of the beam, k and l (vertical displacement and rotation);

$$\left[\left(\frac{EI}{L^3} \begin{bmatrix} 24 & 0 \\ 0 & 8L^2 \end{bmatrix} - \omega^2 \frac{\rho AL}{420} \begin{bmatrix} 312 + 100 \times \frac{420}{\rho AL} & 0 \\ 0 & 8L^2 \end{bmatrix} \right) \begin{Bmatrix} U_{mj} \\ \varphi_{mj} \end{Bmatrix} = 0 \right]$$

b) the first two natural frequencies if $b = 25$ cm;

[265.1 Hz, 1191.7 Hz]

c) the minimum beam cross-sectional dimensions so that the fundamental natural frequency of the system is 20% higher than the operating speed of the motor.

[6.68 cm × 3.34 cm]

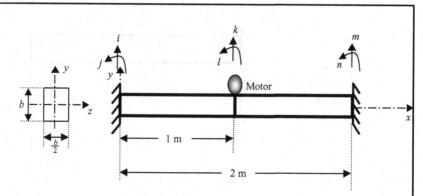

Figure 10.37: Representation of Question 10.12

Q10.13 Solve Questions 10.7 and 10.8 using ANSYS or any other finite-element package. Produce a short report containing the results obtained, an analytical solution and a finite-element hand calculation. Use the following guidelines for the contents of the report and include a list of sections, sub-sections and appendices with page numbers.

1. *Introduction*
   - Introduce the problems you are trying to solve.
   - Outline the objectives of the work.
   - Refer to textbooks where theoretical background can be found.
   - Summarize the content of the report.
2. *Analytical solution*
   - Explain briefly the theoretical background of the analytical solution (Chapter 9).
   - Discuss the advantages and disadvantages of the analytical solution.
   - Calculate the analytical solution for Questions 10.7 and 10.8.
3. *Hand calculation*
   - Outline briefly the finite-element method.
   - Discuss the advantages and disadvantages of the finite-element method.
   - Solve the problems using one finite element.
4. *Finite-element package results*
   - Describe the steps you followed in the finite-element package to model the beams (see Section 10.4).
   - Present the results obtained using the finite-element package.

5. *Comparisons*
   - Compare the results obtained using the analytical solution, the hand calculation and the finite-element package.
   - Comment on the accuracy of the results.
6. *Conclusion*
   - Summarize the work you have done.
   - Include the main points you want to highlight.
7. References
8. Appendices (if any)

APPENDIX A

DAMA and Guidelines for Simulations

A.1 INTRODUCTION

This appendix introduces and explains simulations of the mechanisms and vibrating systems presented in this book. The simulations have been implemented in the LabVIEW software. LabVIEW is a graphical programming language based on connecting icons using wires. The icons represent the operations to be performed, while the wires define the links and data flow between operations. LabVIEW is used in a variety of applications, such as virtual measurement instruments, oscilloscopes, data acquisition instruments, data logging and controlling instruments.

The simulations have been designed in such a way that their layout is consistent, clear and interactive. Almost the same layout template has been used in developing each simulation. There are four boxes in the front panel:

- a red box into which all the input parameters are entered;
- a green box in which all the output numerical values are listed;
- a graphical box in which the animation of the mechanism or vibrating system and the time response are illustrated;
- a white box in which a graphical drawing of the mechanism or vibrating system is sketched and the required input parameters are defined.

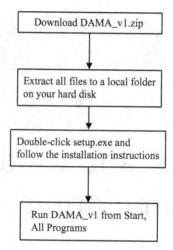

Figure A.1: Installing DAMA on a PC

The simulations are standalone applications that require a LabVIEW runtime engine that is freely available from National Instruments. The software DAMA and the required runtime engine can be downloaded from the book's online resource website (http://eu.wiley.com/WileyCDA/WileyTitle/productCd-0470723009.html). The DAMA installation steps on a PC are shown in Figure A.1.

The DAMA software groups the simulations under five main titles: kinematics of mechanisms, kinetics of linkages, balancing of machines, discrete systems and continuous systems (see Figure A.2). Theoretical background to each group of simulations was explained in the relevant chapter of this book, i.e. kinematics of mechanisms in Chapter 2, kinetics of linkages in Chapter 4, balancing of machines in Chapter 5, discrete systems in Chapters 6, 7 and 8, and continuous systems in Chapter 9.

For all simulations, the input parameters are entered in an interactive way. The effects of changing any input parameter are immediately reflected in the output values and the animation. When possible, the animations are performed in real time so that the student may appreciate the motion of mechanisms and the behaviour of vibrating systems. Descriptions of all DAMA simulations are presented in the following subsections. After the description of each simulation, multiple-choice questions assess the student's understanding of the mechanism or vibrating system. Students can run the simulations to seek the answers to the questions, which will help them to have a deep understanding of the problem.

It should be noted that the scale on the x and y axes of the animation box can be changed by clicking the scale, entering the new scale and pressing Enter. The automatic scale should

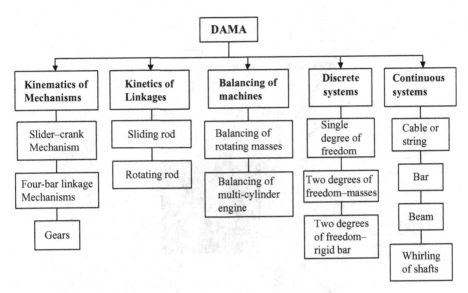

Figure A.2: DAMA configuration

be turned off (by clicking the right mouse button), in order to be able to edit and activate the scale. This will be of great help in adjusting the animation view.

A.2 KINEMATICS OF MECHANISMS

A.2.1 Slider–crank mechanism

Figure A.3 shows the front panel of the slider–crank mechanism simulation, based on Section 2.4.

The input parameters are in the red box:

- the crank angular velocity in revolutions per minute;
- the crank radius OB in metres;
- the connecting-rod length AB in metres;
- the offset distance between the axis of rotation, point O, and the sliding plane in metres;
- the crank angular positions (θ) in degrees at which velocities and accelerations are required.

The velocity and acceleration components for the slider A and a point D on the connecting rod are calculated for all five input angular positions. The position of the intermediate point D can be defined as the ratio between AD and AB (the red box between the two output tables in the green box in Figure A.3). The table on the left of the green box

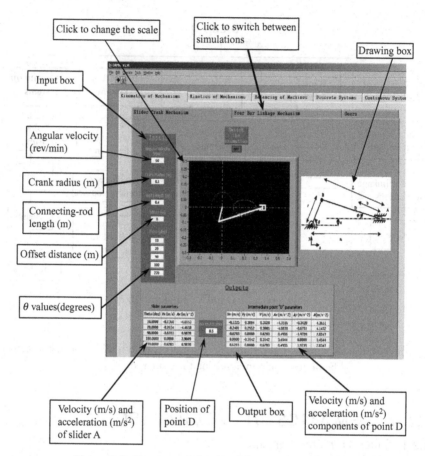

Figure A.3: Front panel for the slider–crank mechanism

contains the velocity and acceleration of slider A and the table on the right contains the velocity and acceleration components in the x and y directions and the magnitudes of the intermediate point D for the five chosen θ values.

The changes in the input parameters are immediately reflected in the animation as well as in the output tables. In the animation box, the dotted circle and the oval are used to guide the user to see the positions of points B and D as a function of time. In the white drawing box, a summary of the input parameters is shown in graphical form.

This simulation assists students in understanding the effect of the crank radius, connecting-rod length and the offset distance on the velocity and acceleration of the slider A and a point D on the connecting rod.

The student can change the input parameter values by selecting a parameter and either typing a new value or using the Up and Down keys. The Enter key confirms that the simulation should use the current value.

Q2.1.1   If the ratio of the crank radius to the connecting-rod length is increased, does the absolute value of the velocity of the slider:
☐ a. remain constant   ☐ b. increase   ☐ c. decrease

Q2.1.2   If the ratio of the crank radius to the connecting-rod length is increased, does the absolute value of the acceleration of the slider:
☐ a. increase   ☐ b. decrease   ☐ c. remain constant

Q2.1.3   If the offset distance for crank angular position equal to zero is increased, does the velocity of the slider:
☐ a. remain constant   ☐ b. increase   ☐ c. decrease

Q2.1.4   If the offset distance for crank angular position equal to 90° is increased, does the acceleration of the slider:
☐ a. increase   ☐ b. decrease   ☐ c. remain constant

Q2.1.5   If the offset distance for crank angular position equal to 180° is increased, does the velocity of the slider:
☐ a. remain constant   ☐ b. increase   ☐ c. decrease

Q2.1.6   If the offset distance for crank angular position equal to 270° is increased, does the acceleration of the slider:
☐ a. increase   ☐ b. decrease   ☐ c. remain constant

Q2.1.7   The absolute maximum velocity of the slider for a zero offset distance and ratio of the crank radius to the connecting-rod length of 0.25 takes place around a crank angular position equal to:
☐ a. 77   ☐ b. 90   ☐ c. 55   ☐ d. 44

Q2.1.8   The absolute maximum acceleration of the slider for a zero offset distance takes place at a crank angular position equal to:
☐ a. 0   ☐ b. 90   ☐ c. 77

Q2.1.9   If the ratio of the crank radius to the connecting-rod length is increased, does the crank angular position in the first quadrant at which maximum velocity of the slider takes place:
☐ a. remain constant   ☐ b. increase   ☐ c. decrease

Q2.1.10   If the ratio of the crank radius to the connecting-rod length is increased, does the crank angular position in the first quadrant at which maximum acceleration of the slider takes place:
☐ a. remain constant   ☐ b. increase   ☐ c. decrease

A.2.2 Four-bar linkage mechanism

Figure A.4 shows the front panel of the four-bar linkage mechanism simulation, based on Section 2.4.

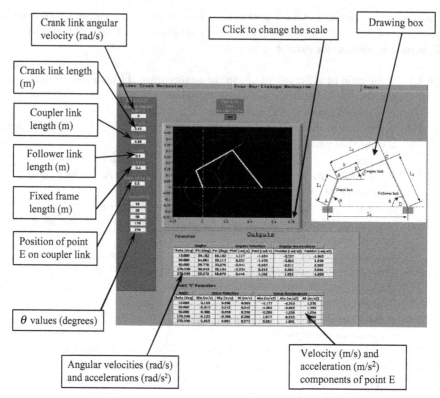

Figure A.4: Front panel for the four-bar linkage mechanism

The input parameters are in the red box:

- the crank-link angular velocity in radian per second;
- the crank-link length AB in metres;
- the coupler-link length BC in metres;
- the follower-link length CD in metres;
- the fixed-frame length AD in metres;
- the crank angular positions (θ) in degrees at which velocities and accelerations are required.

The position of the intermediate point E on the coupler link can be defined as the ratio between BE and BC. The output tables contain the angular position, velocity and acceleration of the coupler link and the follower link, as well as the velocity and acceleration components in the x and y directions and their magnitudes for the intermediate point E for the five chosen θ values.

The changes in the input parameters are immediately reflected in the animation as well as in the output tables. In the animation box, the circle and the arc are used to guide the user to see the positions of points B and C as a function of time. In the white drawing box, a summary of the input parameters is shown in graphical form.

This simulation assists students in understanding the effect of the length of the various links on the angular velocities and accelerations of the coupler and follower links, and on the velocity and acceleration of a point E on the coupler link.

The student can change the input parameter values by selecting a parameter and either typing a new value or using the Up and Down keys. The Enter key confirms that the simulation should use the current value.

Q2.2.1 If the crank-link length is increased, does the absolute value of the angular velocity of the coupler link:
☐ a. remain constant   ☐ b. increase   ☐ c. decrease

Q2.2.2 If the crank-link length is increased, does the absolute value of the angular velocity of the follower link:
☐ a. increase   ☐ b. decrease   ☐ c. remain constant

Q2.2.3 For crank-link angular position equal to zero, if the fixed frame length is increased, does the absolute value of the angular velocity of the coupler and follower links:
☐ a. remain constant   ☐ b. increase   ☐ c. decrease

Q2.2.4 For crank-link angular position equal to 90°, if the fixed-frame length is increased, does the absolute value of the angular velocity of the coupler and follower links:
☐ a. increase   ☐ b. decrease   ☐ c. remain constant

Q2.2.5 For crank-link angular position equal to zero, if the coupler-link length is increased, does the velocity magnitude of the coupler link midpoint:
☐ a. remain constant   ☐ b. increase   ☐ c. decrease

Q2.2.6 For crank-link angular position equal to zero, if the follower-link length is increased, does the magnitude of the angular velocity of the coupler link:
☐ a. increase   ☐ b. decrease   ☐ c. remain constant

Q2.2.7 If the crank-link length is increased for a crank-link angular position in the first quadrant, does the absolute value of the angular acceleration of the follower link:
☐ a. increase   ☐ b. decrease   ☐ c. remain constant

Q2.2.8 For crank-link angular position equal to zero, if the crank link rotates clockwise, the coupler link rotates:
☐ a. Anti-clockwise   ☐ b. Clockwise

Q2.2.9 For crank-link angular position equal to 90°, if the crank link rotates clockwise, the coupler link rotates:
☐ a. Anti-clockwise   ☐ b. Clockwise

Q2.2.10 For crank-link angular position equal to zero, if the crank link rotates clockwise, the follower link rotates

☐ a. Anti-clockwise   ☐ b. Clockwise

Q2.2.11 For crank-link angular position equal to 90°, if the crank link rotates clockwise, the follower link rotates

☐ a. Anti-clockwise   ☐ b. Clockwise

A.2.3 Gears

Figures A.5 and A.6 show the front panels of the gear-train simulations, based on Section 2.3.2.

For the simple gear train, the input parameters in the red box are:

- the radius of the driver gear (gear 1) in millimetres;
- the angular velocity of the driver gear in revolutions per minute;
- the gear ratios for gears 2 and 3 (the driven gears).

The output parameters in the green box are the radii in millimetres and angular velocities in revolutions per minute of gears 2 and 3.

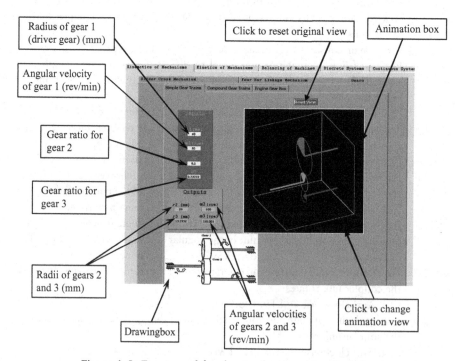

Figure A.5: Front panel for the simple gear-train simulation

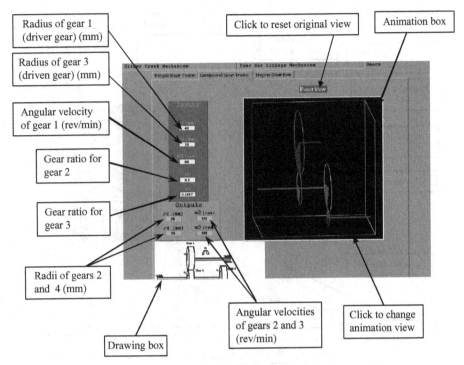

Figure A.6: Front panel for the compound gear-train simulation

For the compound gear train, the simulation also requires the radius of the driven gear (gear 3) as an input parameter. The output parameters for the compound gear train are the radii in millimetres of gears 2 and 4 and the angular velocities in revolutions per minute of gears 2 and 3.

Figure A.7 shows the front panel for the engine gearbox simulation.

Its input parameters are:

- the radius of the driver gear (gear 6) in millimetres;
- the radius of gear 1 in millimetres;
- the angular velocity of the engine in revolutions per minute;
- the gear ratios for gears 1, 2, 3, and 4.

In order to engage gears in the animation box, the "engaged gear" bar can be dragged to the required gear position. The output parameters in the green box are the remaining gears' (i.e. gears 2, 3, 4, 1', 2', 3', 4' and 6') radii in millimetres and the angular velocities, in revolutions per minute, of the wheel shaft and the layshaft.

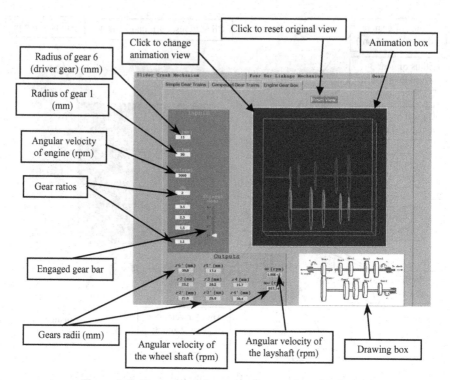

Figure A.7: Front panel for an engine-gearbox simulation

The gear simulations assist students in understanding how gear systems work and the effect of the sizes and ratios of gears on the resulting angular velocity of the driven gears.

Q2.3.1   In simple gear trains, if the radius of the driver gear is increased, does the radius of the driven gear:
□ a. remain constant   □ b. increase   □ c. decrease

Q2.3.2   In simple gear trains, if gear ratio 2 is increased, does the angular velocity of gear 2:
□ a. remain constant   □ b. increase   □ c. decrease

Q2.3.3   In simple gear trains, if gear ratio 2 is increased, does the angular velocity of gear 3:
□ a. remain constant   □ b. increase   □ c. decrease

Q2.3.4   In simple gear trains, if gear ratio 3 is increased, does the angular velocity of gear 2:
□ a. remain constant   □ b. increase   □ c. decrease

Q2.3.5 In simple gear trains, if gear ratio 3 is increased, does the angular velocity of gear 3:

☐ a. remain constant   ☐ b. increase   ☐ c. decrease

Q2.3.6 In compound gear trains, if the radius of the driver gear (gear 1) is increased, does the radius of gear 2:

☐ a. remain constant   ☐ b. increase   ☐ c. decrease

Q2.3.7 In compound gear trains, if the radius of the driver gear (gear 1) is increased, does the radius of gear 4:

☐ a. remain constant   ☐ b. increase   ☐ c. decrease

Q2.3.8 In compound gear trains, if the radius of the driven gear (gear 3) is increased, does the radius of gear 2:

☐ a. remain constant   ☐ b. increase   ☐ c. decrease

Q2.3.9 In compound gear trains, if the radius of the driven gear (gear 3), is increased, does the radius of gear 4:

☐ a. remain constant   ☐ b. increase   ☐ c. decrease

Q2.3.10 In compound gear trains, if gear ratio 2 is increased, does the angular velocity of gear 3:

☐ a. remain constant   ☐ b. increase   ☐ c. decrease

Q2.3.11 In compound gear trains, if gear ratio 2 is increased, does the angular velocity of gear 2:

☐ a. remain constant   ☐ b. increase   ☐ c. decrease

Q2.3.12 In compound gear trains, if gear ratio 3 is increased, does the angular velocity of gear 2:

☐ a. remain constant   ☐ b. increase   ☐ c. decrease

Q2.3.13 In compound gear trains, if gear ratio 3 is increased, does the angular velocity of gear 3:

☐ a. remain constant   ☐ b. increase   ☐ c. decrease

Q2.3.14 In the engine gearbox simulation, if the radius of the driver gear (gear 6) is changed, which of the following parameters is affected?

☐ a. r_6' ☐ b. r_1' ☐ c. All gears' radii

Q2.3.15 In the engine gearbox simulation, if the gear ratio of the driver gear (gear 6) is changed, which of the following parameters is affected?

☐ a. r_6' ☐ b. r_1' ☐ c. All gears' radii

Q2.3.16 In the engine gearbox simulation, if the gear ratio of the driver gear (gear 6) is increased, does the angular velocity of the wheel shaft:

☐ a. remain constant   ☐ b. increase   ☐ c. decrease

Q2.3.17 In the engine gearbox simulation, if the gear ratio of the driver gear (gear 6) is increased, does the angular velocity of the layshaft:

☐ a. remain constant   ☐ b. increase   ☐ c. decrease

Q2.3.18 In the engine gearbox simulation, if the radius of gear 1 is increased, does the radius of gear 1':

☐ a. remain constant   ☐ b. increase   ☐ c. decrease

Q2.3.19 In the engine gearbox simulation, if the gear ratio of gear 2 is increased, does the angular velocity of the layshaft:

☐ a. remain constant   ☐ b. increase   ☐ c. decrease

Q2.3.20 In the engine gearbox simulation, for gear ratio G_6 equal to 2, the angular velocity of layshaft is:

☐ a. twice   ☐ b. half   ☐ c. equal to

the angular velocity of the engine?

A.3 KINETICS OF LINKAGES

A.3.1 Sliding rod

The main aim of the sliding-rod and rotating-rod simulations is to assist students in understanding the application of Newton's second law. Figure A.8 shows the front panel of the sliding-rod simulation, based on Section 4.2.5.

The input parameters are in the red box:

- the rod length AC in metres;
- the distance between the upper sliding plane and the free end of the rod AB in metres;

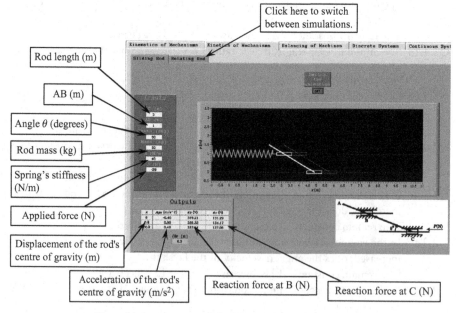

Figure A.8: Front panel for the sliding-rod simulation

- the angle that the rod makes with the horizontal axis in degrees;
- the mass of the rod in kilograms;
- the spring's stiffness in newtons per metre;
- the horizontal force applied at point C on the lower sliding plane in newtons.

The output table contains the acceleration of the centre of gravity of the rod in m/s^2 and the two vertical reactions at B and C in newtons. In the white drawing box, a summary of the input parameters is shown in graphical form.

This simulation helps students to understand the effect of the rod's mass and the applied force on the acceleration of the rod and to understand the effect of the rod length (AC), the position of the upper support B (AB) and the angle that the rod makes with the horizontal axis on the reaction forces at supports B and C.

Q3.1.1 If the rod length AC is increased, does the absolute value of the acceleration of the rod's centre of gravity:
☐ a. increase   ☐ b. remain constant   ☐ c. decrease

Q3.1.2 If the angle that the rod makes with the horizontal plane is increased, does the acceleration of the rod's centre of gravity:
☐ a. increase   ☐ b. decrease   ☐ c. remain constant

Q3.1.3 If the rod mass is increased, does the absolute value of the acceleration of the rod's centre of gravity:
☐ a. increase   ☐ b. remain constant   ☐ c. decrease

Q3.1.4 If the angle that the rod makes with the horizontal plane in the first quadrant ($0° \leq \theta \leq 90°$) is increased, does the reaction force at B:
☐ a. increase   ☐ b. remain constant   ☐ c. decrease

Q3.1.5 If the angle that the rod makes with the horizontal plane in the first quadrant ($0° \leq \theta \leq 90°$) is increased, does the reaction force at C:
☐ a. remain constant   ☐ b. decrease   ☐ c. increase

Q3.1.6 If the rod length AC (AB constant) is increased, does the reaction force at B:
☐ a. increase   ☐ b. remain constant   ☐ c. decrease

Q3.1.7 If the rod length AC (AB constant) is increased, does the reaction force at C:
☐ a. increase   ☐ b. remain constant   ☐ c. decrease

Q3.1.8 If the spring stiffness k is increased, does the acceleration of the rod's centre of gravity at the start of the motion:
☐ a. increase   ☐ b. remain constant   ☐ c. decrease

Q3.1.9 If the spring stiffness k is increased, does the reaction force at C at the start of the motion:
☐ a. increase   ☐ b. remain constant   ☐ c. decrease

Q3.1.10 If the spring stiffness k is increased, does the reaction force at B at the start of the motion:
☐ a. increase   ☐ b. remain constant   ☐ c. decrease

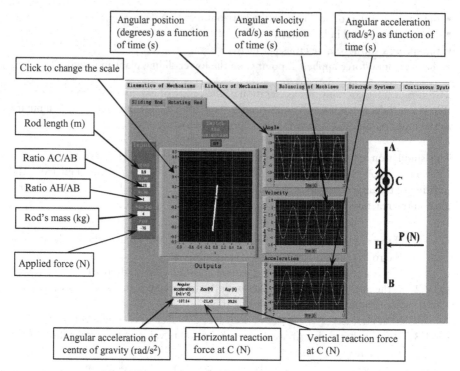

Figure A.9: Front panel for rotating-rod simulation

A.3.2   Rotating rod

For a rotating rod, e.g. a rigid-bar pendulum, the front panel of the simulation is shown in Figure A.9. The rotating-rod simulation is based on Section 4.2.5.

The motion of the rod starts when the rod is in a vertical position as in Example 4.6. The input parameters that appear in the red box are:

- the rod length AB in metres;
- the ratio between the distance from end A and the point about which the rod rotates C (AC) and the rod length AB;
- the ratio between the distance from end A and the point H where the force is applied (AH) and the rod length AB;
- the mass of the rod in kilograms;
- the horizontal force in newtons.

The output table contains the angular acceleration of the centre of gravity of the rod in rad/s^2, the horizontal reaction force at C and the vertical reaction force at C in newtons.

Three additional animation boxes plot the angular position in degrees, the angular velocity in radians per second and the angular acceleration in rad/s^2 as functions of time (in seconds).

This simulation helps students to understand the effect of the rod's mass, the magnitude of the applied load and its position, the rod's length and the position of the point about which the rod rotates on the angular acceleration of the rod and the components of the reaction force at point C.

Q3.2.1   If the rod length AB is increased (when the force is applied at B), does the absolute value of the angular acceleration:
    ☐ a. increase   ☐ b. remain constant   ☐ c. decrease

Q3.2.2   If the rod length AB is increased (when the force is applied at B), does the absolute value of the horizontal reaction force at C:
    ☐ a. decrease   ☐ b. increase   ☐ c. remain constant

Q3.2.3   If the rod length AB is increased (when the force is applied at B), does the absolute value of the vertical reaction force at C:
    ☐ a. increase   ☐ b. remain constant   ☐ c. decrease

Q3.2.4   If the ratio AC/AB (up to 0.4) is increased (when the force is applied at B), does the absolute value of the angular acceleration:
    ☐ a. decrease   ☐ b. remain constant   ☐ c. increase

Q3.2.5   If the ratio AH/AB (with AC/AH < 1) is increased, does the absolute value of the angular acceleration:
    ☐ a. increase   ☐ b. remain constant   ☐ c. decrease

Q3.2.6   If the rod mass is increased, does the absolute value of the angular acceleration:
    ☐ a. decrease   ☐ b. increase   ☐ c. remain constant

Q3.2.7   If the rod mass is increased, does the absolute value of the vertical reaction force at C:
    ☐ a. decrease   ☐ b. increase   ☐ c. remain constant

Q3.2.8   If the rod mass is increased, does the absolute value of the horizontal reaction force at C:
    ☐ a. decrease   ☐ b. increase   ☐ c. remain constant

Q3.2.9   If the force magnitude is increased, does the absolute value of the angular acceleration:
    ☐ a. remain constant   ☐ b. increase   ☐ c. decrease

Q3.2.10 If the force magnitude is increased, does the absolute value of the vertical reaction force at C:
    ☐ a. remain constant   ☐ b. increase   ☐ c. decrease

Q3.2.11 If the force magnitude is increased, does the absolute value of the horizontal reaction force at C:
    ☐ a. remain constant   ☐ b. increase   ☐ c. decrease

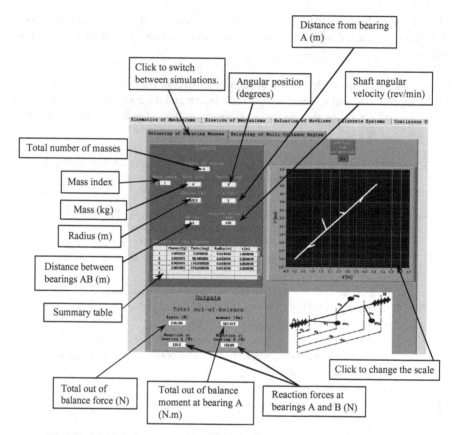

Figure A.10: Front panel for simulation of balancing of rotating masses

A.4 BALANCING OF MACHINES

A.4.1 Balancing of rotating masses

Figure A.10 shows the front panel for the simulation of balancing of rotating masses, which is based on Section 5.2.

The input parameters that appear in the red box are:

- the total number of masses;
- the mass in kilograms;
- the angular position in degrees;
- the radius in metres;
- the distance from bearing A to each plane in metres;
- the distance from bearing A to bearing B in metres;
- the angular velocity of the shaft in revolutions per minute.

A summary of the input parameters is given in a table at the bottom of the red box.

The output table contains the total out-of-balance force in newtons, the total out-of-balance moment at bearing A in newton metres and the reaction forces at bearings A and B in newtons.

The aim of this simulation is to assist students in understanding how to balance the inertia forces generated by the rotation of the masses around the shaft.

Q4.1.1   If the radial positions of the masses are increased, does the total out-of-balance force:
☐ a. remain constant   ☐ b. increase   ☐ c. decrease

Q4.1.2   If the radial positions of the masses are increased, does the total out-of-balance moment:
☐ a. remain constant   ☐ b. increase   ☐ c. decrease

Q4.1.3   If the location of the mass along the shaft (z coordinate) is changed, which of the following output parameters is affected?
☐ a. the out-of-balance force   ☐ b. the out-of-balance moment   ☐ c. both

Q4.1.4   If the angular velocity of the shaft is changed, which of the following output parameters is affected?
☐ a. the out-of-balance force   ☐ b. the out-of-balance moment   ☐ c. both

Q4.1.5   Static balance is achieved when:
☐ a. the total out-of-balance force is zero   ☐ b. the total out-of-balance moment is zero   ☐ c. both are zero

Q4.1.6   Dynamic balance is achieved when:
☐ a. the total out-of-balance force is zero   ☐ b. the total out-of-balance moment is zero   ☐ c. both are zero

Q4.1.7   If static balance is achieved, what do you change to achieve dynamic balance?
☐ a. radial positions   ☐ b. masses   ☐ c. locations of masses along the shaft

Q4.1.8   If the angular velocity of the shaft is doubled, the total out-of-balance force and moment increases by a factor of:
☐ a. 2   ☐ b. 4   ☐ c. 1.414

Q4.1.9   If the distance between the two bearings is increased, does the total out-of-balance force and moment:
☐ a. remain constant   ☐ b. increase   ☐ c. decrease

Q4.1.10   If the distance between the two bearings is increased, does the reaction force at bearing B:
☐ a. remain constant   ☐ b. increase   ☐ c. decrease

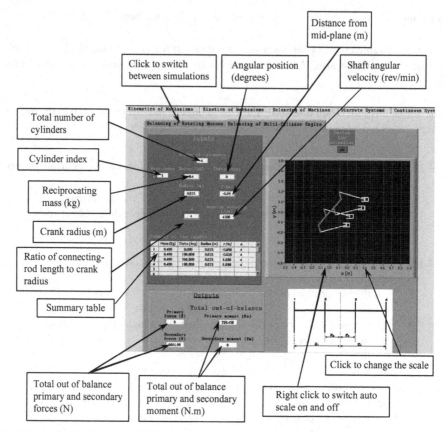

Figure A.11: Front panel for the simulation of balancing of multi-cylinder engines

A.4.2 Balancing of multi–cylinder engines

For balancing of multi-cylinder engines (Section 5.3), the front panel of the simulation is shown in Figure A.11.

The input parameters that appear in the red box are:

- the total number of cylinders;
- the reciprocating mass in kilograms;
- the crank angular position in degrees;
- the crank radius in metres;
- the distance from the mid-plane in metres;
- the ratio between connecting-rod length and the crank radius for each cylinder;
- the angular velocity of the shaft in revolutions per minute.

A summary of the input parameters is given in a table at the bottom of the red box.

The output table contains the total out-of-balance primary and secondary forces in newtons and the total out-of-balance primary and secondary moments at the mid-plane in newton metres.

The aim of this simulation is to assist students in understanding how to minimize the out-of-balance forces and moments of a multi-cylinder engine.

Q4.2.1   Which of the following parameters do you change in order to achieve primary force balancing?
☐ a. cylinder's distance along shaft   ☐ b. ratio of the crank radius to the connecting-rod length   ☐ c. crank radius

Q4.2.2   If the ratio of the crank radius to the connecting rod increases for all cylinders, does the out-of-balance secondary force:
☐ a. decrease   ☐ b. increase   ☐ c. remain constant

Q4.2.3   If the angular velocity of the shaft is doubled, do the forces and moments increase by a factor of:
☐ a. 2   ☐ b. 4   ☐ c. 1.414

Q4.2.4   If the radial positions of the masses are increased, do the primary and secondary forces:
☐ a. remain constant   ☐ b. increase   ☐ c. decrease

Q4.2.5   If the cylinder's location along the shaft (z coordinate) is changed, which of the following output parameters is affected?
☐ a. the primary moment   ☐ b. the secondary moment   ☐ c. both

Q4.2.6   If the ratio of the crank radius to the connected rod is changed, which of the following output parameters is affected?
☐ a. the primary force   ☐ b. the secondary force   ☐ c. both

Q4.2.7   If the ratio of the crank radius to the connected rod is doubled for all cylinders, does the secondary moment:
☐ a. double   ☐ b. halve   ☐ c. remain constant

Q4.2.8   If the ratio of the crank radius to the connected rod is doubled for all cylinders, does the primary moment:
☐ a. double   ☐ b. halve   ☐ c. remain constant

Q4.2.9   A complete balance is achieved for:
☐ a. a three-cylinder engine   ☐ b. a four-cylinder engine   ☐ c. a six-cylinder engine

Q4.2.10 The total out-of-balance force of a multi-cylinder engine is:
☐ a. the primary force   ☐ b. the secondary force   ☐ c. the vector summation of both

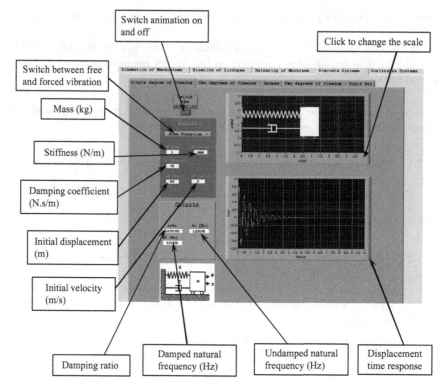

Figure A.12: Front panel for the simulation of the free vibration of a SDOF system

A.5 DISCRETE SYSTEMS

A.5.1 Vibration of systems with a single degree of freedom

Figure A.12 shows the front panel of the simulation of free vibration in a system with a single degree of freedom, based on Sections 6.2, 6.3, 7.2 and 7.3.

The input parameters that appear in the red box are:

- the mass in kilograms;
- the stiffness in newtons per metre;
- the damping coefficient in newton seconds per metre;
- the initial displacement in metres;
- the initial velocity in metres per second.

The output table contains the undamped and damped natural frequencies of the system and the damping ratio. The animation box shows the real-time vibration of the system. An additional animation box plots the displacement as a function of time.

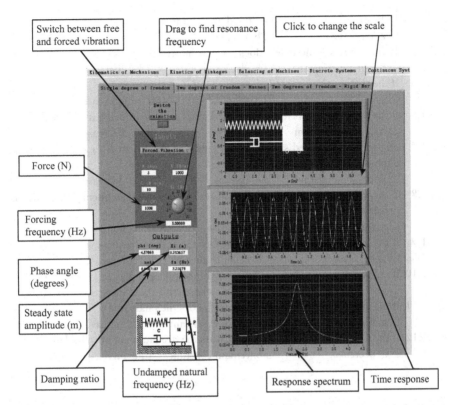

Figure A.13: Front panel for the simulation of the forced vibration of a SDOF system

In the case of forced vibration, as shown in Figure A.13, the simulation requires additional input parameters: the force in newtons and the forcing frequency in Hz.

The output parameters in the green box are the maximum amplitude in metres and the phase angle in degrees. A response spectrum graph shows the displacement as a function of the forcing frequency.

The simulation assists students in understanding the effect of the mass, the stiffness, the damping coefficient, the initial conditions and the external forces on the natural frequency and displacement time response of a system with a single degree of freedom.

Q5.1.1   If the mass is increased, does the undamped natural frequency:
    ☐ a. increase   ☐ b. remain constant   ☐ c. decrease
Q5.1.2   For an underdamped system, if the mass is increased, does the damped natural frequency:
    ☐ a. increase   ☐ b. decrease   ☐ c. remain constant

Q5.1.3   If the mass is increased, does the damping ratio:

☐ a. increase   ☐ b. decrease   ☐ c. remain constant

Q5.1.4   If the stiffness is increased, does the natural frequency:

☐ a. remain constant   ☐ b. decrease   ☐ c. increase

Q5.1.5   If the stiffness is increased, does the damping ratio:

☐ a. remain constant   ☐ b. increase   ☐ c. decrease

Q5.1.6   If the damping coefficient is increased, does the undamped natural frequency:

☐ a. remain constant   ☐ b. increase   ☐ c. decrease

Q5.1.7   For an underdamped system, if the damping coefficient is increased, does the damped natural frequency:

☐ a. remain constant   ☐ b. increase   ☐ c. decrease

Q5.1.8   If the forcing frequency is getting closer to the natural frequency, does the displacement amplitude:

☐ a. remain constant   ☐ b. decrease   ☐ c. increase

Q5.1.9   If the force magnitude is increased, does the natural frequency:

☐ a. decrease   ☐ b. remain constant   ☐ c. increase

Q5.1.10 If the force magnitude is increased, does the phase angle:

☐ a. decrease   ☐ b. remain constant   ☐ c. increase

Q5.1.11 If the excitation frequency is increased, does the phase angle:

☐ a. decrease   ☐ b. remain constant   ☐ c. increase

Q5.1.12 If the force magnitude is increased, does the displacement amplitude:

☐ a. decrease   ☐ b. remain constant   ☐ c. increase

A.5.2 Vibration of masses with two degrees of freedom

Figure A.14 shows the front panel of the simulation of free vibration in a system with two masses and two springs, based on Sections 8.3 and 8.5.

The input parameters that appear in the red box are:

- the masses in kilograms and the stiffness of the springs in newtons per metre; the output table contains the natural frequencies of the system and the displacement amplitude ratios for each mode of vibration; the mode of vibration, phase mode or anti-phase mode, can be selected for animation through the drag tab above the animation box; an additional animation box plots the two displacements as functions of time.

In the case of forced vibration, as shown in Figure A.15, additional input parameters are required: the forces at each degree of freedom and the forcing frequency.

The output parameters are the suppression vibration frequency and the steady state amplitude for each degree of freedom. A response spectrum graph shows both displacements as functions of the forcing frequency. The animation box displays the real vibration of the system under the applied forces and so the drag tab icon is no longer required.

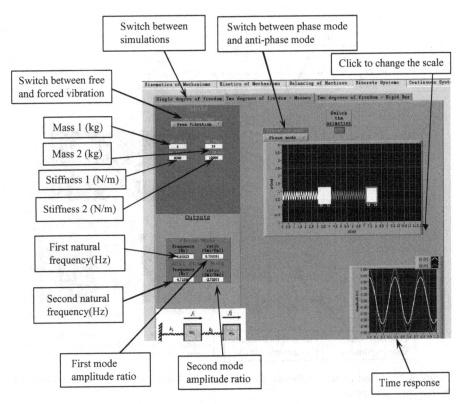

Figure A.14: Front panel for the simulation of the free vibration of a system with two masses and two springs

The simulation assists students in understanding the effect of the masses, the stiffness and the external forces on the natural frequencies of the system, their corresponding amplitude ratios, the suppression vibration frequencies and the steady-state amplitudes.

Q5.2.1   If mass 1 is increased, does phase mode frequency:
  ☐ a. remain constant   ☐ b. decrease   ☐ c. increase

Q5.2.2   If mass 2 is increased, does the absolute value of the amplitude ratio for the anti-phase mode:
  ☐ a. increase   ☐ b. remain constant   ☐ c. decrease

Q5.2.3   If stiffness 1 is increased, does phase mode frequency:
  ☐ a. increase   ☐ b. remain constant   ☐ c. decrease

Q5.2.4   If stiffness 2 is increased, does the absolute value of the amplitude ratio for the anti-phase mode:
  ☐ a. remain constant   ☐ b. decrease   ☐ c. increase

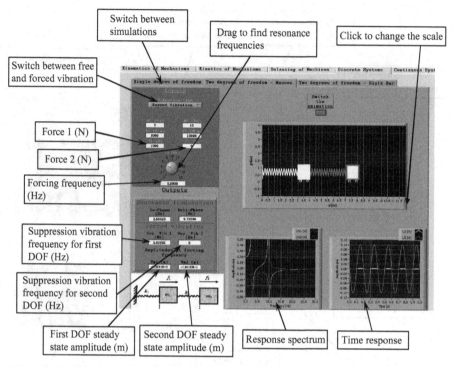

Figure A.15: Front panel for the simulation of the forced vibration of a system with two masses and two springs

Q5.2.5   If the forcing frequency is getting closer to a natural frequency, do the steady state amplitudes

☐ a. remain constant   ☐ b. decrease   ☐ c. increase

Q5.2.6   For Force 1 equal to zero, if Force 2 is increased, does the first suppression vibration frequency:

☐ a. decrease   ☐ b. remain constant   ☐ c. increase

Q5.2.7   For Force 1 equal to zero, if Force 2 is increased, does the second suppression vibration frequency:

☐ a. decrease   ☐ b. remain constant   ☐ c. increase

Q5.2.8   For Force 2 equal to zero, if mass 2 is increased, does the first suppression vibration frequency:

☐ a. decrease   ☐ b. remain constant   ☐ c. increase

Q5.2.9   For Force 2 equal to zero, if mass 1 is increased, does the first suppression vibration frequency:

☐ a. decrease   ☐ b. remain constant   ☐ c. increase

Q5.2.10 For Force 2 equal to zero, if stiffness 2 is increased, does the first suppression vibration frequency:

☐ a. decrease   ☐ b. remain constant   ☐ c. increase

Q5.2.11 For Force 2 equal to zero, if stiffness 1 is increased, does the first suppression vibration frequency:

☐ a. decrease   ☐ b. remain constant   ☐ c. increase

A.5.3 Vibration of rigid bars with two degrees of freedom

Figure A.16 shows the front panel of the simulation of free vibration in a system with a rigid bar and two springs, based on Sections 8.3 and 8.5.

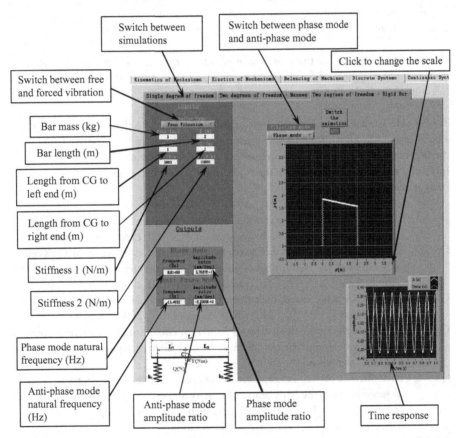

Figure A.16: Front panel for the simulation of the free vibration of a system with a bar and two springs

The input parameters for the simulation are:

- the mass of the bar in kilograms;
- the length of the bar in metres;
- the distance from the bar's left end to its centre of gravity in metres;
- the distance from the bar's right end to its centre of gravity in metres;
- the stiffness of the springs in N/m.

The output table contains the natural frequencies of the system in Hz and the amplitude ratios for each mode of vibration in millimetres per degree. The mode of vibration can be selected using the drag tab above the animation box. An additional animation box plots the displacements as a function of time.

In the case of forced vibration, as shown in Figure A.17, additional input parameters are required: the external force in newtons and the torque in newton metres, acting at the centre of gravity of the bar and the forcing frequency.

The output parameters are the suppression vibration frequency and the steady-state amplitude for each degree of freedom. A response spectrum graph shows both degrees of freedom as functions of frequency.

This simulation assists students in understanding the effect of the mass of the bar, the length of the bar, the stiffness and position of the springs and the external forces on the two natural frequencies, their corresponding amplitude ratios, the mode of vibration (phase or anti-phase) of the system, the suppression vibration frequencies and the steady-state amplitudes.

Q5.3.1 If the mass of the bar is increased, does the fundamental frequency:
☐ a. increase   ☐ b. decrease   ☐ c. remain constant

Q5.3.2 If the length of the bar is increased, for equal distances between the bar's centre of gravity and the two springs, does the fundamental frequency:
☐ a. increase   ☐ b. remain constant   ☐ c. decrease

Q5.3.3 If stiffness 1 is increased, does the fundamental frequency:
☐ a. increase   ☐ b. remain constant   ☐ c. decrease

Q5.3.4 If stiffness 1 is larger than stiffness 2, and for equal distances between the bar's centre of gravity and the two springs, does the higher frequency correspond to:
☐ a. anti-phase mode   ☐ b. phase mode

Q5.3.5 If the distance between the bar's centre of gravity and the left spring is larger than distance between the bar's centre of gravity and right spring, and for stiffness 1 equal to stiffness 2, does the higher frequency correspond to:
☐ a. phase mode   ☐ b. anti-phase mode

Q5.3.6 If stiffness 2 is larger than stiffness 1, and for equal distances between the bar's centre of gravity and the two springs, does the higher frequency correspond to:
☐ a. anti-phase mode   ☐ b. phase mode

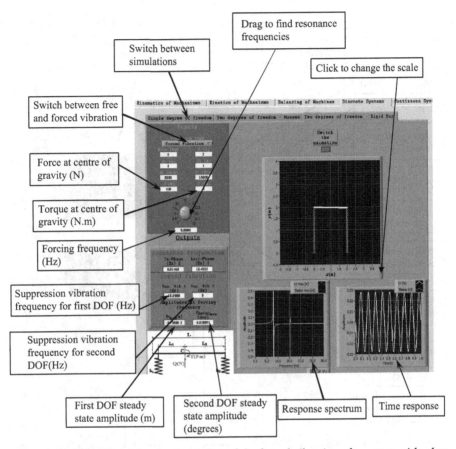

Figure A.17: Front panel for the simulation of the forced vibration of a system with a bar and two springs

Q5.3.7   For external torque equal to zero, if the external force is increased, does the suppression vibration frequency for the first DOF:

☐ a. increase   ☐ b. remain constant   ☐ c. decrease

Q5.3.8   For external force equal to zero, if the external torque is increased, does the suppression vibration frequency for the second DOF:

☐ a. decrease   ☐ b. increase   ☐ c. remain constant

Q5.3.9   If the length of the bar is increased, does the suppression vibration frequency for the first DOF:

☐ a. decrease   ☐ b. increase   ☐ c. remain constant

Q5.3.10 If the mass of the bar is increased, does the suppression vibration frequency for both DOF's:

☐ a. increase   ☐ b. remain constant   ☐ c. decrease

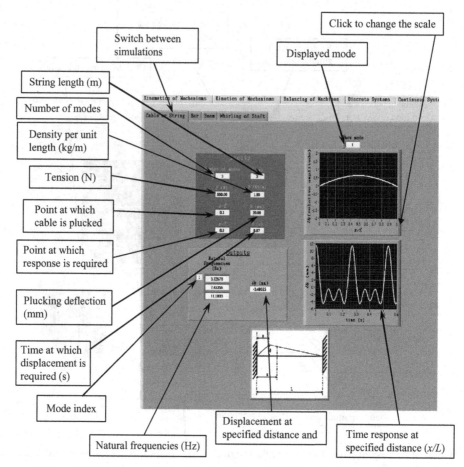

Figure A.18: Front panel for the simulation of the lateral vibration of a string

A.6 VIBRATION OF CONTINUOUS SYSTEMS

A.6.1 Lateral vibration of a cable or string

Figure A.18 shows the front panel of the simulation of the lateral vibration of a string, based on Section 9.2.

The input parameters for the simulation are:

- the number of modes required to be calculated and which will contribute to the displacement time response calculations;
- the cable length in metres;

- the tension in the cable in newtons;
- the density per unit length in kilograms per metre;
- the ratio between the distance at which the cable is plucked and the cable length;
- the amplitude of the plucking in millimetres;
- the ratio between the distance at which the amplitude is required and the cable length;
- the time at which the amplitude is required.

The output table contains the natural frequencies for the required modes as well as the cable displacement at the specified distance and time. In the animation box, the mode shape can be selected using the Show mode icon. The number of modes that can be animated cannot exceed the required number of modes parameter. A box below the animation box plots the displacement as a function of time, at the specified distance (x/L) of the cable.

This simulation helps students to understand the effect of the cable length, cable density, tension in the cable and the initial displacement magnitude and position on the natural frequencies and the displacement time response of the system. It also helps them to understand the contribution of the different modes of vibration to the displacement time response.

Q6.1.1   If the tension in the cable is increased, do the natural frequencies:
   □ a. increase   □ b. remain constant □ c. decrease

Q6.1.2   If the cable mass is increased, do the natural frequencies:
   □ a. increase   □ b. remain constant □ c. decrease

Q6.1.3   If the cable length is increased, do the natural frequencies:
   □ a. decrease   □ b. increase   □ c. remain constant

Q6.1.4   If the tension in the cable is increased, does the displacement amplitude:
   □ a. decrease   □ b. increase   □ c. remain constant

Q6.1.5   In calculating the displacement time response at distance of 1/3 of the cable length, if the cable is plucked at its midpoint, which of the following modes would have a contribution?
   □ a. mode 4   □ b. mode 5   □ c. mode 6

Q6.1.6   The greatest contribution of the fundamental mode to the displacement time response is when it is calculated at a distance of
   □ a. one tenth the cable □ b. half the cable □ c. one third the cable
    length   length   length

Q6.1.7   If the displacement time response is calculated at a distance of 0.1 of the cable length when the cable is plucked at its midpoint, which modes would make a contribution?
   □ a. even modes □ b. all modes   □ c. odd modes

Q6.1.8   If the time response is calculated at a distance of 0.1 of the cable length when the cable is plucked at that point, which modes would make a contribution?
   □ a. even modes □ b. all modes   □ c. odd modes

Q6.1.9   If the cable mass is increased, does the displacement amplitude:
    ☐ a. increase   ☐ b. remain constant   ☐ c. decrease

Q6.1.10 If the cable length is increased, does the displacement amplitude:
    ☐ a. decrease   ☐ b. increase   ☐ c. remain constant

A.6.2 Longitudinal vibration of a bar

Figure A.19 shows the front panel of the simulation of longitudinal vibration of a bar with fixed–fixed boundary conditions, based on Section 9.3.

A drag tab button above the input parameters enable the user to select the boundary conditions: free–free, fixed–free or fixed–fixed. The input parameters for the simulation are:

- the number of modes required to be calculated (for the fixed–free boundary condition, these modes are used to calculate the displacement time response);
- the length of the bar in metres;

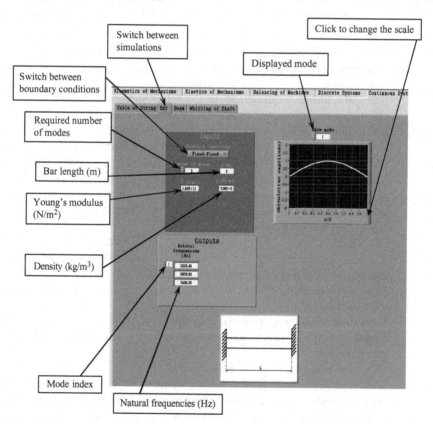

Figure A.19: Front panel for the simulation of the axial vibration of a fixed–fixed bar

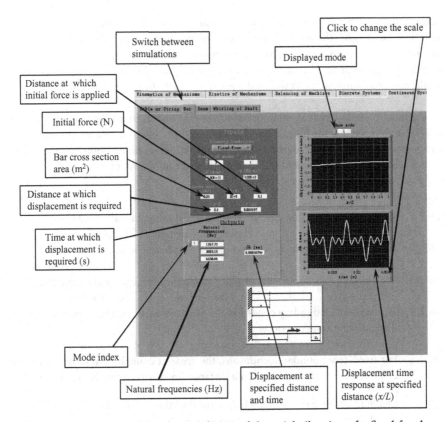

Figure A.20: Front panel for the simulation of the axial vibration of a fixed-free bar

- the Young's modulus of the bar in newtons per square metre;
- the density of the bar in kilograms per cubic metres.

The fixed-free boundary conditions (see Figure A.20) require additional parameters:

- the cross-sectional area of the bar in square metres;
- the initial force in newtons;
- the ratio between the distance at which the force is applied and the bar length (a/L);
- the ratio between the distance at which the displacement is required and the bar length (x/L);
- the time in seconds at which the displacement is required.

The white drawing box shows a diagram that defines the input parameters and the boundary conditions. The output table contains the natural frequencies for the required modes.

In the animation box, the mode shape can be selected using the Show mode icon. The number of modes that can be animated cannot exceed the required number of modes parameter. It should be noted that the animation is for displacement along the bar's longitudinal axis but it is plotted in the vertical direction for clarity. For the free–free boundary conditions, the mode that has zero frequency is the rigid-body mode. For the case of fixed–free boundary conditions, the displacement at the specified distance (x/L) and time is given in the green box. An additional box below the animation plots the displacement as a function of time, at the point at which displacement is required.

This simulation helps students to understand the effect of the bar length, the bar density and Young's modulus, the boundary conditions and the initial displacement magnitude and position on the natural frequencies and the displacement time response of the system. It also helps them to understand the contribution of the different modes of vibration to the displacement time response of the bar.

Q6.2.1   If Young's modulus of the bar is increased, do the natural frequencies:
    ☐ a. increase   ☐ b. remain constant   ☐ c. decrease

Q6.2.2   If the bar density is increased, do the natural frequencies:
    ☐ a. decrease   ☐ b. increase   ☐ c. remain constant

Q6.2.3   Which boundary conditions produce the lowest fundamental frequency?
    ☐ a. Free–free   ☐ b. Fixed–fixed   ☐ c. Fixed–free

Q6.2.4   For fixed–free boundary conditions, if the bar's cross section increases, does the displacement time response:
    ☐ a. decrease   ☐ b. increase   ☐ c. remain constant

Q6.2.5   For fixed–free boundary conditions, the greatest contribution of mode 1 to the displacement time response is at a distance (measured from the fixed end) equal to
    ☐ a. half of the bar length   ☐ b. one third of the bar length   ☐ c. the bar length

Q6.2.6   For fixed–free boundary conditions, when the force is applied at the bar's free end, the modes that contribute to the displacement time response at a distance equal to half the bar length are:
    ☐ a. all modes   ☐ b. even modes   ☐ c. odd modes

Q6.2.7   For fixed–free boundary conditions, if a force is applied at the bar's free end, which modes would make a contribution to the displacement time response?
    ☐ a. even modes   ☐ b. all modes   ☐ c. odd modes

Q6.2.8   For fixed–free boundary conditions, if a force is applied at 0.1 of the bar length from the fixed end, which modes would make a contribution to the displacement time response?
    ☐ a. even modes   ☐ b. all modes   ☐ c. odd modes

Q6.2.9   For fixed–free boundary conditions, if the bar's density is increased, does the displacement amplitude:
    ☐ a. increase   ☐ b. remain constant   ☐ c. decrease

Q6.2.10 For fixed–free boundary conditions, if the bar's Young's modulus is increased, does the displacement amplitude:
    ☐ a. increase   ☐ b. remain constant   ☐ c. decrease

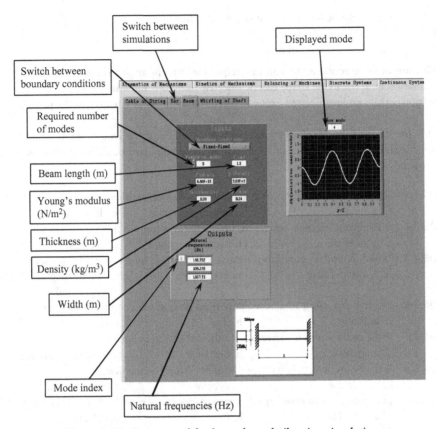

Figure A.21: Front panel for beam lateral vibration simulation

A.6.3 Lateral vibration of a beam

Figure A.21 shows the front panel of the simulation of lateral vibration of a beam, based on Section 9.4.

There are five possible boundary conditions: free–free, fixed–free, fixed–fixed, fixed–simply supported and simply-supported (where both ends are simply supported). You can switch between different boundary conditions by clicking the tab icon. The boundary conditions in the diagram, in the white drawing box, are updated according to the selected icon.

The input parameters for the simulation are:

- the number of modes required to be calculated;
- the length of the beam in metres;
- the Young's modulus of the beam in newtons per square metre;

- the density of the beam in kilograms per cubic metre;
- the width of the beam in metres;
- the thickness of the beam in metres.

The output table contains the natural frequencies for the required modes in Hz. In the animation box, the mode shape can be selected using the Show mode icon. The number of modes that can be animated cannot exceed the required number of modes parameter.

This simulation helps students to understand the effect of the length, width, thickness, density and Young's modulus of the beam and the boundary conditions on its natural frequencies.

Q6.3.1   If the beam thickness is increased, do the natural frequencies:
    ☐ a. increase   ☐ b. remain constant   ☐ c. decrease

Q6.3.2   If the beam length is increased, do the natural frequencies:
    ☐ a. increase   ☐ b. remain constant   ☐ c. decrease

Q6.3.3   Which boundary conditions produce the highest fundamental frequency?
    ☐ a. Fixed–fixed   ☐ b. Fixed–free   ☐ c. Fixed–simply supported

Q6.3.4   For free-free vibration, how many rigid-body modes can be identified?
    ☐ a. None   ☐ b. 2   ☐ c. 1

Q6.3.5   If the beam width is increased, do the natural frequencies:
    ☐ a. increase   ☐ b. decrease   ☐ c. remain constant

Q6.3.6   If the beam density is increased, do the natural frequencies:
    ☐ a. increase   ☐ b. decrease   ☐ c. remain constant

Q6.3.7   For free–free vibration, at what frequency does a rigid-body mode take place?
    ☐ a. 0.01 Hz   ☐ b. 0   ☐ c. 0.1 Hz

Q6.3.8   For simply supported boundary conditions, how many nodes are there in the fifth mode shape?
    ☐ a. 3   ☐ b. 5   ☐ c. 6

Q6.3.9   What is the maximum number of vibration modes that can be calculated for a beam?
    ☐ a. 9   ☐ b. 100   ☐ c. infinite

Q6.3.10 One of the boundary conditions at a free end is:
    ☐ a. displacement is zero   ☐ b. slope is zero   ☐ c. moment is zero

A.6.4 Whirling of shafts

Figure A.22 shows the front panel of the simulation of the whirling of a shaft, based on Section 9.5.

The input parameters for the simulation are:

- the length of the shaft in metres;
- the diameter of the shaft in metres;

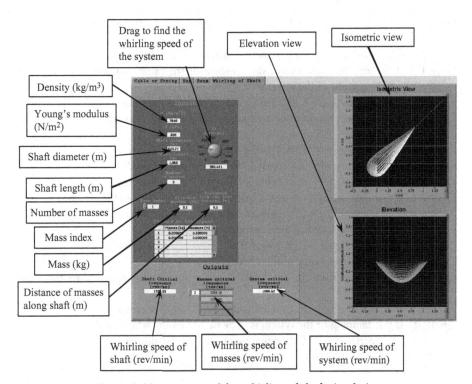

Figure A.22: Front panel for whirling of shaft simulation

- the Young's modulus of the shaft in newtons per square metre;
- the density of the shaft in kilograms per cubic metre;
- the masses that are attached to the shaft in kilograms;
- the distance of the masses along the shaft in metres (measured from the left bearing).

The output table contains the whirling speed of the shaft alone, the whirling speed of each mass and the whirling speed of the system (the shaft and the masses) in revolutions per minute.

This simulation helps students to understand the effect of the length, diameter, density and Young's modulus of the shaft and the attached masses and their positions on the whirling speed of the system.

Q6.4.1   If the shaft diameter is increased, does the whirling speed of the system:
   ☐ a. increase   ☐ b. remain constant   ☐ c. decrease
Q6.4.2   If the shaft length is increased, does the whirling speed of the system:
   ☐ a. increase   ☐ b. remain constant   ☐ c. decrease

Q6.4.3   If the number of attached masses is increased, does the whirling speed of the system:

☐ a. increase   ☐ b. remain constant   ☐ c. decrease

Q6.4.4   If the attached masses are shifted towards the middle of the shaft, does the whirling speed of the system:

☐ a. increase   ☐ b. remain constant   ☐ c. decrease

Q6.4.5   If the Young's modulus of the shaft is increased, does the whirling speed of the system:

☐ a. increase   ☐ b. remain constant   ☐ c. decrease

Q6.4.6   If the shaft density is increased, does the whirling speed of the system:

☐ a. increase   ☐ b. remain constant   ☐ c. decrease

Q6.4.7   If the shaft density is increased, does the whirling speed of the attached masses:

☐ a. increase   ☐ b. remain constant   ☐ c. decrease

Q6.4.8   The whirling speed of the system is:

☐ a. higher than   ☐ b. lower than   ☐ c. equal to

the natural frequency of the system

Q6.4.9   If the shaft's rotating speed gets closer to its whirling speed, does the whirl amplitude:

☐ a. increase   ☐ b. remain constant   ☐ c. decrease

Q6.4.10   If the damping of the system is increased, does the whirl amplitude:

☐ a. increase   ☐ b. remain constant   ☐ c. decrease

APPENDIX B

Properties of Area

B.1 AREA MOMENT OF INERTIA

The area moment of inertia, also called the second moment of area, is a measure of an object's ability to resist bending moment and deflection. For the perpendicular axes x and y, the area moment of inertia, as shown in Table B.1, about the x axis in a general form is defined as:

$$I_{xx} = \int_A y^2 dA \tag{B.1}$$

Similarly, the area moment of inertia about the y axis in a general form is defined as:

$$I_{yy} = \int_A x^2 dA \tag{B.2}$$

B.2 POLAR MOMENT OF INERTIA

The polar moment of inertia (I_P) is a measure of an object's ability to resist torsion and twist. The smaller the polar moment of inertia of a shaft, the more it twists under the action of a torque. For the perpendicular axes x and y, the polar moment of inertia, as shown in Table B.1, in a general form is defined as:

$$I_P = \int_A (x^2 + y^2) dA = I_{xx} + I_{yy} \tag{B.3}$$

Table B.1: Common plane shapes

Shape	Dimensions	Area	Area moment of inertia (I_{xx} and I_{yy})	Polar moment of inertia ($I_p = I_{xx} + I_{yy}$)
Rectangular		bh	$I_{xx} = \dfrac{bh^3}{12},$ $I_{yy} = \dfrac{hb^3}{12}$	$I_p = \dfrac{bh}{12}(b^2 + h^2)$
Circular		$\dfrac{\pi D^2}{4}$	$I_{xx} = \dfrac{\pi D^4}{64}$ $I_{yy} = \dfrac{\pi D^4}{64}$	$I_p = \dfrac{\pi D^4}{32}$
Annulus		$\dfrac{\pi(D_o^2 - D_i^2)}{4}$	$I_{xx} = \dfrac{\pi(D_o^4 - D_i^4)}{64}$ $I_{yy} = \dfrac{\pi(D_o^4 - D_i^4)}{64}$	$I_p = \dfrac{\pi(D_o^4 - D_i^4)}{32}$
Semi-circular		$\dfrac{\pi D^2}{8}$	$I_{xx} = \dfrac{\pi D^4}{128}$ $I_{yy} = \dfrac{\pi D^4}{128}$	$I_p = \dfrac{\pi D^4}{64}$
Quarter-circular		$\dfrac{\pi D^2}{16}$	$I_{xx} = \dfrac{\pi D^4}{256}$ $I_{yy} = \dfrac{\pi D^4}{256}$	$I_p = \dfrac{\pi D^4}{128}$

APPENDIX C

Equivalent Stiffness for Combinations of Springs

C.1 SPRINGS IN PARALLEL

Consider a body of mass m, which is supported by two springs in parallel as shown in Figure C.1. The first spring, denoted as spring 1, has a stiffness k_1 and deflects δ_1 under the action of the body's weight, mg. Similarly, the second spring, denoted as spring 2, has a stiffness k_2 and deflects δ_2 under the action of the body's weight mg. The static deflection of the body, denoted as δ_t, is identical to δ_1 and δ_2:

$$\delta_t = \delta_1 = \delta_2 \tag{C.1}$$

The force in spring 1, F_1, and the force in spring 2, F_2, are related to mg by:

$$F_1 + F_2 = mg \tag{C.2}$$

For linear springs, the following relationships between forces and deflections are applicable:

$$F_1 = k_1\delta_1 \tag{C.3}$$
$$F_2 = k_2\delta_2 \tag{C.4}$$
$$mg = k_{eq}\delta_t \tag{C.5}$$

where k_{eq} is the equivalent stiffness of the system. Substituting Equations (C.3) to (C.5) into Equation (C.2), gives:

$$k_1\delta_1 + k_2\delta_2 = k_{eq}\delta_t \tag{C.6}$$

Figure C.1: Two springs in parallel

Substituting Equation (C.1) into Equation (C.6), the equivalent stiffness of two springs in parallel is obtained as:

$$k_{eq} = k_1 + k_2 \tag{C.7}$$

For n springs in parallel, with stiffness k_1, k_2, k_3, ..., k_n, the equivalent stiffness of the system is given by:

$$k_{eq} = k_1 + k_2 + k_3 + \ldots + k_n \tag{C.8}$$

C.2 SPRINGS IN SERIES

Similarly to the derivation above, consider a body of mass m, which is supported by two springs in series as shown in Figure C.2. The static deflection of the body, δ_t, is equal to the sum of the deflections of spring 1, δ_1, and spring 2, δ_2, and is given by:

$$\delta_t = \delta_1 + \delta_2 \tag{C.9}$$

while the force in spring 1, F_1, and the force in spring 2, F_2, are identical to mg:

$$F_1 = F_2 = mg \tag{C.10}$$

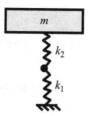

Figure C.2: Two springs in series

The relationships between forces and deflections are the same as given in Equations (C.3) to (C.5). Substituting Equations (C.3) to (C.5) into Equation (C.9) gives:

$$\frac{F_1}{k_1} + \frac{F_2}{k_2} = \frac{mg}{k_{eq}} \tag{C.11}$$

Substituting Equation (C.10) into Equation (C.11), the equivalent stiffness of two springs in series is obtained as:

$$\frac{1}{k_{eq}} = \frac{1}{k_1} + \frac{1}{k_2} \tag{C.12}$$

For n springs in series, with stiffness k_1, k_2, k_3, ..., k_n, the equivalent stiffness of the system is given by:

$$\frac{1}{k_{eq}} = \frac{1}{k_1} + \frac{1}{k_2} + \frac{1}{k_3} + \ldots + \frac{1}{k_n} \tag{C.13}$$

APPENDIX D

Summary of Formulas

Rectilinear motion with constant acceleration

$$v = u + at, \quad v^2 = u^2 + 2as, \quad s = ut + \frac{1}{2}at^2, \quad s = \frac{1}{2}(u + v)t$$

Curvilinear motion

$$v_r = \dot{r}, \quad v_t = r\dot{\theta}, \quad a_r = \ddot{r} - r\dot{\theta}^2, \quad a_t = r\ddot{\theta} + 2\dot{r}\dot{\theta}, \quad v = \frac{ds}{dt} = \dot{s},$$

$$a_t = \frac{dv}{dt} = \dot{v}, \quad a_n = \frac{v^2}{\rho}$$

Slider–crank mechanism

$$\dot{x}_A = -r\omega \left(\sin\theta + \frac{r}{2L} \sin 2\theta - \frac{a}{L} \cos\theta \right),$$

$$\ddot{x}_A = -r\omega^2 \left(\cos\theta + \frac{r}{L} \cos 2\theta + \frac{a}{L} \sin\theta \right),$$

$$\dot{x}_D = \dot{x}_A + \frac{r^2\omega b}{2L^2} \sin 2\theta - \frac{r\omega ba}{L^2} \cos\theta, \quad \dot{y}_D = \frac{r\omega b}{L} \cos\theta,$$

$$\ddot{x}_D = \ddot{x}_A + \frac{r^2\omega^2 b}{L^2} \cos 2\theta + \frac{r^2\omega^2 ba}{L^2} \sin\theta, \quad \ddot{y}_D = -\frac{r\omega^2 b}{L} \sin\theta$$

Four-bar linkage mechanism

$$\dot\psi = -\frac{L_1\dot\theta C_1}{L_2 C_2}, \quad \dot\varphi = \frac{L_1\dot\theta C_3}{L_3 C_2}, \quad \ddot\psi = -\left(\frac{L_1\dot\theta^2 C_4 + L_2\dot\psi^2 C_5 + L_3\dot\phi^2}{L_2 C_2}\right),$$

$$\ddot\varphi = -\left(\frac{L_1\dot\theta^2 C_8 + L_2\dot\psi^2 + L_2\dot\phi^2 C_5}{L_3 C_2}\right)$$

$C_1 = \cos\theta\sin\varphi + \sin\theta\cos\varphi, \quad C_2 = \cos\psi\sin\varphi + \sin\psi\cos\varphi, \quad C_3 = \cos\theta\sin\psi - \sin\theta\cos\psi,$

$C_4 = \cos\theta\cos\phi - \sin\theta\sin\phi, \quad C_5 = \cos\psi\cos\phi - \sin\psi\sin\phi, \quad C_6 = \cos\theta\cos\psi + \sin\theta\sin\psi$

Newton's law of universal gravitation

$$F = G\frac{m_1 m_2}{r^2}, \quad G = 6.673 \times 10^{-11}\,\frac{\text{m}^3}{\text{kg.s}^2}$$

Newton's second law of motion

$$\sum F_x = m\ddot x, \quad \sum F_y = m\ddot y, \quad \sum M = \pm m\ddot y_G\, x_G \pm m\ddot x_G\, y_G + I_G\ddot\theta$$

Work and energy (particle)

$$KE = \frac{1}{2}mv^2, \quad PE = mgh, \quad SE = \frac{1}{2}kx^2, \quad W_e = F.s$$

Work and energy (rigid body)

$$KE = \frac{1}{2}mv_G^2 + \frac{1}{2}I_G\dot\theta^2, \quad PE = mgh_G, \quad SE = \frac{1}{2}kx_G^2, \quad W_e = F_G.s, \quad U_\theta = M_G.\theta$$

Impulse and momentum (particle)

$$mv_1 + \int_{t_1}^{t_2} F\,dt = mv_2, \quad \sum mv_1 = \sum mv_2, \quad e = \frac{v_{B2} - v_{A2}}{v_{A1} - v_{B1}}$$

Impulse and momentum (rigid body)

$$I_G\dot\theta_1 + \int_{t_1}^{t_2} M_G\,dt = I_G\dot\theta_2, \quad \sum H_{G_1} = \sum H_{G_2}$$

Balancing of machines

$$\sum F = \sum mr\omega^2, \quad \sum F_x = \sum mr\omega^2\cos\theta, \quad \sum F_y = \sum mr\omega^2\sin\theta,$$

$$\sum M_x = \sum F_y \times z = \sum mr\omega^2\sin\theta, \quad \sum M_y = \sum F_x \times z = \sum mr\omega^2\cos\theta$$

Secondary forces and moments

$$\sum F_x = \omega^2 \sum \frac{mr}{n} \cos 2\theta, \qquad \sum F_y = \omega^2 \sum \frac{mr}{n} \sin 2\theta$$

$$\sum M_x = \sum F_y \times z = \omega^2 \sum \frac{mrz}{n} \sin 2\theta, \qquad \sum M_y = \sum F_x \times z = \omega^2 \sum \frac{mrz}{n} \cos 2\theta$$

Free Vibration of a System with a Single Degree of Freedom

Undamped system

$$\omega_n = \sqrt{\frac{k}{m}}, \quad x = C\sin(\omega_n t + \psi), \quad C = \sqrt{x_o^2 + \left(\frac{\dot{x}_o}{\omega_n}\right)^2}, \quad \psi = \tan^{-1}\frac{x_o \omega_n}{\dot{x}_o}$$

Overdamped system

$$x = \frac{\dot{x}_o + x_o \omega_n \left(\zeta + \sqrt{\zeta^2 - 1}\right)}{2\omega_n \sqrt{\zeta^2 - 1}} e^{\left(-\zeta + \sqrt{\zeta^2 - 1}\right)\omega_n t}$$

$$- \frac{\dot{x}_o + x_o \omega_n \left(\zeta - \sqrt{\zeta^2 - 1}\right)}{2\omega_n \sqrt{\zeta^2 - 1}} e^{\left(-\zeta - \sqrt{\zeta^2 - 1}\right)\omega_n t}, \quad \zeta = \frac{c}{2m\omega_n}$$

Critically damped system

$$x = (x_o + (\dot{x}_o + \omega_n x_o)t)\, e^{-\omega_n t}$$

Underdamped system

$$x = \left[x_o \cos \omega_d t + \frac{\dot{x}_e + \zeta \omega_n x_o}{\omega_d} \sin \omega_d t\right] e^{-\zeta \omega_n t}, \quad \omega_d = \omega_n \sqrt{1 - \zeta^2},$$

$$\delta = \ln\frac{x_1}{x_2} = \zeta \omega_n \tau_d, \quad \zeta = \frac{\delta}{\sqrt{(2\pi)^2 + \delta^2}}$$

Forced Vibration of a System with a Single Degree of Freedom

Undamped system

$$x_p = \chi \sin \omega t, \quad \chi = \frac{F_o/k}{1 - \left(\frac{\omega}{\omega_n}\right)^2}$$

Damped system

$$x_p = \chi \sin(\omega t - \phi), \quad \chi = \frac{F_o/k}{\sqrt{\left(1 - \left(\frac{\omega}{\omega_n}\right)^2\right)^2 + \left(\frac{2\zeta\omega}{\omega_n}\right)^2}},$$

$$\phi = \tan^{-1} \frac{2\zeta(\omega/\omega_n)}{1 - \left(\frac{\omega}{\omega_n}\right)^2}, \quad T_r = \frac{F_t}{F_o} = \sqrt{\frac{1 + \left(\frac{2\zeta\omega}{\omega_n}\right)^2}{\left(1 - \left(\frac{\omega}{\omega_n}\right)^2\right)^2 + \left(\frac{2\zeta\omega}{\omega_n}\right)^2}}$$

Systems with Two Degrees of Freedom

$$\omega_{1,2}^2 = \frac{-b \pm \sqrt{b^2 - 4ac}}{2a}, \quad a = M_{11}M_{22}, \quad b = -(M_{11}S_{22} + M_{22}S_{11}),$$

$$c = S_{11}S_{22} - S_{12}^2,$$

$$A_1 = \frac{(S_{22} - \omega^2 M_{22})F_{o1} - S_{12}F_{o2}}{M_{11}M_{22}(\omega^2 - \omega_1^2)(\omega^2 - \omega_2^2)}, \quad A_2 = \frac{(S_{11} - \omega^2 M_{11})F_{o2} - S_{21}F_{o1}}{M_{11}M_{22}(\omega^2 - \omega_1^2)(\omega^2 - \omega_2^2)}$$

Vibration of Continuous Systems

Lateral vibration of a cable

$$\omega_n = \frac{nc\pi}{L}, \quad n = 1, 2, 3, \ldots, \quad c = \sqrt{\frac{F}{\rho'}},$$

$$y(x, t) = \sum_{n=1}^{\infty} \sin\frac{n\pi x}{L}\left(C_n \cos\frac{nc\pi t}{L} + D_n \sin\frac{nc\pi t}{L}\right),$$

$$C_n = \begin{cases} \dfrac{2\delta L}{\pi^2 n^2}\left(\dfrac{1}{a} + \dfrac{1}{L-a}\right)\sin\dfrac{n\pi a}{L} \ldots\ldots n = 1, 3, 5, \ldots\ldots \\ 0 \ldots\ldots \qquad\qquad\qquad n = 2, 4, 6, \ldots\ldots \end{cases}$$

Longitudinal vibration of a bar

$$c = \sqrt{\frac{E}{\rho}}$$

For fixed–free bar

$$u_x(x, t) = \sum_{n=0}^{\infty} \sin\frac{(2n+1)\pi x}{2L}\left(C_n \cos\frac{(2n+1)\pi ct}{2L} + D_n \sin\frac{(2n+1)\pi ct}{2L}\right),$$

$$C_n = \frac{8U_o L}{(2n+1)^2\pi^2 a}\sin\frac{(2n+1)\pi a}{2L}$$

Table D.1: Angular frequencies and mode shapes for a bar in longitudinal vibration

End conditions	Boundary conditions	Angular frequencies	Mode shapes
Fixed–fixed	$u_x(0, t) = 0$ $u_x(L, t) = 0$	$\omega_n = \dfrac{nc\pi}{L}$, $n = 1, 2, 3, \ldots$	$U_n = A_n \sin \dfrac{n\pi x}{L}$
Free–free	$\dfrac{\partial u_x(0, t)}{\partial x} = 0$ $\dfrac{\partial u_x(L, t)}{\partial x} = 0$	$\omega_n = \dfrac{nc\pi}{L}$, $n = 0, 1, 2, 3, \ldots$	$U_n = A_n \cos \dfrac{n\pi x}{L}$
Fixed–free	$u_x(0, t) = 0$ $\dfrac{\partial u_x(L, t)}{\partial x} = 0$	$\omega_n = \dfrac{(2n+1)c\pi}{2L}$, $n = 0, 1, 2, 3, \ldots$	$U_n = A_n \sin \dfrac{(2n+1)\pi x}{2L}$

Lateral vibration of a beam

$$\omega_n = (\beta L)^2 \frac{c}{L^2}, \quad c = \sqrt{\frac{EI}{\rho A}}$$

Whirling of shafts

$$\omega_1 = 94.25\sqrt{\frac{EI}{mL^3}}, \quad \omega_1 = 16.54\sqrt{\frac{EIL}{Ma^2(L-a)^2}},$$

$$u_y = \frac{\delta\left(\frac{\omega}{\omega_n}\right)^2}{\sqrt{\left(1 - \left(\frac{\omega}{\omega_n}\right)^2\right)^2 + \left(\frac{2\zeta\omega}{\omega_n}\right)^2}}, \quad \frac{1}{\omega_1^2} \approx \frac{1}{\omega_1^2} + \frac{1}{\omega_2^2} + \frac{1}{\omega_3^2} + \cdots\cdots\cdots \frac{1}{\omega_n^2}$$

Finite-element method

Bar element

$$[m] = \frac{\rho Al}{6}\begin{bmatrix} 2 & 1 \\ 1 & 2 \end{bmatrix}, \quad [k] = \frac{EA}{L}\begin{bmatrix} 1 & -1 \\ -1 & 1 \end{bmatrix}$$

Table D.2: Values of $\beta_n L$ and mode shapes for a beam in transverse vibration

End conditions	Values of $\beta_n L$	Mode shapes
Free–free	$\beta_0 L = 0$ for rigid body mode $\beta_1 L = 4.73$ $\beta_2 L = 7.853$ $\beta_3 L = 10.995$ $\beta_4 L = 14.137$	$U_n = C_n \left(\sin \beta_n x + \sinh \beta_n x \right.$ $\left. + \left[\dfrac{\sin \beta_n x - \sinh \beta_n x}{\cosh \beta_n x - \cos \beta_n x} \right] (\cos \beta_n x + \cosh \beta_n x) \right)$
Simply supported	$\beta_1 L = \pi$ $\beta_2 L = 2\pi$ $\beta_3 L = 3\pi$ $\beta_4 L = 4\pi$	$U_n = C_n \sin \beta_n x$
Fixed–fixed	$\beta_1 L = 4.73$ $\beta_2 L = 7.853$ $\beta_3 L = 10.995$ $\beta_4 L = 14.137$	$U_n = C_n \left(\sinh \beta_n x - \sin \beta_n x \right.$ $\left. + \left[\dfrac{\sinh \beta_n x - \sin \beta_n x}{\cos \beta_n x - \cosh \beta_n x} \right] (\cosh \beta_n x - \cos \beta_n x) \right)$
Fixed–free	$\beta_1 L = 1.875$ $\beta_2 L = 4.694$ $\beta_3 L = 7.854$ $\beta_4 L = 10.995$	$U_n = C_n \left(\sin \beta_n x - \sinh \beta_n x \right.$ $\left. - \left[\dfrac{\sin \beta_n x + \sinh \beta_n x}{\cos \beta_n x + \cosh \beta_n x} \right] (\cos \beta_n x - \cosh \beta_n x) \right)$
Fixed–simply supported	$\beta_1 L = 3.926$ $\beta_2 L = 7.068$ $\beta_3 L = 10.21$ $\beta_4 L = 13.351$	$U_n = C_n \left(\sin \beta_n x - \sinh \beta_n x \right.$ $\left. + \left[\dfrac{\sin \beta_n x - \sinh \beta_n x}{\cos \beta_n x - \cosh \beta_n x} \right] (\cosh \beta_n x - \cos \beta_n x) \right)$

Beam element

$$[m] = \frac{\rho A L}{420} \begin{bmatrix} 156 & 22L & 54 & -13L \\ 22L & 4L^2 & 13L & -3L^2 \\ 54 & 13L & 156 & -22L \\ -13L & -3L^2 & -22L & 4L^2 \end{bmatrix},$$

$$[k] = \frac{EI}{L^3} \begin{bmatrix} 12 & 6L & -12 & 6L \\ 6L & 4L^2 & -6L & 2L^2 \\ -12 & -6L & 12 & -6L \\ 6L & 2L^2 & -6L & 4L^2 \end{bmatrix}$$

Index